A CANADIAN CLIMATE OF MIND

A Canadian Climate
of Mind

*Passages from Fur to Energy
and Beyond*

TIMOTHY B. LEDUC

McGill-Queen's University Press
Montreal & Kingston • London • Chicago

© McGill-Queen's University Press 2016

ISBN 978-0-7735-4761-2 (cloth)
ISBN 978-0-7735-4762-9 (paper)
ISBN 978-0-7735-9871-3 (ePDF)
ISBN 978-0-7735-9880-5 (ePUB)

Legal deposit second quarter 2016
Bibliothèque nationale du Québec

Printed in Canada on acid-free paper that is 100% ancient forest free
(100% post-consumer recycled), processed chlorine free

This book has been published with the help of a grant from the Canadian
Federation for the Humanities and Social Sciences, through the Awards
to Scholarly Publications Program, using funds provided by the Social Sciences
and Humanities Research Council of Canada. Funding has also been received
from the Faculty of Social Work at Wilfrid Laurier University.

McGill-Queen's University Press acknowledges the support of the Canada
Council for the Arts for our publishing program. We also acknowledge the
financial support of the Government of Canada through the Canada Book
Fund for our publishing activities.

Library and Archives Canada Cataloguing in Publication

Leduc, Timothy B., 1970–, author
A Canadian climate of mind: passages from fur to energy and beyond /
Timothy B. Leduc.

Includes bibliographical references and index.
Issued in print and electronic formats.
ISBN 978-0-7735-4761-2 (cloth). – ISBN 978-0-7735-4762-9 (paper). –
ISBN 978-0-7735-9871-3 (ePDF). – ISBN 978-0-7735-9880-5 (ePUB)

1. Human ecology – Canada. 2. Native peoples – Ecology – Canada.
3. Human beings – Effect of climate on – Canada. 4. Climatic changes –
Canada. 5. Climatic changes – Psychological aspects. 6. Climatic
changes – Social aspects – Canada. I. Title.

GF511.L43 2016   304.20971   C2015-908342-7
                                           C2015-908343-5

This book was typeset by Interscript in 10.5/14 Sabon.

Contents

Illustrations

Unless otherwise specified, the photographs found in this book were taken by the author. Apart from various works in the public domain that are referred to below, I want to give specific thanks to the following individuals and organizations for allowing me to include their images in this book: Barbara and David Clark, Kevin Gordon, Jamie Leduc, and City of Toronto Museums and Heritage Services.

Foreword

WILLIAM WOODWORTH RAWENO:KWAS

The profound understanding with which *A Canadian Climate of Mind* weaves together what might seem at first disparate narratives conveys to us a healing cultural integration perhaps unique in the literature around our relationship with nature. The sensitive life-encounters so compassionately recounted in this book are suffused with the spirit of a great elder in the Haudenosaunee (Iroquois) tradition. Jacob Ezra Thomas was a traditional chief of the Six Nations Confederacy. He carried the chiefly duties of the Snipe Clan of the Cayuga Nation with the title Deyawenhethon (He Is So Real in a Double Way). Jake was my teacher, and through me he became a teacher to Tim Leduc as he considered the complex challenges to minding a passage through our climate of change in these lands we call Canada.

To me, Leduc's book is the culmination of our mutual journeying, friendship, and respect for all those who have come before us. Our kindred spirits coalesced around the mixed ancestries of our relationship – Native, British, French, and the incredible natural world of southern Ontario. As the following pages attest, this is a region imbued with a rich natural history, vibrant pre-historic Native settlement, and complex colonial narratives that are evident even in the dense urbanity of a place like Toron:to – the place where our relationship and this book begins. All of our work together seemed surrounded by our mutual ancestors, and particularly the ever-appearing spirit of Jake Thomas as I shared with Tim the many teachings and experiences of my time with Jake at Six Nations an hour west of Toronto.

The great Iroquois Confederacy that was the source of Jake's teachings was conceived almost a thousand years ago.[1] It practised a form of spiritually grounded democracy in the varied landscape of what is now upstate New York. The recession of the last Ice Age carved hills and valleys, mountains, gorges, and long lakes among which the benevolent Carolingian forests thrived. My ancestors, and Jake's, came from this homeland to the Grand River as refugees from the American Revolution. Under the leadership of Joseph Brant, a remnant of people from each nation wisely decided to remain the friends and allies of the crown of Great Britain, as we had been since 1710 when our chiefs visited the court of Queen Anne. Brant was able to negotiate a vast new territory in which to resettle members of the Six Nations, including at first the Mohawk (the majority) and then the Cayuga, Onondaga, some Oneida, and a few Seneca. Refugees of settler genocide from other Native nations joined them, including the adopted Tuscarora, thus completing the original six nations. Today members of thirteen Native nations make their home at Six Nations of the Grand River.

Something about this dark past makes me think of the dystopic visions we often hold of ourselves today, encompassing mass extinctions, over-population, the seeming irreversible human disturbances of ecological and climate systems, and so on. At the same time, new forms of human communication are often suffused with narratives of dysfunctional human history and futures as we search for some alternative route. Last year a team of NASA-funded mathematicians created a complex mathematical model for the future and concluded that the utter collapse of human civilization (as we know it) will be difficult to avoid.[2] Such apocalyptic stories are a significant dimension of the climate of change that Leduc also considers, but he does so from the hopeful view – consistent with Jake's teachings – that we are being given an opportunity to heal long-dysfunctional ways of relating with each other and the world.

Jake's story[3] bears sharing here by way of an apt entrance into the realms of Leduc's engagement with other ancestral teachers, beginning with the Canadian eco-thinker John Livingston and the first Indigenous saint, Kateri Tekakwitha. Personal and cultural strife like the climate of change discussed in this book sometimes return us into the arms of Mother Earth, and we as tired children find solace and renewed purpose there. Jake was no exception. He was born in 1922 between two great world wars in a still pastoral traditional Native community largely untouched by the Indian Act. All this was about to change abruptly. In October 1924 a harmonious council of traditional hereditary chiefs was ousted at gunpoint by the RCMP, followed by a hurried and weakly supported community election of a municipal-styled band council, organized and funded by the Canadian government. The incident is typical of the many colonial severances by *rolling heads*[4] that inform Leduc's powerful range of perspectives and places which he has chosen to weave in *A Canadian Climate of Mind* as he contemplates the deep cultural roots of our current challenges.

Despite the violent ousting at Six Nations, traditionalists continued to meet in a parallel council and hold longhouse ceremonies in the Native languages and according to ancient forms of governance defined by the Great Law of the ancestors. Jake's mother, a Cayuga speaker, and father, an Onondaga, were among them. In the traditional matrilineal lines Jake was a Cayuga, embraced and raised into manhood by his father, David Thomas, and his maternal grandfather, David Skye. Following numerous healings facilitated by the medicine societies from early childhood, he took up carving and learned the songs and dances of the longhouse. During the Second World War, many Six Nations young men went to war; however, Jake was held back to learn the ceremonies and speeches of Thanksgiving and Condolence, the events of the ceremonial cycle and creation, the Great Law, and the Code of Handsome Lake – one of the few left to absorb these deep traditions. In 1943 he began a family with Doris Keye, a member of the Onondaga Wolf Clan, and in succession they had fourteen children. Two of his sons died in infancy, and another was killed in an automobile accident when he was twenty-one. Jake took up farming in the traditional food cycles of the three sisters (corn, beans, and squash), supplementing the family sustenance by raising livestock and fishing in local streams. Devastatingly, in November 1964 Doris died of cancer, leaving Jake alone to raise a huge family. A two-year descent into alcoholism finally led him to seek a traditional *medicine friendship* with Raymond Spragge, bringing him out of the confusion.

Beginning in 1971 when Jake was fifty-one years old, a series of events began to mark a significant consolidation of his identity – his father, known as one of the finest orators of the Six Nations, died, a loss that was followed by the deaths of Jake's mentor, Howard Skye, a faith-keeper of the Wolf Clan, and two other prominent speakers in his Onondaga longhouse; he was raised a chief; and he remarried.

With the deaths of his father, Skye, and two other Onondaga whom he respected greatly, Jake found himself largely on his own and was eased gently into the role of ceremonial leader and elder. In the spring of 1973 he was chosen by his Snipe Clan mother, Ida Sky, to be elevated as a hereditary chief. In October 1973 he married Yvonne Hill, a member of the Seneca Snipe Clan. He had met Yvonne on his appointment as curator at the Woodland Indian Cultural Centre in Brantford. His chiefly duties and curatorial responsibilities evolved into several significant educational appointments culminating in September 1976 in an assistant professorship in the newly formed Department of Native Studies at Trent University in Peterborough where he taught popular courses in Iroquoian culture and languages. He retired in 1991 and began teaching language and culture classes at Six Nations. In 1993 he started the Jake Thomas Learning Centre from his home. This is where I first met Jake in 1995 when I attended a week-long teaching of the cycle of the traditional ceremonies. This experience changed my life focus and began a deep and respectful affection for Jake.

During a sabbatical from my architectural practice in Toronto, precipitated by an economic downturn and a personal crisis of loss in the AIDS epidemic, I had assumed a teaching position in the architectural faculty of Lawrence Technological University outside Detroit. My mother, a Mohawk from Six Nations, and my father, a British/American, had met in Detroit where my maternal grandfather had taken the family to avoid the Canadian residential-school system. Over time I became acutely aware of my Native origins and determined to explore these connections more deeply in a doctoral program offered by the California Institute of Integral Studies in San Francisco. There I was asked to return to the traditions of Six Nations, which opened the door to a relationship with the highly reputed elder and teacher Jake Thomas.

Jake gave me the responsibility of learning the Thanksgiving address, of assisting in his recitation of the Great Law in 1996, and of becoming a travelling companion in his duties across our traditional homelands and beyond, work that included retracing the locations and ceremonies mentioned in the story of the Peacemaker. I witnessed Jake's teaching and participated in his Condolence ceremonies, his carving of sacred Hadoui masks, and his interpretation of the wampum belts, while also attending gatherings in his home of visiting Native traditionalists and researchers. I spent many hours simply sitting with him, sharing meals with him and Yvonne, helping out in the garden and sugar bush, and assisting him in the gathering of medicine plants and spring water. In this way, the circle of a long journey was completed, and I surrendered my time and life to assisting and accompanying Jake in his work until his death.

In the midst of this work, I occasionally asked him who would carry on the practices and knowledge of our culture when he was no longer with us. He would invariably respond, "You're here!" Mentioning the great white pine of the Great Law, Jake felt the burden of holding up the tree, but he trusted that we, his students, would help raise this tree once again. Jake passed away suddenly at his home in August 1998 at the age of seventy-six. Then I too was left on my own, as he once was before me. As I returned to active life as an advocate of traditional Native culture and practising architect, I felt the estrangement that has been the hallmark of the Indigenous experience in the West. These were timeless days which continue to nurture my present mindfulness.

Slowly, new openings to the work I needed to do emerged, including a teaching assignment in the architecture program at the University of Waterloo and my serendipitous meeting with Tim Leduc. Jake's call for help in holding up the white pine became central to our work together, something that can be seen reverberating through *A Canadian Climate of Mind*. The wisdom of the ancestors must continue to be respected and passed on to future generations. This is how I mind my work and the writing of Leduc in these pages.

In the diverse stories of Native-settler relations shared in this book, I began to see the grand climatic, ecological, ancestral, and cosmological context for my own life. I hope that everyone who partakes of these narratives, like myself, will see their story too. In this critical moment of change in how we mind relations, Leduc reminds us, compassionately and compellingly, of our common human inheritance for peacefully seeking fresh paths toward adaptations that have continually allowed us to live sustainably, nurtured by our Mother, this Earth.

Acknowledgments

A Canadian Climate of Mind is about the role of Thanksgiving in living in a good and sustainable way, and consequently it offers recognition and thanks throughout to various significant ancestors, elder teachers, animal and tree presences, places, and cosmological beings. In a sense, the writing reflects all whom I want to acknowledge in supporting this work and my life. That said, there are some specific people I want to highlight here. First, the teachings of William Woodworth Raweno:kwas have become like a beacon that continually opens me out to the depths of these lands, and thus through him I must also thank his teacher Chief Jacob Thomas. Various people read drafts of chapters and provided invaluable advice, including William Woodworth, Barbara Clark, Mark Dickinson, David Banerjee, Ray Rogers, Christina Lessels, Stephen Scharper, Mora Campbell, the two anonymous external reviewers, and my editor, Mark Abley. My children, Étienne and Iona, have provided an overall familial climate of energetic joy, mischievousness, and, at times, turbulence that could not help but influence how I approached each day of writing. I am thankful that they have both found their way into being symbolically represented in the passages of this book. My partner, Christina, offered comments and encouragement throughout the writing and I am grateful for her support. Finally, I give thanks to all my relations, who have taught me what it means to mind relations in a good way, even during times of great climatic change.

A CANADIAN CLIMATE OF MIND

Oh! Stranger …

Oh! Stranger look with reverence.
Man, Man! Unstable Man,
It was thou who caused the severance.

Above my head a great blue heron flies in a southerly direction from the wetlands a few hundred yards to the northwest as I make my way up the oak-, sassafras-, and staghorn-sumac-lined ravine paths that weave their way to one of my favourite spots in Toronto. For more than ten years I have stopped at this high ravine ridge in the city's west end to take in its view of the mouth of the Humber River entering Lake Ontario. It is not simply natural beings that stop me in my tracks here, for a stark human presence also strikes me, particularly during dusk of a winter evening. At this time of year, dusk arrives just as the rush hour home fills the roads, as can be seen a hundred yards below me in the sixteen lanes of traffic, two streetcar tracks, and railway lines that energetically hug the lake with streaks of light. The shadowy bush around me is clearly severed from the lake's dark glittering waters by an urban necessity driving through it.

A black iron fence separates me from the tomb of John and Jemina Howard, the nineteenth-century colonists who donated much of the land that is today the 151-hectare High Park in Toronto's west end. My current view is quite different from the mid-1800s sightline that John Howard looked upon when building this tomb. His paintings show only a country path and railroad intervening between the lake and the Howards' Colborne Lodge behind me. The name High Park was given to this place because of its view over the lake and the mouth of the Humber River just to the west.[1] In deeding this land to the city

1 *Howard Tomb in 1874–75*, by John George Howard. It is at this tomb, a century and a half later, that the author began to reflect on some of the book's central concerns.

in 1873, the Howards gifted to us a remnant sense of the Humber Plains, a habitat running along Lake Ontario from the Humber River to the eastern edge of this park and stretching nine kilometres north.[2] For me, it has become a place of natural repose in an urban reality, a place where I often feel a sense of reverence well up from its contours and presences.

But it is not simply an ecological gift that has been passed down to the present, for the colonial period of the Howards also inaugurated the industrial and urban changes to human, ecological, and climate relations that are so apparent in the traffic below. The "Oh Stranger" epitaph that is etched on the gate before the Howard tomb speaks to a Christian original sin that informed nineteenth-century Toronto, yet it also resonates with the fast-paced severance below. Beyond the lights that can be seen are the greenhouse gases of an urban metabolism that

2 Two Row Wampum held by William Woodworth Raweno:kwas (spring 2015). Picture was taken along the Humber River or what is introduced as the *passage de taronto* in chapter 2. The Two Row Wampum is engaged in this book as means for minding Canadian-Indigenous relations in a climate of change.

intimately connects my ways of living in a city like Toronto with today's climate changes. I see this in the surrounding blackness not because it is observable on this cool night but because I have been contemplating the human role in such changes as they manifest in northern warming for much of the time that I have been coming here.[3]

Melting glaciers and ice, shifting lands and water, changing animal patterns, and impacts on Inuit culture have become established features of life in a warming north. Elsewhere around the planet some regions are being challenged by drought, others by the uncertainty of extreme hurricanes, and island states and vulnerable coastal areas by the threat of flooding deluges. Within this urban place that has become my home, unprecedented ice storms, flash floods, and record summer heat waves have begun to impress on us that our modern drive is moving the planet toward an uncertain future. While these significant changes are most easily observable out in the world, it is becoming clear that the most important locus of change needs to be in our ways of relating *with the world*. As Naomi Klein succinctly puts it in her popular account of climate change, we have been given "a civilizational wake-up call" that is speaking powerfully "in the language of fires, floods, droughts, and extinctions."[4] And what is it saying? We need to renew our ways of minding relations by reimagining "the very idea of the collective, the communal, the commons, the civil, and the civic."[5] Our planetary climate is teaching us about the social nature of what we live within, and the need for us to change in response to these lessons.

There is an odd beauty to my view on this speeding trail of red surrounded by darkness that draws forth from me feelings of reverence, though our climate of change colours them with more of a numinous tinge than I like. While an awe-inspiring experience of something holy tends to confirm our understanding of where we fit in the world, the numinous takes the form of a prodigious energy that breaks in upon our beliefs, rationalizations, and expectations with an unsettling impact. That which manifests numinous energy "can only be evoked, awakened in the mind; as everything that comes 'of the spirit' must be awakened."[6] In our case, we are being awakened to the realization that we are part of something much bigger than our modern minds have been able to appreciate. It is not simply that the surrounding world is becoming more uncertain today, but that those changes are asking for significant shifts in how we mind relations. That is what the darkness beyond the perpetual light of urban living evokes in me as I stand on this ravine ridge and attend numinous presences beyond the sensible.

The epitaph's call for the "stranger" before it to "look with reverence" leaves me with the uneasy feeling that there is more to our interconnected climate, ecological, energy, economic, and social crises than that which is suggested in the common debates of our day. Watching the greenhouse-gas-emitting spectacle drive by me brings to the fore questions about the modern love affair with technology and the continuing faith that it can solve our current crisis in ways that require less than substantive cultural changes. Intensive development of Alberta's oil sands, fracking and offshore drilling, the optimistic promise of carbon sequestration, proposals for renewable energy – in all of this the predisposition seems to be maintaining current ways of living through technological advances and efficiencies. Underlying this tendency is the endless search for economically viable responses that can allow the economy to grow endlessly. The economic crisis that reared its head in 2008 has made it apparent that modern nations like Canada have little desire to live without the kind of linear growth that has become an expected norm since the Industrial Revolution of Howard's time. Even in a context of so much uncertainty, the movement toward re-minding modern ways of living has largely been limited to tinkering around the edges.

The interconnected issues of climate change and energy use are central to what follows, but they are not the overarching social context to which the book's title, *A Canadian Climate of Mind*, refers to. Of broader interest is climate as a kind of social and cultural milieu that informs how we, particularly Canadians, mind relations with an increasingly numinous surround. Beyond meteorological phenomena, climate also refers to a "prevailing trend of opinion"[7] or milieu that has both physical and subjective connotations.[8] The approach I am taking is also consistent with a climate science that is not only revealing an interactive web of physical relations but also highlighting the increasing capacity of humanity, since the Industrial Revolution, to evoke planetary responses.[9] A climatic milieu has always surrounded humanity, and this is again becoming clear to the modern world as we witness the impact of "our inadvertent and unwanted agency."[10] The current crisis that rushes past us everyday is not just climatic or ecological but rather envelops the modern ways of minding our energetic relations with each other and presences beyond the human.

Standing on this high ridge, I am impressed with the need today for making our drive less a "stranger" to the living world. We need to renew a reverence and humility that can ground our great scientific knowledge, economic policy options, and technological innovations. This book is about that social challenge and whether Canadians and, more broadly, humans can devise radical, courageous, and creative ways of being mindful of our many relations. Reflecting on this moment of change, Bruno Latour writes that "we had never been modern, but now we are even less so – fragile, frail, threatened – that is, back to normal, back to the anxious and careful stage in which the 'others' used to live before being 'liberated' from their 'absurd beliefs' by our courageous and ambitious modernization."[11] Yet, despite all the signs of why radical cultural change is required, "we seem to cling with a new intensity to our idols, to our fetishes, to our 'factishes,' to the extraordinarily fragile ways in which our hand can produce objects, and over which we have no command."[12] Our modern "factishes" are the deep source of continuing debates around climate change, fuelling their endless twists and turns, producing restrained responses to a numinous energy that is only beginning to bear down on us. It is this social climate of change that introduces itself to me each time I look out from this spot at the Howard tomb.

Untold rituals often inform the writing of a book. I am referring to practices and experiences that help move thought in directions that often seem unthinkable at a computer screen in an enclosed room but that usually receive little, if any, mention. For some it is walking or jogging, others have a drink at the pub, some practise art, others congregate in ceremonies, there is meditation, and then there are the anecdotes of problems solved in dreams. In the history of environmental thought, there is also a long tradition of engaging nature as a source of inspiration. From Henry David Thoreau to Aldo Leopold to Rachel Carson to Richard Louv's more recent *Last Child in the Woods,* land experience is consistently touted as essential to sustainable ways of thinking and living.[13] It is an old environmental ritual that has even more ancient colonial and Indigenous ancestries in Canada's diverse lands, and it is an experiential approach that grounds the telling of *A Canadian Climate of Mind* in acts of walking, jogging, contemplating, and writing in wooded ravines and savannahs, along river and lake shorelines, and at urban-nature intersections like the lighted one that drives past the Howard tomb.

Though it is generally assumed that more remote wilderness is where one goes to inspire ecological reverence, urban settings can inspire thought in much the same spirit as Thoreau's Walden Pond or Carson's coastal ecologies. This was something that the deceased Canadian naturalist John Livingston knew well. Before heading the Audubon Society of Canada in the 1950s, being a founding producer of the CBC's *The Nature of Things* in the 1960s, teaching at Canada's first Environmental Studies program starting in the 1970s, and writing various classics in Canadian environmental thought, he was initiated into the life of a naturalist by the ravines along the Don River and the marsh ecology of Ashbridges Bay in Toronto's east end. Reflecting on the marshes of his childhood, Livingston wrote there was no other world for him, "nothing beyond shimmering light on water, smooth clean muck, green plants, trickling sounds, flickering tadpoles, living, being."[14] These early experiences not only instilled a sense of reverence in a young boy but also inspired his later insights on what is needed to sustainably mind ecological relations in our fast-paced time.

The words of Livingston often reverberate in my mind as I traverse the ravine paths of the Humber Plains and come across great blue herons, ruby-crowned kinglets, goldfinches, foxes, coyotes, an occasional deer, and other beings in the midst of Toronto's urban reality. A central premise of his writing is that coming into relation with ecologies like the Ashbridges Bay of his childhood or the ravines surrounding the Howard tomb can, even in this city, transform and heal modern ways that are enmeshed in our emerging crisis. In *Rogue Primate*, which won the 1994 Governor General's Literary Award for non-fiction,

Livingston described these essential experiences as "the dissolution of the ego-centred self, as when one was drawn close, ever closer and at last into the gold-flecked eye of a toad."[15] Wildlife conservation and environmental studies were for him not necessarily activities but "state(s) of mind."[16] From the marshy flight of a heron to immersion in a ravine forest, relations with the beings in our presence and their ecological contexts can transform our minds. It is a view that has precursors in the likes of Leopold and, at the dawn of the environmental movement, Gregory Bateson's influential *Steps to an Ecology of Mind*, which proposed that "the mind is not limited by the skin."[17]

The immersive sensibility of Livingston's thought frames chapter 1 with a Canadian focus, though I bring him into dialogue with others and a responsive climatic milieu that inspires the book's title. We start by considering thought on an ecology of mind because that is where environmental thinkers began attempting to reconnect modern ways of minding relations with a broader milieu. But our sense of the human relation to climate change and energy has, we will see across the book's chapters, added dimensions and insights. The climate is not only the context of ecologies and their many biological relations but is also deeply interconnected to changes in ice cover, ocean temperature and circulation, the shape of land, and, as we are increasingly coming to understand, human cultures. With the dawn of the Industrial Revolution a short two centuries ago, a modern globalizing culture was born that has increased its capacity to energetically communicate with the global climate in disruptive ways that were for most people unthinkable fifty years ago. Our greenhouse-gas emissions are evoking numinous responses from the surrounding planetary climate, and we are now being asked to respond in turn.

At the Howard tomb, our climatic milieu is further defined by the way in which the dimensions of Livingston's ecology of mind resonate with the epitaph's haunting second stanza: "Man, Man, Unstable Man! It was thou who caused the severance." His childhood initiation was triggered not only by relations with "hundreds of kinds of birds, plants, butterflies and other forms of life" but also by the shock of having those relations suddenly erased by Toronto's urban need for a sewage- and water-treatment project.[18] What he witnessed was the latest development in what was then North America's largest lake-filling project. Looking back, he wrote of being in a paradoxical position where the "things I value – such as birds – are being destroyed by other things I also value: human life."[19] What troubled him even more was the feeling that our allegiance must always be with the modern human as felt relations with other beings are severed. Speaking of a similar dynamic that was silencing birds, Rachel Carson concluded that the "'control of nature' is a phrase conceived in arrogance, born of the Neanderthal age of biology and philosophy, when it was supposed that nature exists for the convenience of man."[20] Livingston's paradox remains unresolved after five decades of environmentalism and over a century of conservation, and the reason proposed in these pages is that the roots of our crisis run deeper than what many modern "factishes" allow us to contemplate.

Shaping the vehicular passages that drive by me are a lake, ravine ecologies, and cultural histories that follow major regional rivers like the Humber, Don, Rouge, Credit, and their related creeks from the Oak Ridges Moraine north of the city. Waters like these have etched the development of Toronto, southern Ontario, and Canada in diverse ways, though, beyond anecdotes, they are rarely considered as the context for our present climate of change. The fast-pace asphalt of Lake Shore Road and the Gardiner Expressway that hugs Lake Ontario in front of me can lead us "to forget that we are riding on the ancient 'road' system of the Indians, the *coureur de bois*, and the traders."[21] Winding roads like Davenport along the ridge of former Lake Iroquois follow the historic meanderings of old Indigenous paths. A few streets east of High Park is the former trail today known as Indian Road. Meanwhile, names like *Toron:to* (Haudenosaunee/Wendat), *Ontar:io* (Haudenosaunee), and *Kanatha* (Haudenosaunee/Wendat) speak to the original influence of diverse Indigenous cultures on the city, province, and nation, all of which are contemplated in the chapters that follow as inherited reminders of urban life's power to sever us from the past.

In 2012 the Mohawk teacher and architect William Woodworth Raweno:kwas wrote of a vision of Toronto's skyline that came to him while at Ashbridges Bay some eight decades after Livingston – one that introduces us to the colonial dimensions of severance and reverence that are of concern in this book. On one of his regular walks along the Lake Ontario shoreline during which he offered Thanksgiving to the creation, Woodworth looked across the water and saw the city's core in a different light. The Canadian National (CN) Tower that illuminates the night sky to the east of where I stand was reimagined as "an Aboriginal sending and receiving device, a kind of prayer staff reaching out for the aid of the unseen spirits who assist us, even unconsciously."[22] Below it were the "circular gathering place that opens to the sky field, the great informer of the Ancestors, formerly named SkyDome," and to the east four skyscrapers at the heart of the financial district that conform to the colours and placing of the Haudenosaunee medicine circle.[23] While Woodworth offers a uniquely Haudenosaunee perspective, Victoria Freeman writes that his "orientation to the past is broadly discernible" in the unique cultural interpretations of the Anishinaabe and Wendat who also have Indigenous roots to Toronto.[24] We will come to see that the land's ancestors are present despite the modern drive to progress, and this is as true for Toronto as it is for anywhere else in Canada.

What I find striking in Woodworth's re-visioning of the city as I stand at the Howard tomb is the paradox it seemingly shares with Livingston's Ashbridges Bay experience, though of a different cultural sensibility with longer historic relations to these lands. He speaks not only of an unconscious impressing of the Indigenous past on the forms of contemporary Toronto but also of the simultaneous violent disregard the city displays for its ancestors.[25] It is a view he expresses when describing the Humber River site of the Haudenosaunee village of Teiaiagon, northwest of this tomb, "as a sacred place, sanctified by the horror of what happened there," with the surrounding English houses built over the past in a way that epitomes colonization.[26] From the Humber Plains in the west end to Toronto's CN Prayer Tower and Medicine Wheel Skyscrapers to Ashbridges Bay in the east, the land's Indigenous ancestors underlie a present that at best has forgotten and, at worst, actively denies their continuing presence. In the words of Freeman, this city "is a place where the colonial past and the people affected by it often appear to be completely absent, as if colonialism never happened – or perhaps as if it had been completely accomplished."[27] Such critical

Indigenous perspectives enlighten another dimension of the severance that we will contemplate, a darkness of mind that informs our drive as we speed past any sustained awareness of Indigenous, colonial, ecological, and climate presence.

Rattling its way between and past the cars is a passenger train that draws my attention to Canada's historic industrial track that accelerated these colonial dynamics and moved us toward our climate of change. The book's subtitle, *Passages from Fur to Energy and Beyond*, foreshadows a concern with the way in which Canada's relation to energy is intimately connected to the nation's colonial resource tradition. This is an issue not simply of politics and economics but also, extending the work of Harold Innis, of travel, communication, and culture. Innis made the point that Canada was moulded by a staples economy which "organized transport over wide areas especially adapted for handling heavy manufactured goods going to the interior and for bringing out a light, valuable commodity."[28] Our institutions have been centred on the transport of resources, symbolically epitomized in the CN Tower to my back. The trans-Canada railroads followed old waterways like Lake Ontario and the Humber River that once brought fur and other resources from the interior to European and then American markets. It was one of these iron tracks that Howard painted from this view at a time when it and a path were all that separated lake and ravines. From canoe to railway to pipelines, the process and the resource have changed while the overall "factishes" have remained steady. Returning to Klein, our climate of change challenges not only "capitalism, but also the building blocks of materialism that preceded modern capitalism, a mentality some call 'extractivism.'"[29] The never-ending debates around Alberta's oil sands and national responses to climate change are connected to Canada's historic enactment of colonizing severances, to which I will add dimensions in each chapter.

Though severances are a central theme in what follows, the book's primary focus is on the re-emerging sense of ecological, climatic, and social milieus that *have* in the past and *are* currently calling for more reverential ways of minding relations. This is what the Ashbridges Bay experiences of Livingston and Woodworth both point toward, though from two unique cultural positions. Informing Woodworth's vision is a Haudenosaunee Good Mind tradition that is introduced in chapter 2 and offers both a critical understanding of colonial processes and dutiful ways to reverence like the Thanksgiving ceremony. In a spirit resonant with Livingston's ecology of mind, his tradition teaches that people should be humble in recognizing that "real intelligence isn't the property of an individual – the real intelligence is the property of the universe."[30] From an Anishinaabe perspective, Deborah McGregor similarly explains that, in her culture, knowledge and "stories are gained from animals, plants, the moon, the stars, water, wind, and the spirit world."[31] The lived experiences of Livingston and Woodworth embody differences that are vital for contemplating various dimensions of *A Canadian Climate of Mind*.

In many ways, my approach is inspired by the Kaswén:ta or Two Row Wampum teaching of Woodworth's elder teacher and now ancestor, the former Cayuga Peace Chief Hadajigrenhta (he makes the clouds descend) Jacob Ezra Thomas, who died in 1998. A treaty dating back to 1613, the Two Row represents two beaded rows of blue in the midst of white as symbols of an Indigenous canoe and European mast ship going side-by-side down their common river. We need to respect the path of each while simultaneously recognizing their grounding in the common river of life.[32] Colonial dynamics have disrupted the canoe by actively attempting to force everyone to live the way of the modernizing ship. At the same time, the increasing power of this ship has also disrupted the common waters, land, and climate upon which respectful relations depend. The traditions represented by Livingston and Woodworth have recognized the relation of these changes to the ship, one from within the ship and the other from the position of the canoe. They offer different perspectives on the root causes of our situation and what is needed to mind relations differently, but they also come together in vitally important ways. By bringing the wisdom of these traditions into dialogue, they offer unique insights on practices for a good minding of relations. My goal is not to reduce difference but, as with the Two Row, respect diverse perspectives on our common waters while recognizing points of commonality where a dialogue can bring us into more just relations.

Our climate of change is asking us whether it is possible to find a way back to sustainable relations with each other and our common source. As such, we also contemplate in this book those signs in Canada's history of a third approach that is situated between the ship and canoe and yet has the same Two Row intent of respecting our common waters, land, and climate as a reflection of respect for each other. In many ways, I will be building upon the reimagining of Canada's colonial history offered by people like John Ralston Saul when he describes the nation as "a métis civilization" of people, institutions, and sensibilities profoundly influenced and shaped by Indigenous nations, despite appearances.[33] Though the first three centuries of contact were marked by conflict and missionary conversions, colonists and Indigenous peoples often came together with a sense of respect and equality.[34] Mediating these relations were the sublime ecologies and climates of Canada that could not be conquered in the same way that those of Europe and the United States were. Rather, Saul writes that settlers often survived by compromising "with the Native population and with the demanding requirements of a society which was dominated by the difficult place in which it was found."[35] Woven in the common river of the Two Rows was the potential of a métis or braid, a relational possibility that informed the historic emergence of a distinct Métis Nation.[36] It is a potentiality that is arising again as the climatic waters become more turbulent and the modern ship less sure of itself.

Something across the traditions of Woodworth and Livingston coalesced into calls for me to slow down, walk, sit, and be more aware of the possibility that the ecology and climate of places can think us in phenomenal ways; that reverence for our surrounding relations can change our minds. This is what Woodworth seems to do when the CN Tower metamorphisizes from simply a symbol of colonial Canada into a Prayer Stick for renewing relations. The power of our cultural drive requires us to renew ways of living in places that have often been significantly disrupted, changed, and severed.[37] It is, one hopes, starting to become clear that minding relations means more than the thought inside our heads; it is about how our worldly relations are in a sense transforming how we think and live with each other. Very different questions emerge from this perspective: Can we discern a climate of thought from the places we live in? And if we can, what can we learn about good and sustainable ways of minding relations?

Questions of this nature inform all seven chapters, though each is centred on particular places that offer unique insights on broader national and global dynamics underlying today's climate of change. While the first two chapters begin in Toronto with Livingston in Ashbridges Bay and then Woodworth along the Humber River, subsequent chapters move beyond Lake Ontario to uncover potential footholds where ecology of mind and Good Mind can be braided for a deeper sense of what social change looks like. Each place will offer interdisciplinary perspectives on what it is to mind land and climate in Canada,[38] scholarly thought on the interdependence of mind with broader surrounding ecologies and climates, and Indigenous understandings. Throughout there is a concern with the way in which nature can inspire, even in milieus of ever-intensifying severances. Any of us can consider this tension in the places we inhabit if we slow down long enough to become intimate with our surrounding relations, while also remembering to reflect on our participation in the severing changes observed at the Howard tomb. A sense of reverence and severance needs to be held in our being as we engage the world at this time of change.

It is the striking presences of each place that guide the passages of *A Canadian Climate of Mind*, which on this high ridge include greenhouse-gas-emitting cars, the Humber River, Lake Ontario, an urban forest ravine, a heron overhead, the railroad, a CN Tower Prayer Stick, and the Howard tomb epitaph – all of which are contemplated in the following chapters. The places we become familiar with are filled with human, ecological, climatic, ancestral, and cosmological presences that are calling us to different ways of relating. To say with Livingston that ecology is a state of mind is to recognize one's participation in a larger mystery that must be recognized in our thought and being, what is referred to in these pages as minding relations. This is also what Woodworth was doing when he had his Toronto vision while practising the Haudenosaunee ceremony of Thanksgiving, a feature of the Good Mind that recognizes that "we need each other" and owe thanks to the Earth, the "grasses, waters, trees, plants, winds, the moon, the stars, the sun, the universe, the whole thing."[39] Acting with thankfulness and reverence is vital to recognizing that our "intelligence is the property of the universe," that our energy for change comes from far beyond the resources dug up from the earth. To energize good ways of minding relations, we may have to relearn how to be responsive to ever-changing climates that remind us of our ancestral relations and duties.

Though I did not start writing this book with the explicit intention of bringing an ecology of mind into dialogue with the Haudenosaunee Good Mind, the latter tradition gradually took on more and more importance for me. The Ashbridges Bay vision of Woodworth arose from the guidance of his teacher Jacob Thomas, and it is this particular ancestral lineage as embodied by Woodworth and other students of Thomas that I follow and to which I am indebted. That said, the places found herein have diverse Indigenous ancestries, and as such we do at times consider Anishinaabe, Wendat, and other understandings as described by people situated in those traditions. In my recognition of the central role of Indigenous voices, it is not my intent to appropriate. Following the respectful spirit of the Two Row, I engage Indigenous voices and terms in relation to these places as means of disrupting common modern factishes about what it is to mind relations in our climate of change. In this regard, a glossary of both Indigenous and braided terms is provided at the end of the book as an aid to readers. The hope is that these terms can deepen our sense of the change that is before us, and thus suggest other passages toward a sustainable and good climate of mind.

As I refocus my eyes on the tomb's epitaph, I cannot help but contemplate the presence of death and ancestors on this darkening ridge and in this book, even as our lighted urban ways rush past such an awareness. Thinking about the vital role the dead have historically played in cultures across the planet, Robert Pogue Harrison writes that dialogue with the ancestors must be ongoing so as to "keep open the possibility of a 'reciprocative rejoinder' that never simply denies but freely avows or disavows the will of the ancestors."[40] It is a view resonant with Woodworth's valuing of ancestors, who serve to remind us of duties like Thanksgiving that need renewal in the place and time we occupy. From a slightly different perspective, Livingston writes that his "interpretation of 'self-actualization' is death and recycling ... the sensation of release from life-long mental slavery."[41] It is as though the severance at the core of the modern ship is reacquainting us with those dark spectres that eventually subsume everything. If we can renew a dialogue, the ancestors may still have much spirited wisdom to offer us on the painful nature of our climate of change, and the question of how to navigate these numinous passages to a good future.

1

When the Pain Started

There was behind my parents' house a city ravine, with a little stream
running through it ... There was no world whatever, outside that world ...
nothing beyond shimmering light on water, smooth clean muck, green
plants, trickling sounds, flickering tadpoles, living, being. That was when
the pain started.

John A. Livingston (2007/1981), 129

With the approach of spring in Toronto's west-end ravines, I feel
closer to those wild places where John Livingston "sought, and
found; when one relinquished, and was free."[1] At this time of year,
the ability to see with any depth is increasingly obscured by green
growth, especially as the season progresses and one moves within the
forest-covered areas above the restored wetlands of Grenadier pond
in the Humber Plains. Around its northwestern marsh, the golden
reeds of fall and winter backdrop April's emerging green shoots at the
open water edge. The red-winged blackbirds sing their high pitched
conk-la-ree calls, and up the ravines a layer of green buds on the
younger small trees begin to provide a low-level cover. When I return
a few weeks later in May, the budding leaves offer more visual and
auditory buffers from the surrounding hum of traffic. The oriole adds
a new tune to a bird and insect chorus that draws my awareness
deeper into a minding of the marsh below as the sound of the urban
severance recedes to the south – except for the sporadic but common
piercing of frustrated horns and relentless construction from around
its edges.

In early-twenty-first-century Toronto there are few natural spaces
where the seeming primacy of human activities and ideas feels secondary to surrounding ecological and climate presences, and this ravine

3 Toronto marsh with a great blue heron (summer 2015). It is places like this
restored west-end wetland that help me mind the experience of John Livingston
in the marshes of Ashbridges Bay and the ideas that flowed from it.

on the edge of a wetland is one of those rare places where I experience
fleeting moments of, in Livingston's words, "nothing beyond shimmering light on water."[2] I am not alone in feeling called out to green
spaces like this as spring makes its presence felt. With winter releasing
its grip, the land comes back to life and people share in the energetic
renewal, as witnessed in the spring of their step and their joy in being
outside. Suddenly places that were hushed for months are filled with
human conversation and recreation, and surrounding it all is a burst
of green and birdsong. Though we may not all have natural spaces
where we go regularly, many of us do have places where we seek
renewal – from community temples to recreation centres to our bed-
rooms. Here the stress of our modern rush can be released and we can
feel a little less bent out of shape by the social pressures around us.
There is something about being in relation with ecological presences
that is particularly soothing to our driven lives, even if we are only
half aware of why we feel such a desire to get out of our homes and
experience something bigger than our human communities. What I
learned from the life of Livingston was that we can find these spaces

of renewal in an urban place like Toronto; we just need to take some time to become familiar with what is around us.

While Livingston can be called a conservationist, environmentalist, teacher, and author, he primarily defined himself as a naturalist.[3] Of particular interest to him were, in the words of his friend and Canadian author Graeme Gibson, "the birds: he obviously loved them best."[4] It was a love that started early and was then sealed when he witnessed the rare least bittern, a small heron, driven from the Toronto marshes of Ashbridges Bay when a much-needed sewage treatment plant was built.[5] Experiencing the yearly visits of bittern, great blue heron, egret, various species of duck, red-winged blackbirds, and a host of other marsh birds allowed him to feel imprinted on this place. The eventual loss of these relations felt like something being cut "out of my stomach," as if the extracted ecology was a part of him. In describing this experience in two of his books, *One Cosmic Instant* and *The Fallacy of Wildlife Conservation*, Livingston highlights it as a defining moment in his life, work, and thought. The stark paradox was, in his words, that "things I value – such as birds – are being destroyed by other things I also value: human life."[6] It was a loss without and within. This breaking down of internal and external worlds would come to inform his later definition of wildlife conservation as "a state of mind," one that connects the conservation of ecologies to a mind deeply affected by life in an unsustainable society – a modern mind in need of a spring-like renewal.

The twentieth century saw many conservationists and then environmentalists have similar agonizing experiences in far-removed ecologies. Around the same time that Livingston underwent his initiation, the American conservationist Aldo Leopold was reflecting back on an experience he had while working with the American Forest Service years earlier. In the essay *Thinking like a Mountain*, he writes of watching "a fierce green fire" die in the eyes of an old wolf he had shot, and in that moment seeing something that was "known only to her and to the mountain."[7] He had been participating in the then common activity of attempting to eradicate wolf populations. With the dying green fire of one wolf etched in his mind, the subsequent years of deer increases due to wolf population declines and defoliation of mountain ecologies led him to realize that unless we

4 Traffic along a sunny Lake Ontario (summer 2015).

learn "to think like a mountain" we will inevitably "have dustbowls, and rivers washing the future into the sea."[8] It was experiences of this nature that inspired him to promote a "land ethic" that situates the human in the community of "soils, waters, plants, and animals, or collectively: the land."[9] Though the ecology and beings of a mountain are clearly different from those of a marsh on the edge of urban expansion, their common message is that a change in human ways of thinking and being is needed.

Just as the ideal of managing a mountain without wolves could not keep up with reality, Livingston's experience arose in the midst of

5 Traffic along a dark Lake Ontario (summer 2015). It was the encompassing
drive of a predominant technological experience that concerned John Livingston's
ecology of mind, something that I can appreciate when viewing or participating
in this kind of drive along Lake Ontario. The alternative future path of renewable
energy like wind is seen in the background, something that is talked about in later
chapters in relation to the issues discussed here.

calls for better water management. The Toronto population expanded
from 31,000 in 1850 to 400,000 in 1912, with Ontario's city dwellers becoming the provincial majority in 1911.[10] An urban reality was
expanding around the city's waterways at the turn of the twentieth
century, as about ten tonnes of untreated sewage were dumped yearly
into Lake Ontario through streams.[11] This practice prompted the

outlawing of dumping in 1908 followed by a series of improvements. Between 1912 and 1918 most of the increasingly polluted Ashbridges Bay wetland was filled by excavated material from Toronto's industrializing expansion to create the port lands.[12] The progression of this work was the source of Livingston's painful paradox as unsustainable ways of minding water relations led to more extreme management of the water system and reduction of biodiversity. His wound was the result of being intimately situated between two communities, human and marsh. The tension initiated a life focused on the need for a different "state of mind," one that could balance human needs with an awareness of broader relations.

We all have comparable experiences of loss that leave us raw on the outside, deflated within, and searching for the meaning of it all. The death of a loved one, an uncontrollable move from our home, a colonial severance from land and culture, an intimate possession stolen – we can each populate such events with our own personal and cultural remembrances, if the feelings of hollowness and tear-singed eyes have not been buried in denial. How do we mind relations that are lost to us and yet paradoxically deeply a part of who we are? Such losses can become a kind of ghostly presence that comes to inform our ways of interacting with the world, even when long forgotten. This is what seemed to happen to Livingston as the Ashbridges Bay wound initiated his life work. As we will see, the ecology of mind tradition that emerged in the latter half of the twentieth century is in many ways a response to the losses accrued by the unsustainable ways of the modern ship on the Two Row. Guided by Livingston, this tradition is a vitally good place to begin learning about some of those practices of a good Canadian climate of mind that are concerned with finding our way through the pain, not around it.

Less than a century after Livingston's experience, I am part of the more than 80 per cent of Canadians who now live in urban centres, with the Greater Toronto Area (GTA) population of five million being the nation's largest. Surprisingly, in the midst of its ever-expanding human presence, power structures, concrete, asphalt, and rising greenhouse-gas emissions, a different ecological place grabbed hold of my being. Following Livingston's example, marshes, ravines, and significant urban-ecology intersections like the one highlighted at the

Howard tomb gradually became places where I attempted to ground a driven urban life whose pace and intensity often feels unsustainable. Rivers, forested ravines, parks, a lake, extensive trails, and other natural areas continue to give Toronto "some vestiges of its original wildness."[13] That said, this place has clearly experienced significant progressive change over the past two centuries. From noise and air pollution, to reduced habitat, to human-induced smog and climate changes, the urban rush continues to expand and quicken since Livingston's 1930s initiation. We live in a time etched by wildly painful paradoxes that are intensifying and pulling us into more intimate relations with the creation around us. These paradoxes have the potential to move us toward a more sustainable ecological and climate consciousness, but only if we can look at our role in the pain. Our introductory guide to this reminding of relations is Livingston, and we begin in Ashbridges Bay before edging out later in the chapter to a consideration of what this discussion means for our current climate of change.

Why would anyone want to live in Toronto? This question reverberates through my mind not only because of Livingston's Ashbridges Bay initiation but because it is tied up with my earliest sense of this city. For someone who grew up in rural eastern Ontario, not far from the Saint Lawrence River and near the Quebec border, the question not only speaks to my general childhood impression of Toronto but is also a variation on comments I still hear from family when I return to where my life began. I grew up between Montreal and Ottawa, both of which were a short hour drive away compared with Toronto's four, though even these short trips were rare as a child. When we occasionally made our way toward Toronto, it was a place to drive around and not into. The city seemed abstract, concrete, impersonal, too fast, distant, and estranged from any kind of nature, particularly the cedar bushes, crab apple tree groves, corn fields, and ponds that marked my childhood. It was a place of financial and political power with people who, at least from the stereotypical view of my hometown, saw themselves as the centre of the nation. Such were my rural biases when I reluctantly moved to Toronto for school in the mid-1990s with a hope that my time there would be short. An unanticipated two decades later, I have a very different feel for this place along the beautiful sparkling waters of Lake Ontario.

The flight of a white egret over my head on the rocky shores of today's Ashbridges Bay carries me to reflections on the love Livingston had for birds. It was a feeling that took hold of him as a child in these once avian-rich marshes and along the ravines of the Don River. In these ecologies he learned not only to attach "names to the hundreds of kinds of birds, plants, butterflies and other forms of life" but also to realize that the behaviour of these beings have a certain predictability.[14] In his last book, *Rogue Primate*, he recalls learning "the song of the wood thrush in my childhood," perhaps in one of the Don ravines, and how after that "the bird became my familiar and my friend."[15] Continuing this love letter, he writes: "My childhood experience of the bird, and its lifelong annual reinforcement, ensure that each spring's recognition of our relationship is more satisfying than the last. What I celebrate is not merely the existence of the wood thrush, however; it is my connection to him. My bond to him. My self in him."

With familiarity, the patterns of these beings can transform our thought. But with the ecological insight "that one must go to marshes to find bitterns, rails, swamp sparrows," Livingston also came to the devastating realization that when "some favourite location for a species is changed or eliminated by human activity, there are issues at stake."[16] The birds, in their presence and then absence, initiated him into the environmental paradox of our time: that to be human in a modern world is to participate in the destruction of that which nurtures and fosters life, even our own. It was clear to him there is a need to mind relations differently.

In the early decades of the twentieth century, the marshes of Ashbridges Bay spanned five square kilometres and had for centuries been one of the largest and richest marshlands for migrating birds and local wildlife in eastern Canada.[17] It was a wonderfully rich place for experiencing a myriad relations. One of the earliest written accounts of the area was by United Empire Loyalist surveyor Joseph Bouchette in 1793. He wrote of "the untamed aspect which the country exhibited when first I entered the beautiful basin," the "dense and trackless forests" lining the lake's shores, luxuriant foliage that sheltered two

Mississauga families, and close-by marshes like Ashbridges Bay that "were the hitherto uninvaded haunts of immense coveys of wild fowl."[18] The natural abundance of this bay was described in 1834 by writer Anna Jameson as "the haunt of thousands of wild fowl, and of the terrapin, or small turtle of the lake; and as evening comes on, we see long rows of red lights from the fishing boats gleaming across the surface of the water, for thus they spear the lake salmon, the bass and the pickereen."[19] A century later the young Livingston experienced a marsh still "full of wild animals, especially during spring and fall bird migrations,"[20] though its waters had undergone significant change.

In an appreciation following Livingston's death in 2006, Graeme Gibson remembered being puzzled at first by his passion and enthusiasm for birds. The common view in the 1960s, when they first met, was that birding was not a masculine activity, and yet Livingston "was an assured and forceful man."[21] A key moment in Gibson's change of mind about the importance of birds came "when a very large hawk unexpectedly soared out from beneath the Bloor Street viaduct" that goes over the Don River and ravines.[22] From this experience, he followed Livingston's example and began exploring local ravines and the lakeshore where he was humbled "to find that I'd lived more than thirty years without ever seeing the extraordinary vitality and beauty of the bird life around me." Others had similar remembrances after Livingston's death, such as George Archer, who recalled being led as a twelve-year-old by a teenaged Livingston into the ravines along the Don and Humber rivers on birding excursions from which he "developed a lifetime interest in birds and nature."[23] Gibson and Archer were neither the last nor first to be transformed by the teachings of this Canadian naturalist and by presences within these urban ecologies.

About a half-century before Livingston, another young boy by the name of Ernest Thompson Seton found a place in the same Don ravines where he "learned quickly that in the woods, 'the silent watcher sees the most'; drawing from nature forced the eye to see and the spirit to accept."[24] The eventual co-founder of the Boy Scouts of America, wildlife artist, and nature writer was initially dismayed when his family moved to Toronto in 1870 because of the seeming absence of nature in the city. But he soon found a ravine bordering the Don River that "wound its way, unseen, from the hills to the north, through the middle of Toronto, and south to Lake Ontario."[25] As with Livingston, the young Seton went to the ravines whenever possible, and what he experienced there became central to his thought and life. Reflecting on a particular crow he named Silverspot in his book *Wild Animals I Have Known*, he asks us to consider the importance of becoming familiar with wild animals and the rarity of such an experience at what was then the dawn of the twentieth century. He does not "mean merely to meet with one once or twice, or to have one in a cage, but to really know it for a long time while it is wild, and to get an insight into its life and history."[26] For both Livingston and Seton, the birds along Lake Ontario became central familiars and teachers about what it is to be human within something much larger.

Though minding the birds was central to the education of Livingston in a reverential sensibility, his Canadian ecology of mind was also deeply etched by the painful severing of the Ashbridges Bay marsh. This part of his experience was foreshadowed in a set of events that occurred around the same time that Seton was entering the Don ravines. Along the Taddle Creek that historically flowed through the University of Toronto campus before entering Lake Ontario just west of the Don River's mouth, a controversy arose in the 1880s about what to do with its increasingly polluted waters. A mere two and a half decades earlier, the picturesque Taddle Creek had been damned, creating McCaul Pond to the east of the campus.[27] The flooded ravine at first created an inspiring body of water, but that changed quickly. A rising population added pressure to the sewage system which was then centred on privy-pits connected to surface street drains that subsequently emptied into waterways like the Taddle. Acts of fishing, skating, and contemplating were soon replaced with "an evil odour" of raw sewage which went downstream and festered in the pond.[28] Creek and pond were increasingly seen as public nuisances, and the university decided to bury the creek and drain the pond.

The burial of urban waterways historically arose in the latter half of the nineteenth century amidst an increasing recognition of the relation between health and the water that in urban centres was being filled with human sewage. Many welcomed the university's participation in addressing the Taddle Creek problem. Before these changes and the addition of chlorine to drinking water, average yearly death rates in Toronto due to typhoid and related diseases ranged between 22 and 44 people per 100,000 – a number that subsequently dropped to 1 by 1915 and that is now negligible.[29] Some years were worse than others. The typhoid spectre "carried off over eleven hundred souls" in 1847–48, and cholera outbreaks led to 273 deaths in the early 1830s and 420 in 1849.[30] With the "scientific discoveries about the contagion routes of cholera, typhoid, and other infections in the 1880s, western European and North American cities began building thousands of filtration plants to purify domestic water supplies."[31] Toronto's construction of buried sewer lines began in the 1870s, and by 1891 there was approximately ten tonnes of untreated solid waste entering Lake Ontario on a yearly basis.[32] The 1908 by-law stemmed the dumping, and by 1930 there was over 1,084 kilometres of sanitary sewers, 104 kilometres of storm and relief sewers, a new pumping station, and a filtration plant.[33] A solution to human impacts on Toronto's water was pushed toward Lake Ontario, eventually requiring the modern transformation of the Ashbridges Bay that so inspired the young Livingston.

On the eve of Taddle Creek's 1884 burial below today's Philosopher Walk on the northern edge of the University of Toronto, some words were written by a graduate student that Livingston would have appreciated: "No more shalt thou behold the varied scenes upon thy banks, nor thy mighty influence. No more shall the sight of thee inspire noble thoughts. But thy work has been accomplished. Thou goest down to thy grave, unknowing and unknown."[34] Another anonymous student wrote that "here by the banks have dwelt a little space, And known and loved this mem'ry-haunted place," adding that these days have ended now that the city and university have decided to "make thee – Ichabod! – a common sewer! Taddle."[35] These words marked the end of a debate at the university on the value of maintaining Philosopher's Walk along the meandering Taddle Creek and McCaul Pond. At its moment of burial, the loss of the once inspiring stream and pond was seen by some as having the potential to degrade their thought, despite its increasingly polluted and potentially deadly state. The severance of the marsh and birds from Ashbridges Bay a few decades later similarly led Livingston to ask questions about the kind of ecological experience that is needed to sustainably mind relations with the world.

These were questions that arose from within the ecological fissures of an expanding urban system and way of living. What was humbling for Gibson was the extent to which, prior to his friendship with Livingston, he had scarcely paid attention to the birds in his midst. The structure of everyday urban experience had left him barely aware of "one of the great and ancient lineages of life."[36] Through the birds, Gibson came to realize his grounding in "an elaborate cultural construct that had very little to do with nature" – an idea that, he points out, is central to Livingston's thought and his Ashbridges Bay initiation.[37] Such a concern with the increasing disconnection of urban living to ecological relations has recently been popularized in Richard Louv's *Last Child in the Woods* as a kind of "nature deficit disorder" that is on the rise. His diagnosis is based on synthesizing research on the human costs and impacts of being alienated from nature, including "diminished use of the senses, attention difficulties, and higher rates of physical and emotional illness."[38] Decades earlier, Livingston similarly wrote about the way urban and technological developments are increasingly abstracting children from surrounding relations, with potentially delusional impacts for them, Canada, and the planet.

As one strolls along Philosopher's Walk today, it is easy to be oblivious to Taddle's watery inspiration, which now rushes through a concrete storm sewer below. For Livingston, this is not only because the creek and marsh have been removed from our sensory awareness, but also because the fast-paced technological realities of urban living further buries its flowing inspiration on an internal scale. Drawing upon Canadian philosopher George Grant, he situates the ecological crisis in a "technological imperative" that makes it seem abundantly natural to be constantly reshaping nature to human concerns.[39] His primary concern is not with technology in the form of tools that we use, but rather a technological minding of our lives that is making us more machine than human, more cog in an ever-expanding technological reality than participant in the song of a wood thrush. Despite these concerns, it was in the midst of Toronto's expansion that the birds offered him an opportunity to mind human-nature relations in a different way. What Livingston learned through these familiar relations

was the need for humans, modern or otherwise, to come to a "compliant acceptance" of our place in a world of broader life forces. The
modern ship needs to do nothing less than reconcile "with that long-
forgotten quality that is the nature of being."[40] Such a change is of a
cultural and spirited scale that has significant implications for how we
mind relations with the world during a climate of change like ours.

———

In spite of my initial negative impressions of life in Toronto, I came to
this city with a desire to learn about social and environmental issues,
first at the University of Toronto and then through Canada's oldest
Environmental Studies program at York University where Livingston
taught for three decades. Though he was no longer present when I
arrived in 1999 for graduate work, his writings were still in the air. In
my first term of study, I was given a guidebook to environmental
studies that consisted of a series of essays co-authored by Livingston.
One passage struck me: "The concept of 'environment' may be seen
not only to mean the phenomena which are studied, but also to mean
an inherent way of studying – environmental thinking. The notion,
relatively undeveloped [in modern culture], is that subject matter has
an intrinsic form of thought as well as content, and that the environmental context of substantive concerns may be used to determine
form of thought."[41] Our engagement with the world can transform
our ways of thinking, not simply the content we study. Surrounding
environments are less objects of research than an adjective that modi-
fies what "is to be studied and also the manner in which it is to be
studied."[42] It is a sensibility consistent with the emerging insights of
the early environmental movement, though particularly influenced by
Livingston's sense of conservation "as a state of mind" and environment as not simply ecology but as the varied "contexts" we live and
think within – from ecological to technological.[43]

   The environmental dimension of minding relations crystallized in
Livingston's writing around the same time as he arrived at York
University, which coincided with the emerging environmental movement. Others in the late 1960s and early 1970s were making similar
connections. In 1967 Paul Shepard wrote his seminal essay describing

"ecological thinking" as the self extended into "the landscape and ecosystem."[44] He later described Livingston's beloved birds as beings that externalize our minds, with the visible habitat being "the outward form of the whole space of the mind" and "its unseen distance like the unconscious" that the birds fly into.[45] A few years following Shepard's intuition, Gregory Bateson's influential *Steps to an Ecology of Mind* concluded that "the mind is not limited by the skin."[46] These insights arose in response to increasing concern about human impacts on the environment. As Bateson writes: "You forget that eco-mental system called Lake Erie is a part of your wider eco-mental system – and that if Lake Erie is driven insane, its insanity is incorporated in the larger system of our thought and experience."[47] Intersecting local and global environmental issues were spurring a growing chorus of voices on the need to renew our minding of worldly relations in much the way Leopold decades earlier highlighted the importance of *Thinking like a Mountain.*

Even when terms like an ecology of mind were not explicitly referenced, as with Rachel Carson's 1962 publication of *Silent Spring*, there was a sense that modern awareness was being challenged by surrounding relations. The spectre of spring's ever-cycling bird song being silenced was an image Carson employed to remind people that we have intimate relations to surrounding natural realities, and the potential loss that can occur if our ways did not change. It was a powerful message that Gibson remembers as an undercurrent in his discussions with Livingston and the subsequent hawk experience.[48] The writings of Carson on the ecological impacts of DDT are grounded in a sense of human responsibility. She adds dimensions to the knowledge of an ecological science that is often focused on the internal biology of organisms and their community interactions.[49] In contrast, Carson highlighted the way in which complex ecosystems encompass the human and thus require a capacity to understand how specific processes "work together, reading them as responses, adaptations, resistances to places and circumstances whose local detail and connections with other locations contribute to how organisms, whether human or other, can be."[50] Acts such as overusing DDT, eradicating a wolf population, burying a creek, or replacing a marsh have untold impacts that can haunt the way people mind relations long after their severance.

Just as Carson lamented the potential loss of something unimaginable, the ever-returning birds of spring, Livingston's initiation into a different way of thinking arose from a love for birds and the loss of those relations to ever-expanding urban processes. The idea that environmental contexts like a marsh ecology or urban form can shape our thinking was new to me, but also oddly familiar. Though I did not know Livingston's work prior to arriving at York University, as a Canadian I was aware of one of his popular co-creations, the CBC television show *The Nature of Things* and the thought of its long-time host David Suzuki. Following Livingston's death, Suzuki and other Canadian environmentalists like Gibson, Monte Hummel, and Farley Mowat spoke of the influential role he played in Canada's early environmental movement.[51] His de-centring of human concerns within a broader bio-centric perspective was something that Suzuki states "took a long time for me to understand ... and it was a very important part of me coming to understand the environmental movement in a much deeper way."[52] Bio-centrism is, he writes elsewhere, the bravest position an environmentalist can take because it offers a radically different view on "the roots of our destructive path."[53] As with Livingston, Leopold, and many other environmentalists, Suzuki has come to advocate an experiential and thoughtful engagement of natural surrounds, a recurring theme in his show *The Nature of Things* and popular documentaries like *A Planet for the Taking* (with Livingston) and *The Sacred Balance*.[54]

An ecology of mind is grounded in the experience of finding oneself a part of surrounding presences, such as when Livingston refers to himself as one with the wood thrush, bittern, or Ashbridges Bay marsh. But in our modern context this emerging sense of unity is more often highlighted by its loss, such as the potential hushing of spring birdsong. This is also discussed by Suzuki. Though he had many significant nature experiences from childhood to early adulthood, he writes that one significant "moment of enlightenment" occurred when he was a young father hiking with his two children on the edge of an ancient forest in British Columbia.[55] There they witnessed a magnificent "forest shaped by the forces of nature for ten thousand years, a community of life where death gives birth to new life in an endless recycling," and beside it a land cleared by industrial logging and awaiting the replanting of a much simpler forest.[56] The stark contrast of a mature ecology beside a powerful human severance awoke in Suzuki, as it did in Livingston, Leopold, and Carson, a deeper sense of the environmental crisis and the need to mind our relations differently. Thinking environmentally was something Suzuki primed me for, and Livingston's thought opened my mind further.

What at first grabbed my attention in the idea of "environmental thinking" at York University was the social implication that it requires an interdisciplinary array of knowledge. Even more importantly, this broad-ranging search for knowledge is not simply something that people have decided to entertain out of intellectual curiosity alone; rather, we are being prodded into doing so by our worldly milieu. Interacting climate, ecological, and human contexts are more complex than our thought, and thus our ways of minding relations have been called to be more flexible by issues that reflect a responsive world. In contrast to the disciplinary tradition of demarcating particular foci of knowledge, an ecology of mind is grounded in places, issues, phenomena, or even "texts" whose reading requires us to cross many human abstractions.[57] It must be noted that this sense of engaging surrounding relations as "texts" to be read itself reflects a predisposition of the academic. The bittern and other birds of Livingston's initiatory experience were presences that first opened his mind to various texts that included natural histories, conservation thought, and the science of marsh ecologies. When these same presences became severed from his experience, other texts came into play that helped him critically mind the political, economic, and cultural assumptions underlying his pain. The point is that the broad interdisciplinary array of academic inquiries arises from the felt experience and needs to be brought back into the intimate minding of relations.

Though there is some truth to the statement that "environmental thinking" is "relatively undeveloped" in the West and particularly Canada, intellectual approaches dating to prior to the modern era can help us think through ways of fostering this change of the modern mind. One can potentially go as far back as Plato's founding of his school outside Athens' walls in the grove named after the Greek hero Hekademos or Academos. While the exact motivation for situating his school in a grove some 2,400 years ago is difficult to discern with certainty, it has been suggested that it was perhaps Socrates's conviction and death that led Plato to seek some detachment from the political centre of power.[58] This aim for institutional objectivity was a tactical rather than ideological decision, for Hekademos also "had a well-defined political mission: to turn the future counsellors of rulers, if not the future rulers themselves, into philosophers" so that political action would be grounded in metaphysical foundations.[59] This strategic positioning became a pattern in the history of Western European and American academia as many of these institutions were created with the aim of promoting objectivity on what were once predominantly semi-autonomous garden-like campuses.[60] Engaging with the world, and doing so in a thoughtful way, requires a level of detachment not from the world itself but from one's dominant cultural and political mode of thought.

Such a historic ideal was recognized at the University of Toronto when the institution came into being in 1827 and created Philosopher's Walk along the banks of Taddle Creek going north of the campus. Natural settings outside dominant political, economic, and cultural forces were seen as important places for thinking. There is some consistency here with an ecology of mind that simultaneously engages human institutions in a process of transformation but yet is grounded in these relations that surround the human. That said, the differences are also important, as indicated when Livingston writes that "the 'academic' problem is one in which there is little emotional investment in the solution but very much in the method," and as such ecological preservation often tends to be in conflict with academic ways of conceiving problems.[61] The focus of academia on that which can be rationalized tends to marginalize the felt experience of ecological presences like bittern, and these relations are for him central if we are to return to a sustainable way of minding relations. It is a limiting predisposition that he historically connects to Plato.

From Hekademos's position between the ancient city reality of Athens and its surrounding Mediterranean ecology, Plato would, in the words of Livingston, bewail "the disappearance of the forests from the slopes of Greek mountains."[62] Despite being aware of this deforestation and its local impacts, all indications are that his school spent more time philosophizing about the politics of life in Athens and

Greek civilization than thinking about the metaphysical, cultural, and political implications of such ecological change. For Livingston, Plato, Aristotle, and the Greek tradition influenced a historic succession of anthropocentric Western philosophers who considered humanity to be the pinnacle of creation and natural realities simply there for human uses.[63] Others have supported his view on the ecological impacts of a still present, though transmuted, Platonic tradition that revolutionized Western consciousness by fostering a "turning away from the sensuous (sense-based understanding) to the intelligible (concept-based, discursive understanding) as the basis of knowing and acting."[64] In highlighting this critique, I do not intend to dismiss the great knowledge and practice that Plato fostered, including the search for an objective place from which to mind relations critically and independently. Rather, the concern is with an imbalance that privileges rational thought and its products over felt ways of coming to knowledge, an idealistic pattern that is deeply entrenched in the modern ship.[65] The ancestral legacy of Plato will continue to arise in later chapters as a figure of mixed importance, offering both critical insights on limiting tendencies and forgotten alternatives.

Livingston's emphasis is on a bird's-eye view that recognizes the limits to knowledge in a complex reality that is intimately interconnected to our being. A common critique of such a broad interdisciplinary approach is that it can undermine the focus of disciplines and promote a "multidisciplinary illiteracy" marked by superficial knowledge.[66] So where is the grounding of a bird's-eye view to be found? It is places like a marsh or creek, issues like climate change, and particular presences like bittern that give an ecology of mind a contextual focus. This way of thinking has many strengths. Yet, arising from within the modern ship's institutions like the university or a national economy, it tends to be entangled with the assumptions that are fuelling our changes. As Carson writes, living in "an era of specialists, each sees his own problem and is unaware of or intolerant of the larger frame into which it fits."[67] Though calls for interdisciplinarity are common today, initiatives often do not go beyond a quantitative sensibility that skirts the difficult power-laden questions of how to consider qualitative relations between different kinds of knowledge.[68] There is little discussion about the balance between science, economics, and technological innovations, on the one hand, and what are often conceived as the less rigorous social sciences, humanities, and religious studies, on the other.

The Ashbridges Bay experience of Livingston instilled in him a deep concern about the potential narrowing influence of our modern context on how we mind environmental issues. Similar concerns are being raised around climate change today. In the words of Mike Hulme, it is "the latest project over which human governance, control and mastery is demanded."[69] In contrast with a modern period that assumes a certain detachment of the human from the world, human impacts on local and planetary life systems are calling for a more impassioned minding of relations. The rational intellect so central to Hekademos, contemporary climate research, and environmentalism generally is, for Livingston, merely one important part of the interdisciplinary knowledge that is needed. An ecology of mind is a situated way of coming to understand relations with the surrounding world. The approach is consistent with Donna Haraway's description of "situated knowledges" as reflecting partial views that emerge from a particular embodied relation with a surrounding context; the contradictions and paradoxes of a living knowledge are functions of the fact that they are "views from somewhere."[70] Recent writings on ecological and climate thinking have individual nuances but commonly point to the social nature of recognizing our place within that which we are coming to know.[71] Such a situated and experiential approach further deals with the potential of a bird's-eye interdisciplinarity to be shallow by grounding us in particular relations like that of a marsh.[72] It does so while also maintaining our awareness on narrowing institutional, cultural, and national predispositions in any minding of relations.

Though the familiar bird relations of Livingston in Ashbridges Bay followed by their severance are particular to him, there are vital resonances across many ecological thinkers. This is because of the common collective experience of severed natural relations in the context of an ever-expanding modern ship. A responsible minding of relations is passionately aware of our paradoxical position and yet attempts to foster more compliant relations with the surrounding world. In the remembrance of many of Livingston's students, he promoted an "emotional and visceral, life-long commitment to all things living."[73] Passionately taking interdisciplinary thought outside the modern Hekademos, or perhaps paradoxically deeper within its fissures, is a central feature of what he begins to guide us toward. A good Canadian climate of mind needs not only to find space beyond the walls of a city like Toronto and an Hekademos like York University but also engage an outside world that is situated profoundly within their painful cracks.

The rural eastern Ontario vision of Toronto that informed my early life participated in a romantic predisposition to perceive urban realities as starkly disconnected from anything wild – just as Gibson was shocked by Toronto's bird life. Human environmental impacts have tended to radiate out from highly urbanized and industrialized places where there is an energetic human presence, and an awareness of these processes has given birth to a popular and environmental predisposition toward wilderness experiences distant from city life.[74] Valuing experiences in northern forests, mountain ranges, or isolated rivers has been a tendency that I am familiar with from my childhood experience of family camping and fishing trips north of Ottawa. In light of Livingston's urban initiation, it is not surprising that he critiqued such a view as romantic nostalgia for an untouched nature. It "does just as much to separate man from his environment as does the cult of the cold intellect."[75] In contrast to such a pure separation, we have seen that an ecology of mind has often arisen from the painful witnessing of modern impacts, from a severed marsh to a silent spring to a buried creek. Though climate change is an intensification of these dynamics, I was only semi-conscious of these issues and thus largely began with the assumption that the starting point for my questions was in a wilderness far from human and urban disturbances. My climate of mind thus left the city and the Hekademos where I was situated and headed north to engage Inuit *silatuniq*, wisdom.

While "environmental thinking" can be conceived as a "relatively undeveloped idea" in the West, at least since the modern period, my experience as a social worker in Indigenous communities followed by research into Inuit understandings of climate change made it clear to me that other cultures do have long traditions of minding environments and climate. The north has been described as the global climate system's "canary in the mine," with researchers and Inuit documenting quick changes to flora, fauna, ice, and *sila*.[76] At the time of my work in the north, this latter term was coming up in climate research as a synonym for weather, but, as I talked with the Inuit philosopher and policy worker Jaypeetee Arnakak, I began to realize that there was much more to sila. His thought was not solely focused on a physical meteorological system. Rather, he described sila as an ever-moving and imminent force that surrounds and permeates Inuit life, with it most often experienced in the weather. What emerged from our dialogue was a complementary Indigenous view on what is needed to wisely mind the relation between a quickly warming north and the modern drive of urban places like Toronto.

Almost a year after we began talking, Arnakak asked me, "Why don't researchers ever ask us about wisdom?"[77] For him, silatuniq is of central importance to anyone concerned with sila's northern warming in a context of global changes. I brought to our dialogues an awareness of sila's presence in two largely divided academic disciplines. On the one hand, as noted, contemporary climate research was defining sila as weather. At the other extreme, ethnographies from the first half of the twentieth century described sila as the spirit of the air, upholder of the weather, and breath source of all life. It was while discussing this disciplinary divide that Arnakak asked me his question and brought up silatuniq as a way of inquiring into "the context and consequence of applying knowledge and/or how our interacting with the surround affects that surround."[78] Such silatuniq seemed relevant to northern warming and global climate changes that are at root planetary responses to the modern ship's exhalation of greenhouse gases; the breath source of life is being disrupted by particular human actions. It gradually dawned on me that there also seemed to be a resonance between this definition of wisdom and Livingston's sense of environment as contexts that, because they relationally situate us, require something beyond academic rationality to mind.

Though Indigenous thought is not significant in Livingston's writings, it was recognized by Paul Shepard as he began conceiving ecological thinking in the 1960s. Consistent with Inuit silatuniq, he writes that Indigenous approaches often have means to "carry us beyond ourselves, pursuing the nature of thought as the thought of nature."[79] The long Palaeolithic history of hunter-gathering had led the human mind to pattern itself after the animals, ecology and climate it depended on. Since Shepard, there has emerged a more detailed sense of the role that changing climates have played in the way people mind relations.[80] Research in paleoclimatology, human evolution, and cognitive biology is suggesting that many of humanity's evolutionary shifts coincided with major cyclical climate changes toward and away from recurring Ice Ages. These human changes seem to coincide with unsettled extreme periods of drought and flood rather than during gradual temperature changes or stability. Our evolving minds, tool-making, and social culture may well have emerged in "the decade-long transitions between states, not the steady-states (warm-wet and cool-dry) themselves," as the former created enough disruption to manifest niches where alternative ways of relating became possible.[81] This goes from broad cultural ways of living in diverse climates to the kind of initiations that would bring whole groups and individuals to a silatuniq for living with uncertain climatic contexts, a potentially very old ancestral wisdom of the human species that we more deeply consider in chapter 6.

While academic research has been offering more details on Indigenous ways of reciprocally minding ecological and climate relations,[82] Indigenous thinkers have been culturally and regionally situating such a silatuniq. From Canada's west coast, Nuu-chah-nulth hereditary chief and scholar E. Richard Atleo (Umeek) describes the term *Tsawalk* as a theory of context that includes various units of physical existence from the individual to community to local ecology to climate to ancestors and other non-physical powers that go beyond what can be bodily seen.[83] From Africa, Tsepo Mokuku describes the Lesotho concept of "*lehae-la-rona* as a process of engaging people both rationally and emotionally about" the interconnectedness of all living and non-living things in the cosmos.[84] Other perspectives can be added to these, and starting with the next chapter the book more fully engages Haudenosaunee, Wendat, and some Anishinaabe and pre-modern Western ways of minding relations as they emerge from particular places. There are cultural, ecological, and climate differences across these diverse cultures, but there are also important commonalities. In the words of Cree scholar Shawn Wilson, there is an increasing "awareness of the similarities of experience of Indigenous peoples worldwide" that has resulted in the term Indigenous being a reference to a kind of knowledge system that inclusively cuts across traditions without collapsing them.[85] It is in this sense that the term Indigenous is used in this book.

One of those points of commonality is the role that initiatory pain can play in opening up niches on personal, cultural, or species scales for coming to wisdom, silatuniq. In the 1920s, not long before Livingston's Toronto experience to the south, an Inuit shaman by the name of Igjugarjuk recounted an initiation into a silatuniq that "lives far from mankind, out in the great loneliness, and it can be reached only through suffering."[59] While Livingston's pain was an unintended result of his Ashbridges Bay relations in a period of modern expansion, Igjugarjuk's suffering arose from a conscious ritual directed by his elders that entailed "being left alone on the tundra in a small hut just big enough to sit cross-legged." For thirty days he sat in the cold, fasted, and thought only of the Great Spirit that underlies all life, and in that state he explained that he "sometimes died a little."[62] What Igjugarjuk died to were the everyday things and experiences of his human life as experienced in his Inuit culture. A social cosmos requires finding ways of going beyond what is known to hear what ecology, climate, and sila are saying, and ideally the shaman served this role for family and community, which does not mean that this ideal was always achieved.[86] This northern path to silatuniq was that of "dying a little," just as Livingston's initiation required dying to modern assumptions, beliefs, and practices. The niche for minding climate and ecological relations on a regional scale arose in the midst of a culturally imposed stress, academic language for a painful change in perspective. An initiation into this wild and felt nature may be essential for the modern ship if it is to relearn how to mind relations wisely.

From a "shimmering light" on Ashbridges Bay to a northern "great loneliness," wildness is a phenomenal experience that gives us some distance, not absolute disconnection and objectivity, from dominant ways of minding relations in an urban Hekademos. I was interested in sila, but, as with many climate researchers who went north, I originally focused on Inuit traditional ecological knowledge or TEK. That was the case until Arnakak shifted our discussion from this well-worn path of the modern mind to the more distant consideration of silatuniq.

Indigenous knowledge is of increasing interest to environmental researchers because of its potential applicability in climate models and adaptation, but critiques are being raised that it is a narrow version of Indigenous knowledge. As Arnakak explains, the dominant engagement of Indigenous knowledge like *Inuit Qaujimajatuqangit* is largely "a thinly veiled corporatist agenda regarding environmental and resource development."[87] A focus on TEK tends to decontextualize observations that can be used in models from a broader silatuniq that can challenge modern conceptions and practices. Just as silatuniq calls for an expanded view, Atleo writes that *Tsawalk* does not "call into question the methodologies of the physical sciences" but rather implies that the relational network they consider requires ways of also attending the spiritual dimensions of existence through such things as ceremony, storytelling, and gift-based economics.[88] An intercultural response to climate change requires going beyond the dominant disciplinary discourses of "science, economics and politics,"[89] opening interdisciplinary inquiries out to post-colonial studies, anthropology, religion, and more. It also further clarifies that the wisdom of interest to this book is a situated, not universal, way of minding relations, though it has certain Indigenous resonances across places, cultures, and times.

While there is consensus across many cultures and time periods about the value of engaging something like a wild "great loneliness" as a means for expanding our sense of worldly relations, the unintended experience of Livingston suggests that this can also occur in an urban place like Toronto. He advocated bonding to the natural places in our midst and moving beyond "the wilderness experience" and toward an experience of wildness.[90] In contrast to the external focus of wilderness preservation, wildness was for Livingston the "sweet bondage ... in which one is an autonomous organism, yet bonded and subsidiary to the greater whole."[91] This could be experienced anywhere, though it seemed to require some kind of move beyond human-centred relations and stimulation. His experience and thought is in some ways consistent with William Cronon's classic critique of the wilderness concept in environmentalism as a duality that keeps us focused on some impossibly untouched elsewhere while disconnecting us from local wild realities.[92] It is a romantic mistake that "often means not idealizing the environment in which we actually live,"[93] which for most of us today are urban realities like Toronto.

In following Livingston into Toronto's marshes and ravine paths, I was surprised to find that this place's ecological and geological contours can still be sensed in remnant ecologies like those found along the Don and Humber rivers.[94] Twentieth-century conservation efforts increased the city's green space from 67 hectares in 1954 to 3,161 in 1974, beginning with a largely recreational focus and then shifting in the 1970s and 1980s toward regenerating ecological systems and promoting watershed conservation.[95] As such, many of the bird, tree, plant, animal, and insect species that inspired Livingston almost a century ago can still be found here.[96] The naturalist Wayne Grady writes of a pair of coyotes he tracked in the "ravine off the Don Valley where Ernest Thompson Seton built a makeshift cabin when he was ten years old."[97] At the same time, many other species that were once here are no longer or are much diminished, marginalized by human activities that transformed the land and introduced new plants and animals. To this is being added the challenge of urban smog and climate change, as well as the alternate possibility of partially responding to these issues through expanding urban forests.[98] Severance and reverence sit closely together in urban places, as Livingston experienced.

While Igjugarjuk's small deaths brought him silatuniq for northern living, Livingston's pain was situated in a modern ship which taught him that our unsustainable ways go far beyond water pollution, habitat fragmentation, greenhouse gases, and climate change. The nature of his marsh initiation led to a deep concern with the growing modern power of a "technological imperative," an issue that manifests itself externally in the historic severances of Ashbridges Bay, Don River, Taddle Creek, Humber River, and other waterways from the Oak Ridges Moraine north of the city to Lake Ontario, the Saint Lawrence River, and the Atlantic Ocean beyond. The waters that shaped Toronto are now part of a highly managed and designed human system of flood control, with the river valleys specifically taking on "non-natural urban functions, such as carriers of traffic, transportation, and information."[99] More than that, the city's overall approach to water management and use now also comprises about one-third of its "energy bill and produces one-tenth of our greenhouse gas emissions."[100]

These rivers and ravine valleys have become "urban cyborgs," a complex mixture of machine and ecology.[101] What Livingston's initiatory insight into the technological imperative clarifies is that the ever-quickening pace of this subsuming machine reality is not only replacing an external ecology but also has the powerful potential to narrow our interiority and thus restrain – rather than situate – the bird's-eye view from a marsh.

Resonating with the rushing car lights that I often look upon from the Howard tomb, Livingston writes that the "urban sensory assault" is quantitatively overloading our being with experiences that are coming from a single source of stimulation – that which is manufactured by humans.[102] Offering a similar thought, Suzuki writes that our ever-expanding cities are fostering a kind of technological thinking focused "on standardization, simplicity, linearity, predictability, efficiency and production."[103] These central assumptions of the modern mind are exactly what severs us from the silatuniq of a great loneliness, for the external urban cyborg is gradually enfolding us, Livingston states, within "its own conceptions of instrumentality, neutrality and purposiveness."[104] We are being subsumed in a cyborg system that, returning to Haraway's thought, reflects the dark potential of a planetary "grid of control" rather than the optimistic possibility of balanced kinship between organism and machine. The ideal for her is to situate ourselves critically in scenarios that range from creative to destructive extremes, revealing "both dominations and possibilities unimaginable."[105] Despite the similarities in these views, it is clear that the pain of Livingston's initiation led him to advocate ways that aim to recontextualize the always expanding machine side of our cyborg nature, not simply live with the tension. The difficulty is that, in embodying the modern drive of the urban cyborg, "we may enact patterns of domination in our own selves, working when our bodies are telling us 'they' need to sleep or eat; doing this to ourselves, we do it to others."[106] We then participate in the proliferation of urban severances and resulting delusions.

The past decade has seen a rise of curiosity in the relation of mental health to nature experiences and urban life that in many ways confirms the experiential insights of Suzuki, Carson, Leopold, and Livingston. Urban stimuli tend to result in more pronounced activity of the brain's amygdala and anterior temporal pole, areas that activate in the presence of dangers like environmental threats.[107] It produces an overactive state of mind that prioritizes memories of negative experiences, thus leading to a vicious cycle wherein the intensity of urban living is experienced as more threatening and potentially depressing. This seems to be one of the reasons that the risk for anxiety and depression can be 40 per cent higher for those who live in urban rather than rural areas.[108] From an ecopsychological perspective, the prevalence of depression reflects "a revolt of nature, our bodies saying 'no' to the crushing demands and abuses of modern life"[109] – the painful internal companion of the external revolt by ecology and climate against human abuses. Interestingly, one way of responding to the internal ramping up of the amygdala seems to be by engaging ecologies where the human feels situated in something bigger than itself.[110] It is not only that we need to "bring nature back in" to the urban rather than always subordinating ecology for human needs;[111] we must also recognize that such an external act also has internal dimensions in our healthy minding of relations. Reversing Livingston's initiation is clearly a central challenge for our increasingly technologized urban environments and ways of living, and is also key to the good climate of mind explored in this and subsequent chapters.

Beyond the walls of Hekademos, we are coming to see climate change and related environmental issues as being deeply connected to who we are and not primarily as external problems to be managed. Drawing upon Livingston, his Toronto colleague and eco-philosopher Neil Evernden explains that too often environmental discussion positions us outside the world as an objective diagnostician and manager who forgets that "the source of the environmental crisis lies not without but within, not in industrial effluent but in assumptions so casually held as to be virtually invisible."[112] This silatuniq is also being recognized in some climate research; for example, Hulme suggests that the interconnected nature of climate change means it may be better to contemplate this issue "as an idea of the imagination rather than as a problem to be solved."[113] In this sense, our climate of change is less an issue to be managed than a numinous mystery that is pulling us into differently situated and yet interdependent ways of minding relations. A mystery is a kind of

"problem that encroaches upon its own data, that invades the data
and thereby transcends itself as a simple problem," and thus reveals
our interconnectedness.[114] The presences of this world are pulling us
toward a wiser and more reverential way of minding relations; that
is, if we are willing to engage the pain of this mysterious, numinous,
transformation. Perhaps we are dying a little.

We are not so much called to walk outside Hekademos as being
drawn into the wild cracks that are today mysteriously emerging in a
multiplicity of places, here in Toronto, the north, Canada, the planet,
and our personal being. A change of mind is engulfing the modern
ship in a way that follows a pattern consistent with the kind of
heightened sensibility one often experiences when attempting to enter
a far-removed "great loneliness." The difference is that we are not
entering an untouched wilderness for inspiration, but rather our climate of change is breaking in upon us. From this numinous uncertainty has emerged interdisciplinary and global forums of knowledge
sharing, international political debates, ever-intensifying technologies
for remote monitoring and modelling of risks, and heightened public
concern about more extreme events that increasingly haunt the
world. Fissures are appearing all over the world, and within them
socially engaged perspectives and approaches for minding something
more powerful than the human can possibly arise. It seems that we
are, returning to Bruno Latour's words, being brought "back to normal, back to the anxious and careful stage in which the 'others' used
to live before being 'liberated' from their 'absurd beliefs' by our courageous and ambitious modernization."[115] To give up some rational
control and allow a world of increasingly numinous presences to
mind us will likely be a painful initiation for the modern ship, and yet
may be the central mystery we need to wisely renew on our way to a
sustainable future.

———

Toward the end of his life, Livingston recounted for Farley Mowat
the experience of "weeping with rage, anger and frustration" when
he found out what was to happen to the marshes of Ashbridges
Bay.[116] His love for its beings brought to the fore questions that, at

least in an industrializing modern culture, only a ten-year-old mind can contemplate: "How can I warn the frogs and toads and newts? Can I get them out of there, take them away somewhere?"[117] As we look back on Toronto's urban transformation, it is clear that his experience is merely the tip of an iceberg of losses.[118] How much more stomach-wrenching pain has been experienced, forgotten, and/or not told by children, youth, adults, and elders? This is a question that goes beyond any individual loss, asking us to contemplate a kind of accruing historic pain that our drive leaves us unconscious of. That is, until our climate of change begins to rock the modern ship, thus reminding us to slow down and take a closer look at how we relate with the many presences surrounding us. Each chapter that follows will offer us different perspectives on the historic depth of these severed relations, as well as potential practices like the bird's-eye interdisciplinarity of this chapter for renewing a good Canadian climate of mind.

At the core of Livingston's initiatory pain is the incommensurability of being bonded to both human and natural communities during a modern period that is decreasing the space for seeing ourselves as one with the wood thrush, bittern, blue heron, or marsh. The paradox of his ecology of mind is that wild experiences are considered vital to sustainable ways of thinking and living, and yet such experiences seem to be receding further from our modern realities. Human centrality appears unquestionable. But, as we have seen, there is a more nuanced reading of Livingston's paradox, and that is to see the wild as increasingly impinging on our "urban delusions" in a modified form that is becoming more numinous with each passing moment of inaction. That is what Hurricane Sandy, a Toronto ice storm, Red River floods, melting glaciers, extreme temperatures, and so much more signify. Any wild experience ultimately entails some degree of pain related to the way in which we are brought into something larger and more encompassing. For, as with Igjugarjuk's experience in sila's great loneliness, these experiences are about dying a little to our assumptions, beliefs, and factishes about being human in such a numinous mystery. Our climate of change is bringing that loss of control home for a modern ship that has long pushed this sensibility to its margins. If we cannot find ways of consciously entering such an awareness on individual and cultural scales, then it seems it will be brought to us.

We do not need to go anywhere today for inspiration, for wild realities are surrounding us with numinous inspirations about the need to re-mind the place of humanity within the creation wherever we are, from the Inuit north to urban Toronto. Wherever we live, there are places close by where we can sense this call for change if we find ways of slowing down, becoming familiar, and digging below our modern drive. This is what Livingston's bird's-eye view pointed me toward in its outline of a lovingly situated engagement of the beings in our midst as teachers, guides to a broad range of knowledge that needs to be unhinged from the disciplines of Hekademos and fac-tishes of urban life. From this position, a good sustainable climate of mind has to embed rational interdisciplinary thought in a passion to preserve a relational possibility beyond the machine, to recognize that thought can arise within the climatic fissures of academic, urban, and national walls. At the same time, the experience of becoming more familiar with Livingston's marshy flights and the silatuniq of Inuit led me to appreciate how little ecological knowledge, cultural under-standing, and felt connection I had to the urban place where I live.

Not being from Toronto and for most of my time here working on the assumption that I would leave in a year or two, or five at most, it seemed superfluous to become familiar with this place. In contrast to the long ancestral line of ecological knowledge and cultural sto-ries that embeds Inuit silatuniq and ways of living, I had no cultural grounding in this place's ecology, climate, or cultural stories. Were a particularly snowy winter, wet spring, or turbulent summer storm unique events, representative of past patterns, or some variation in between? What birds were here and at what times of year? In many ways, I was a stranger to where I was living, thus adding resonance to the Howard tomb's call to "Oh! Stranger ..." But then, in the midst of my reflections on climate change, this place gradually pulled me into acts of journal-keeping, contemplating, imagining, and minding relations within an urban reality I thought I wanted noth-ing to do with. My knowledge of Toronto changed as my familiarity with its interdisciplinary dimensions increased. What has been more interesting is the way in which these experiences deepened that knowledge with felt connections. Something in the climate of these ecological fissures within the city began teaching me a different way of minding relations.

Spending time in the Humber Plains of Toronto has over the years made it clear that experiences like Livingston's joyous reverence and agonizing severance continue to reverberate through the land and climate, from Canada's north to Lake Ontario. Its ravines, waterways, and ecologies also highlighted for me the continuing presence of earlier Indigenous ways of minding this place, that neither the wild nor Indigenous peoples have been severed by urban, industrial, and, as we begin considering in the next chapter, colonial developments. As William Woodworth and other thinkers native to this place clarify, ancestral ways of minding relations are marked throughout this urban environment in such things as street names, paths that have become roadways, and the names of the city, province, and country I live in. Through following Woodworth's Ashbridges Bay vision toward a Haudenosaunee Good Mind, I was given a different passage into our mysterious climate of change from a tradition with a longer ancestral relation to Lake Ontario and Toronto than this chapter's ecology of mind. We now stretch the stomach-wrenching loss of Livingston into Toronto's and Canada's colonial period as we attempt to step deeper within the historic fissures of the ship's modern walls.

Meeting Place in the Forest

When the greatest of these trees was struck by lightning or fell tired with
age it becomes an honoured "landmark" – a meeting place in the forest
intervening in the trail. This decomposing Grandfather, returning once
again to the Mother earth, transforms into home for wild life, shelter
for those along the trail, and eventually a place of ceremony.

William Woodworth Raweno:kwas (2012)

The Lake Ontario skyline between the Don and Humber rivers is
today defined by the CN Tower, a landmark that rises rocket-like
above the surrounding skyscrapers, buildings, and seemingly endless
roadways. Its size and symbolic significance in early-twenty-first-
century images of the city makes it difficult to imagine a time when
something more natural towered along the shores of this lake and in
the minds of some inhabitants. That is the Haudenosaunee vision
William Woodworth Raweno:kwas offers, for in his oral tradition
the name of the city derives from the Mohawk *Toron:to (dolon-do)*,
"the place where the great tree fell – the damp log."[1] The "grandfather" he refers to in this chapter's opening quote is a great white pine
or *onerahtase'ko:wa* that came to mark his Haudenosaunee sense of
Toron:to as a "meeting place."[2] Long before Livingston's childhood
experience, these trees grew near the shores of Lake Ontario to a
height of sixty to ninety metres with girths up to a metre wide.[3] In his
vision, the decomposition of such a great one provided shelter for
wildlife and people traversing this place along the old portage trail
known as the *passage de taronto*. More than that, the mythic importance of the onerahtase'ko:wa in the Haudenosaunee Good Mind
signified it as "an enduring sacred place," a good place to stop and
meet others.

With a wet spring mist hanging in the air above the fast-moving Humber River, I am easily taken to a time before Livingston's initiation or a Taddle Creek Hekademos when the ravine and trail system that surrounds this river was known as the passage de taronto. The ancestral meaning of this name stretches beyond the mist of colonial documentation in French maps from the early 1680s that first mention the *lac* and passage de tkaronto or taronto during the period of Seneca and Mohawk habitation here.[4] The *lac* refers to Lake Simcoe, 125 kilometres north of the city, and the passage is the portage trail and Humber River that traverses down this southern side of the Oak Ridges Moraine toward Lake Ontario. For five years I brought a graduate class to this site known today as Etienne Brûlé park to consider our climate of change in relation to the Haudenosaunee teachings of Woodworth.[5] Woodworth himself was present at these sessions, and, with seagulls flying overhead, a white egret reposing in the river, and a warm fire centring our gathering, he often began by opening us out to the deep temporal layers of this place that are often severed from the city's contemporary sense of itself.

On the ridge to the northeast is the site where the Seneca village of Teieiagon was situated in the latter half of the seventeenth century, and below it, closer to the river, there was a smaller Mohawk village at that time.[6] The bones and artifacts of Woodworth's Haudenosaunee ancestors are in the land, and occasionally they come to the surface in the course of excavations in the yards of today's wealthy Baby Point residents. To make this clear, he shows us a replica of a comb made of moose antler and the bones of two women which were discovered when natural-gas lines were installed. They had been buried in the 1680s, and Woodworth was asked to participate in their reburial in a more secluded and safer space along the river. This, he explains, was a delicate process since the comb was adorned with spiritual symbols that signified its owner's status as a medicine woman. It depicts a bear morphing into a panther amidst other symbols, all of which give it a potential association with Mishipizheu, "the chief manitou of the underwater realm."[7] In his tradition, the ancestor who held this comb was someone who could engage these spirit helpers to transform herself while healing or, alternately, cursing others. Recognizing this potential power, the Six Nations

6 Fallen onerahtase'ko:wa near passage de taronto (winter 2015). A fallen white
pine reaches about twenty metres across a ravine near the passage de taronto.
In spring this ravine fills with water, fostering a green growth that reminds me
of Woodworth's description of a fallen dolon-do.

community, west of Toronto, asked Woodworth to help in ensuring
that these women and their belongings were reinterred properly and
respectfully, to protect the peace of their ancestors and those involved
in the task.

    Below in the ravine where we formed our circle, the park's dedica-
tion to Étienne Brûlé speaks to a period preceding Teieiagon when the
Wendat nation known as the Tahontaenrhat for centuries lived along
this passage. It was Wendat who guided Brûlé along this river in 1615
to its mouth on Lake Ontario. The Wendat maintained a strong presence on the passage until the mid-1600s when their confederacy
unravelled in the face of complex colonial dynamics related to
French and Haudenosaunee, something we detail in later chapters.
Their violent ousting from the region opened space for the Seneca
and Mohawk to build Teieiagon here. By 1700, the Haudenosaunee

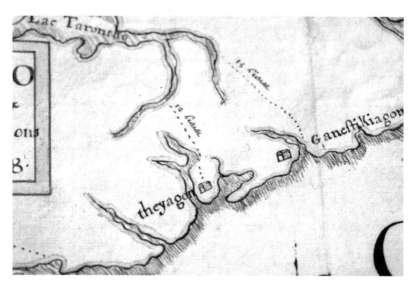

7 *French Map of Lake Ontario, 1675,* by Father Pierre Raffeix, courtesy of Map and Data Library, University of Toronto (original with Bibliothèque Nationale in France). This French map, which was published in 1688, shows Lac Taronto (Lake Simcoe) and to the south the Seneca/Mohawk village of Theyagon (Teieiagon) on a stretch of the *passage de taronto* (Humber River) just north of Lake Ontar:io.

were forced south of Lake Ontario, leaving the passage de taronto to the Mississauga or Anishinaabe, who referred to the waterway as *cobechenonk*, meaning "portage" or where we "leave the canoes and go back."[8] The French would also have a sustained presence for a half-century with a trading post and then fort. From within this complex Indigenous and colonial history arises the reference to Toron:to, its meaning and cultural origin somewhat shrouded in an uncertain mist of change.

It is during the time of Woodworth's Haudenosaunee ancestors at Teieiagon that Canadien maps first make reference to the *lac* and passage de taronto. The most common view is that *taronto* derives from the Iroquoian language, either Mohawk or Wendat, and means "trees standing in the water."[9] This reference is possibly to a meeting place at the narrows between Lake Simcoe and Lake Couchiching where Wendat and Anishinaabe drove stakes in the water to create fish weirs that were in use for four thousand years. Another suggestion is that the term derives from the Wendat *toronton*, meaning both a

place of meetings and a land of plenty,[10] which also has consistencies
with the abundance of the fishing-weir meeting place. These most-
cited interpretations commonly suggest some kind of meeting place,
a view that is also consistent with Woodworth's Mohawk sensibility.
That said, according to the oral tradition held by Mohawk Peace
Chief Jacob Thomas, it was not a fishing weir made of logs but rather
a fallen tree, or *dolon-do*, perhaps an old onerahtase'ko:wa, that
marked this portage route close to Lake Ontario, or *Ontar:io (ondar-
io)* in Mohawk, the "beautiful waters."[11]

Modern Toronto has been shaped by the passages of its past, even
if that past is often severed from our present consciousness. The city
was situated here in the 1790s because of the importance of the lake
and rivers for controlling the region's colonial economic networks,
and as such it continued "patterns that had been in place for thousands of years."[12] The significance of this place is reflected in the
changing relations it has witnessed, including those between United
Empire Loyalists and Anishinaabe starting in the 1790s; Canadiens
and Anishinaabe throughout the eighteenth century; Anishinaabe,
Canadiens, and Haudenosaunee over the latter half of the seventeenth
century; Haudenosaunee, Wendat, and Canadiens in the preceding
half-century; and the ancestors of the Wendat and Anishinaabe in the
pre-colonial period. The more recent global migrations that have
made Toron:to one of the most culturally diverse cities in the world is
following a deep pattern, one that can be seen as having a sacred
grounding in the onerahtase'ko:wa meeting place.

Listening to Woodworth's Haudenosaunee sense of Toron:to as the
river flows past, we are offered passage to a different ancestral way of
minding the urban rush that pervades life today. Places of this nature
can be found wherever we are in this country if we take time to dig
below the modern surface of asphalt and concrete. These waters are a
good place to begin contemplating the deep colonial severances that
fuel Canada's role in our current crisis. Urban lives are no more disconnected from the colonial and pre-colonial ancestries of the places we
live in than they are from those wild ecological and climate relations
highlighted in Livingston's ecology of mind. Though this chapter is
situated in relation to Woodworth and various Haudenosaunee writings, it is not my intent to suggest a definitive Haudenosaunee sense of

what Toron:to means. On this misty passage de taronto, I prefer to stay
in the mystery of multiple interpretations that overlap and diverge
across this place's changing Indigenous histories, affirming Thomas
King's insight that we are the stories we tell.[13] That said, the chapter is
situated in the ancestries that Woodworth and I relationally embody in
this place, with the goal being to temporally deepen, interculturally
broaden, and lovingly situate the interdisciplinary knowledge of a good
Canadian climate of mind.

One rainy spring day Woodworth and I deepened our conversation
on the ancestral presences of Toronto as the venue changed from a
class along the river to a downtown conference at the University of
Toronto. A couple blocks south of the buried Taddle Creek, we co-
presented a paper entitled "Recovering the Tribal Identity of Toronto:
Ancestral Conversations" that was concerned with land compassion,
colonialism, and Indigenous Toron:to.[14] While our focus was on the
"conscious amnesia of the deep cultural roots and history" that
informs modern ways of living in the city, the dialogue was grounded
in our ancestries. As descendants of Haudenosaunee, British, and
Canadien cultures, we embody various historic relations that had significant meetings along these waters.[15] My starting point was
Livingston's marshy flights of mind beyond Hekademos and the rising awareness of my ghostly way of walking through Toronto as if I
did not live here, as if it was a place on the way to somewhere else.
The ancestral tenor of Woodworth's words clarified for me the colonial dimensions of this stance by displaying the quality of a reverential Good Mind as it engages the meeting place where the great
onerahtase'ko:wa fell.

The Ashbridges Bay vision that Woodworth had of the Toron:to
skyline arose while he was performing a Thanksgiving ceremony along
the lake shore. It is a ceremony he was instructed to carry out by his
elder Thomas, who had taught Native studies at Trent University for
fourteen years and lived in Six Nations, about one hundred kilometres
west of Toronto, until the end of his life in 1998. It was in Thomas's
last years that Woodworth came to learn from him Haudenosaunee

teachings on the human place and role in creation, also known as the Good Mind. In his book *Teachings from the Long House*, Thomas describes the Good Mind as being based in a code of "peace, power, and righteousness," a discipline that conscientiously must confront "forces of war and destruction" focused on undermining good ways of minding relations.[16] The lessons Woodworth learned were grounded in this ancestral wisdom while not being stuck in that history, for these teachings were also concerned with present realities such as "healing the profound rift between the indigenous and the Western mind."[17] The way to do that is, according to Thomas's instructions to Woodworth, by resuming this land's traditional duties. Foremost of these duties "is Thanksgiving, which is carried out with the assistance of tobacco burning, and is comprised of gratitude for the many forms of Creation beginning on the earth and ending in the sky."[18]

It was at Ashbridges Bay that Woodworth began regularly leaving thankful offerings of tobacco for the Creator while asking for guidance from the land's ancestors about the duties he was to fulfill.[19] The act of performing this duty began having a profound impact on Woodworth's sense of the city he was living in. As he writes, he began "to receive the messages held in the modern landscape of our ancestral homeland," messages that transformed how he saw the buildings at Toron:to's core and understood the city's name.[20] It was not simply the Thanksgiving ceremony that inspired Woodworth's vision of the fallen dolon-do as an onerahtase'ko:wa, for he had also discussed the Haudenosaunee oral sense of Toron:to's name with his teacher. He was taught that this term is rooted in the Mohawk word dolon-do, which refers to a damp log and not sticks. Another student of Thomas, Brian Rice Natoway, likewise interprets the word as referring to logs, not sticks placed in the river.[21] But Thomas further stressed to Woodworth that there had to be more to this reference. Why would a place be named simply after damp logs unless there was something significant about it? By following the duty of giving thanks, Woodworth was instructed by his elder that he could learn more about this mystery from the place itself. It was in doing this that he saw the dolon-do as a great onerahtase'ko:wa which, because of its significance in the Haudenosaunee Good Mind, made it an important marker for stopping on the watery passage today known as the Humber River.

The Haudenosaunee teachings of Thomas suggested that there is a way to consciously give the ancestors and land space to help us reimagine who we are, where we are, and where we are going, and that is through performing duties traditional to the shores of Lake Ontario. As Woodworth explains, his tradition assumes that all acts of imagination and creativity "are, in the most 'real' ways, acts of memory, deep remembering, ancestral memory."[22] A connection between creative knowledge and a sustained relation with ancestors is common to many Indigenous and pre-modern Western traditions.[23] Years earlier I had learned that Inuit knowledge and silatuniq is described as being "passed down to us by our ancestors, things that we have always known, things crucial to our survival, patience and resourcefulness."[24] Even in an urban reality like Toronto, the presence of the ancestral past in the land is seen as vital to the many Indigenous peoples who currently live here. From the Haudenosaunee to the Anishinaabe and Wendat, the Indigenous past and sacred are "one and the same, still existing at very specific sites but also experienced as being everywhere in the city in a largely invisible but unbounded way."[25] This is also of concern to Indigenous people whose ancestors are buried far from this city and who desire to connect their past to local Indigenous histories.[26] Such a genealogical focus on ancestry adds dimensions to the situated ecology of mind tradition that often has a kind of autobiographical dimension in its focus on the individual's relationality, as reflected in the telling of Livingston's initiation in the previous chapter.[27]

The Indigenous duty to remember the ancestors that Woodworth performs arises from a particular Haudenosaunee way called the Good Mind. Also learning under Thomas, Roronhiakewen (He Clears the Sky) Dan Longboat clarifies that central to this tradition is a sense of imagination that significantly differs from the view on the modern ship. It is not a disconnected fancy but rather "a homing device for finding a way into the sacred unity of time, mind, spirit, and place," a way of going beyond the limits of knowledge.[28] This is consistent with what Haudenosaunee scholar Marlene Brant Castellano identifies as three knowledge sources: "traditional knowledge (from generation to generation); empirical knowledge (gained from observation); and revealed knowledge (acquired through spiritual origins and recognized as a gift)."[29] It reminds me of a discussion I had with Jaypeetee Arnakak around Inuit silatuniq and wise knowledge of land and sila relations. At one point he explained that he was intrigued by Buddhist meditation and other contemplative practices he had read about since they reminded him of the trance-like states that he enters when fishing or hunting, a focused and yet de-focalized consciousness.[30] The cultural and ecological knowledge about the animal, water, or land at some point drops below awareness when one is in a good state of mind for hunting, and a different intuitive or imaginal understanding guides knowledge and actions. This is central to silatuniq, and it seems also to be vital to a Good Mind.

Being an architect with extensive empirical knowledge in this area, Woodworth's practice of the Thanksgiving ceremony took him into an imaginal space where the memory of the ancestors bubbles up from the unconscious throughout the city's built form. This ranges from the CN Tower Prayer Stick to the Medicine Wheel Skyscrapers to features of the land near the former village of Teieiagon above the Humber River. But what he suggests in outlining this vision is not simply a conscious Haudenosaunee approach to the ancestors through Good Mind ceremonies, for he also seems to recognize an unconscious past that continues to shape the city's creation. Inasmuch as there is room for creative inspiration to inform the art of any architect, an internal space is opened where unconscious influences can make impressions on the mind. Though these are often of the individual's cultural or historic grounding, Woodworth proposes that the ancestral presences in the land can also inspire those who open such vital spaces even if not directly engaged, even if actively forgotten. Going one step further, he believes that the conscious evocation of ancestral ceremonies can create larger fissures from within which the Good Mind comes to a very different vision of Toron:to's skyline as colonized and yet holding open the possibility for renewal.

For Woodworth, revisioning the city's name around a fallen dolon-do is one symbol that he imagines rising from the place's ancestral origins. There are still onerahtase'ko:wa groves on the Humber Plains that, when I enter, immediately slow my walk down and take me into a distinctly different Toronto. The tallest trees reach straight into the sky about twenty-five metres with their evergreen branches, the bare lower portion of the trunk allowing one to see farther into the distance. Covering the forest floor is its distinct cluster of fine needles that, in summer, create a unique and inviting aroma. These trees are descendants of those that lined both shores of the lake prior to the colonial and then urban deforestation of southern Ontario, and despite their smaller size they give a peaceful feeling when one comes into their presence. It is to this place that my mind goes when I listen to Woodworth speak of a Toronto named after a fallen dolon-do. His onerahtase'ko:wa-inspired Good Mind has deep resonances with the bird's-eye view of Livingston as it takes me into felt relations with surrounding presences and yet clarifies other dimensions like imagination, the role of ancestors, and ceremonies like Thanksgiving.

While experience is of primary importance in both these traditions, John Mohawk makes it clear that experience is not on its own sufficient in the Haudenosaunee minding of relations. In the same spirit as Thomas's teaching about the Good Mind's imagination, he asserts that "on the one hand there are dreams and visions, and on the other hand there's a responsibility to maintain a clear version of reality, and those two streams of thoughts and reaction have to live cooperatively."[31] Similarly, recent thought on ecological thinking highlights the importance of becoming attuned to the complex realities that surround us. Attunement is based on sustained relations with the land as an integral "mode of thought that, among other things, synthesizes more specialized modes of thought."[32] We are thinking here not about ecology but rather about ecological realities that require us to imagine ways of synthesizing multiple disciplines of knowledge. The situated interdisciplinary knowledge of an ecology of mind needs to be supported with an experience that can offer other felt perspectives, a sensibility consistent across traditions. It was a recognition of these similarities that led Mohawk to propose that the Haudenosaunee Good Mind is "headed into a new space" where it may have "vast allies in the industrialized world," a thought he was surprised to have.[33]

At the same time, the high reach of the onerahtase'ko:wa into the sky and its evergreen quality are symbolic of the Good Mind's participation in a broader cosmological reality that is rarely contemplated in the ecology of mind tradition. As Longboat explains, the Good Mind is "that part of the human mind, that tradition tells,

Shonkwaya'tison (the Creator) took part of His mind and placed inside the minds of the first human beings."[34] Its spirited roots can be traced further back to the Creator's grandmother, Sky Woman, who initiated the creation when she fell from the Sky World to Earth, or Turtle Island as it is referred to by Haudenosaunee. According to Rice's telling of this myth, which is more fully examined in the book's last three chapters, there was a great onerahtase'ko:wa at the centre of her Sky World where her uncle's body rests. In his telling, this tree connects "the different worlds of light" that radiate pure light energy and life from the worlds above to those below.[35] Most versions of the story refer primarily to a great Sycamore tree of lights that is uprooted, leaving a gaping hole through which Sky Woman falls toward Turtle Island. Though Rice's addition of the onerahtase'ko:wa is unique, it can be seen as an extension to two trees of what is an archetypal *axis mundi* symbol in the human imagination. The *axis mundi* is the creative centre that energetically connects all the worlds, often conceived as a tree, mountain, or some other pivotal landmark.[36]

From the Sky World to Turtle Island is gifted a reverential way of relating that humans can manifest through minding the onerahtase'ko:wa's high reach and performing the ceremonial duties of a Good Mind. These actions are meant to foster "love, respect, honesty, caring, compassion, sharing, happiness, health, peace, love," all of which connect people to something more than a narrow self-interest.[37] A central ancestor in this teaching is the Peacemaker, who, according to Thomas and his students, returned the message of peace to the hearts and minds of the Haudenosaunee people during a time of turmoil.[38] Prior to the colonial period, the Peacemaker brought the original Five Nations of the Haudenosaunee Confederacy together after much war and violence. With the help of Ayenwatha, the two of them approached each nation with the aim of healing the pain of loss from years of war, as well as offering an alternative good way of relating to each other. In the words of Thomas, the Peacemaker taught "that when you work together, you exemplify peace, which in turn results in power of mind."[39] This is epitomized in a rising Sun that "will get brighter and brighter and bring warmth and comfort to the people." It is the energetic source of the onerahtase'ko:wa's skyward reach and its deep white roots that stretch out in the four directions as a symbol for the connecting peace of all nations.

The Good Mind talked of by Woodworth participates in renewing peace by performing duties that symbolically remind people of everything their life depends on. During a period of turmoil, the Peacemaker helped the Haudenosaunee see, in his words, their real light nature, and thus "reanimated their self perceptions with the memory which returned them to their innate 'good mind.'"[40] As a continual reminder of this transformation from conflict to peace, the onerahtase'ko:wa tree of peace was planted in the middle of the Confederacy, with the "Four Great White Roots of Peace spreading to the four directions" and the destructive weapons buried underneath.[41] This peace is symbolized in the five-needle cluster of the tree, the Ayenwatha Wampum Belt that tells of the Confederacy's origin, and the nature of a Good Mind whose aim is to spread the peace.[42] The Peacemaker, Sky Woman, and Thomas are some of the ancestors with whom Woodworth engages in imagining the significance of a fallen dolon-do on this watery passage that became the city of Toronto. This tradition will offer a way of ancestrally grounding Livingston's bird-eye view as we come toward a braided minding of what underlies our current climate of change.

While Woodworth remembered the Indigenous ancestry of Toron:to in our University of Toronto talk, I pondered the human role in climate change and the need for fostering something akin to the silatuniq of Inuit. Each return flight from northern research or a related conference concluded with a view of Toronto's seemingly neverending urban grid that surrounds the airport, the Humber Plains, and my home. These recurring vistas raised questions about Livingston's Ashbridges Bay initiation into a technological imperative, and its relation to sila's warming. In many ways southern Ontario "dominates the rest of Canada from the perspective of transport,"[43] and it has done so for the better part of the past century. Its international airport is one of the world's busiest, moving around twenty-eight million passengers per year, and thus energetically defines the region.[44] On the ground, there is in the early twenty-first century millions of daily car, truck, and rail trips within the city's borders.[45] Before engaging Woodworth, what invariably struck me on my return to Toronto was not the onerahtase'ko:wa or its built version the CN Tower Prayer Stick, but rather the sheer extent of a transportation hub that recapitulates older patterns of the passage de taronto. I land in the midst of a portage route that "functioned as the pre-contact equivalents of Highway 400,"[46] one that gives us a way of further situating Livingston's painful severance in an earlier colonial period.

Sila's northern warming as a particular manifestation of global climate changes is very much connected to issues of power in long-established systems. On the largest scale, more than 80 per cent of human-based greenhouse-gas emissions, since the Industrial Revolution, have arisen to the south of the Inuit in Canadian, American, and Western European centres of power. Despite playing a limited role in these processes, the Inuit are experiencing a changing north that is affecting animals, the land, ice, sila, hunting, and communities.[47] The climate's global scale positions everyone in the issue in much the way that Indigenous thinkers conceive of "insider research" as an approach that, in contrast to the modern scientific ideal of objectivity, recognizes the importance of our experiencing the consequences of how we think and live.[48] I came to understand myself as an "insider" situated in a particular nation and city that is globally intertwined with sila's warming. All the intensive movement highlighted by the lights surrounding this hub emits significant greenhouse-gas emissions that not only fuel climate change but, following Livingston, also reshape our ways of minding these relations. What became apparent as I brought his bird's-eye view into dialogue with Woodworth's Haudenosaunee Good Mind is that the colonial dynamics at the root of climate change can be seen in Toronto as much as in the north.

As I stand by the flowing river to the south of the airport, my mind is taken back to a colonial portage that reached north over the Oak Ridges Moraine to Lake Simcoe, or Lac Taronto as it was originally called on a 1675 French map.[49] It is a fifty-kilometre hike from the river's mouth on Lake Ontario to the moraine that is the source of headwaters for "sixty rivers and major streams flowing south" that include the Humber, Don, Rouge, and Credit.[50] The moraine itself is an "environmentally sensitive ridge of rolling hills, wetlands, kettle lakes, and woodland" that has been recognized as southern Ontario's rain barrel, an integral part of the city's ecology and thus a focus of conservation.[51] Portaging northward over the moraine takes one into what was up until the 1650s the heart of Wendat territory, another Iroquoian Confederacy of five nations that included the Attignawantan, Arendarhonon, Attigneenongnahac, Tionnontaté, and Tahontaenrat.[52] The last nation had villages around the passage de taronto until the early 1600s, and, considering their historic geographic proximity to the Seneca who were across Lake Ontario, it should not be surprising that their language had the strongest linguistic connections with this Haudenosaunee nation. Within the Humber valley have been found "dozens of Indigenous villages and campsites,"[53] and the continuing appeal of the river came to effect the colonial founding of Toronto.[54]

Not only was the city physically etched by flowing rivers like the Humber and Don, but the European settlement of Toronto was built upon millennia of cultural patterns centred on water.[55] It is from this history that the most common etymology for the city's name as taronto, the meeting place "where the trees are standing in the water," derives.[56] These "trees" are situated in the narrows between today's Lake Simcoe and Lake Couchiching where Wendat and Anishinaabe drove stakes into the water to create fishing weirs that were used for more than four thousand years.[57] Long before the arrival of the Wendat's Iroquoian ancestors to southern Ontario some two millennia ago,[58] the Algonquin-speaking ancestors of the Anishinaabe hunted in this region and today their descendants have stories that tell of these important weirs as a yearly meeting place for harvesting fish. In 1615 Samuel de Champlain was brought there to meet Wendat and, perhaps, Anishinaabe who were gathering as a war party to attack the Haudenosaunee to the south.[59] Given the French origin of the maps and New France's close alliance with Wendat and Anishinaabe throughout the seventeenth century, there are many reasons for believing that this seasonal meeting place is connected with the naming of Lac Taronto and then the associated portage toward Lake Ontario.

The fishing-weir interpretation of taronto also resonates with the colonial focus of French traders, British settlers, and then the Canadian nation on resources from the interior, even though the primary intent during this period was on securing beaver furs. Waterways like the passage de taronto, the Great Lakes, and the Saint Lawrence River were central to the development of a nation focused on changing staples, from fish to fur to forests to minerals to our current focus on oil. In the words of Harold Innis, cheap and accessible "water transportation favoured the rapid exploitation of staples and dependence on more highly industrialized countries for finished products."[60] Escalating hostilities between the Wendat and Haudenosaunee eventually led to the dispersal of the former from central and southern Ontario, all of which occurred in a context of colonial powers vying for furs.

Along the passage de taronto, a French colonial presence begins with Brûlé in 1615 and continued with "a parade of French explorers, traders, and soldiers" for about one hundred and fifty years,[61] and over that time the Indigenous presence changed significantly from Wendat (until 1650) to Haudenosaunee (1650s–1700) to Anishinaabe. Though the first French map to describe "this region as the 'Country of the Missesagues' appears in 1755," it was around 1702 that the Anishinaabe began reoccupying southern Ontario's two most important trading sites at present-day Toronto and Kingston.[62] This was a return of the Mississauga to a land their ancestors had traversed before the arrival of their eventual Wendat allies.[63] The term Mississauga is Anishinaabe for eagle clan,[64] and according to Margaret Sault, director of lands at New Credit First Nation, it is a colonial word that distorts how they referred to themselves, namely as Anishinaabe, which means "human beings."[65] It is a naming issue they share with the Wendat, who were called Huron by the French owing to a perceived similarity between the bristly hairstyle of male Wendat and a boar's head or huré. During the Anishinaabe reoccupation of Toronto, the French presence also increased with the building in 1720 of a Magasin Royal at the passage's mouth, followed by construction in 1751 of Fort Rouillé near where the British would build Fort York.[66] By decades end, the fort had been burnt down as the French retreated eastward from British forces to their eventual defeat in Quebec on the Plains of Abraham.

The move toward modern Toronto dawned in the 1790s with the arrival of British Lieutenant-Colonel John Graves Simcoe. When his surveyor general, Joseph Bouchette, arrived at the "untamed" passage on board the ship *Mississaga*, he wrote of "dense and trackless forests" lining Lake Ontario and the "ephemeral habitation" of "wandering savages," who then consisted of two Anishinaabe families, and the French trader Jean-Baptiste Rousseau.[67] On the boat with Bouchette were a hundred Queen's Rangers and Elizabeth Simcoe, the colonel's wife. Upon their first approach on the morning of 29 July 1793, she wrote: "As no person on board had ever been at Toronto, Mr. Bouchette was afraid to enter the harbour till daylight, when St. John Rousseau, an Indian trader who lives near, came in a boat to pilot us."[68] The following day, as the soldiers cut "wood to enable them to pitch their tents," she traversed the "beautifully clear and transparent" water of the bay, disembarked, and then "walked thro' a grove of oaks, where the town is intended to be built."[69] Looking across the bay from near where Fort Rouillé once stood, Lady Simcoe thought that the wooded peninsula broke "the horizon of the lake" and thus "greatly improves the view, which indeed is very pleasing."[70] This spit that became the Toronto islands after a nineteenth-century storm was a traditional stopping place in the eighteenth century for the Anishinaabe because of the abundant fish and herbs for healing.[71]

It was the importance of the spit to the Anishinaabe that led the French to situate Fort Rouillé where the Simcoes stood, as indicated in a 9 October 1749 letter to officials in Quebec: "On the advice that we have had that the Indians of the North ordinarily passed at Toronto on the west coast of Lake Ontario … we have felt that it would be advisable to establish a post in this place."[72] For Lady Simcoe's husband, the bay and peninsula were ideally positioned for similar reasons. Following the grooves of the past, he saw the possibility of using the passage de taronto to access Lake Huron and thus bypass the American presence at Detroit and retain British control of the fur trade.[73] At this time, Upper Canada had a sparse population of about ten thousand living amidst forests that were still "best penetrated by the rivers."[74] Guiding his first exploration of this passage was the French trader Rousseau. Portaging a big swampland near the lake that now bears Simcoe's name, they crossed paths with an Anishinaabe hunter who informed them of an alternate winding trail that eventually became Yonge Street.[75] His explorations were secondarily concerned with settlement and primarily with "strategic roads of military significance," and this trail served that purpose.[76]

While Yonge Street would link the town with Lake Simcoe and Georgian Bay, Dundas Street was created to bypass Niagara with the goal of eventually linking east with Montreal.[77] From the bay's peninsula to these main streets to the eventual grid of roadways, Toronto was built with a military mind at the helm.

A related concern for Simcoe was replacing the French and Indigenous references to this place. For a while Toronto became York before reverting when it was incorporated in 1834, Cataraqui was permanently changed to Kingston, and the passage de taronto was renamed the Humber after a river near Simcoe's estate in Devonshire, England. His approach resonates with Northrop Frye's elucidation of a "garrison mentality" that he saw as rooted in a fear "of being silently swallowed by an alien continent."[78] Long before the direct route of airports, the colonial immigrant to Canada edged, Frye would write, down the Saint Lawrence River into the land "like a tiny Jonah entering an inconceivably large whale."[79] The wild setting they arrived into was nothing like the Europe they left. In 1810 Simcoe's provincial capital had the relatively small population of six hundred settlers, mostly government and military personnel, due to a reputed isolation that was "palpable, and daunting."[80] With its weather fluctuating between icy winter and hot humid summer, one British sentry declared that "only a strip of paper separated the land from the fires of damnation."[81] Another new arrival described the passage de taronto's forests as "a perpetual gloom of vaulted boughs and intermingled shade, a solemn twilight monotony."[82] It was such sentiments that led Frye to argue that a garrison mentality influenced Canadian urban development as "an arrogant abstraction, the conquest of nature by an intelligence that does not love it."[83]

Other features of this colonial mentality arose in a dispute about the Anishinaabe land transfer to the British. For the Anishinaabe, "the common-sense boundary" of the emerging settlement was the Credit River, while Simcoe's military surveyors "insisted on a straight line" that disregarded "the numerous streams flowing southeastward into Lake Ontario."[84] The "arrogant abstraction" of a road grid was to define the city and region rather than its ecological contours. These issues actually began earlier, in 1787, when the Indigenous peoples' cession of the area that became Toronto was negotiated at the

Carrying Place on the Bay of Quinté – the birthplace of the Peacemaker of the Haudenosaunee Good Mind tradition. The initial agreement with three Anishinaabe chiefs was for an imprecisely defined land in exchange for £1,700 worth of kettles, mirrors, guns, tobacco, and rum.[85] Following the boundary disputes, a more thorough negotiation with eight Anishinaabe chiefs in 1805 created the Credit River agreement. It occurred in the context of an increasing settler population that was encroaching on hunting lands and resulted in the ceding of "392 square miles from Ashbridges Bay to Etobicoke Creek."[86] The Anishinaabe eventually moved west in the 1850s to New Credit near the Haudenosaunee ancestors of Thomas at Six Nations who were facing similar issues. As Longboat relates, Simcoe insisted that "there could be no direct leasing by Indians to Crown subjects," a policy that led to a shrinking territory due to permanent land sales, surrenders, and "successful court actions on the part of squatters."[87] According to the Six Nations, a 950,000-acre territory was reduced by 1850 to the 46,000 acres that currently remain. The overall impact of the garrison mentality was that Indigenous people became "part and parcel of a subdued land and very much relegated to the past," or so it seemed.[88]

Though the city recognized its Indigenous past by including "a Native person on the Coat of Arms" in 1834, Anishinaabe scholar Rodney Bobiwash points out that "the person portrayed is obviously in Plains and not Mississauga dress" and thus further signifies the progressive and romantic alienation of an Indigenous presence to somewhere else in time and place.[89] The romantic view of wilderness discussed in chapter 1 as areas relatively uninhabited by humans is, in North America, built upon a colonial history of removing Indigenous peoples from those places in much the same spirit as Toronto's ancestral amnesia. They are two sides of the same coin. To manifest this multi-dimensional severance in southern Ontario required removing 80 per cent of the region's original forests and distancing Indigenous from non-Indigenous populations. In such a setting, wilderness then becomes a place always devoid of the human, while the urban simultaneously enacts a violent removal of Indigenous presence as "names are changed, and a steady Europeanization" transforms the land.[90] These are the outward processes that aim at

furthering the amnesia of a garrison mentality by anxiously repress-
ing and severing the ancestors.[91] As Sault laments, "if we were closer,
we could keep reminding people of our presence. 'Out of sight, out
of mind.'"[92]

With the mid-nineteenth-century Industrial Revolution, Toronto
built roads, railroads, and factories upon the colonial foundations of
Simcoe's military grid as its population expanded. By the early twentieth century, Toronto's relation to resources from the Canadian Shield
and western provinces made it an important financial centre, with its
stock exchange surpassing Montreal's in the 1930s and the city becoming the nation's largest in the 1970s.[93] These changes partook in the
increasing urbanization of Ontario and Canada, with the trend continuing into the twenty-first century with approximately 80 per cent
of Canadians now living in cities (six million in the GTA alone).[17] This
urban growth was paralleled by a sixfold increase in personal vehicles
between 1945 and 1975, as well as an additional "40,000 kilometres
of highways and roads in a system centring on the Toronto region."[94]
But it is not only that roadways and automobiles have continued to
rise in number in Toronto and worldwide; our drive has also intensified with vehicle-kilometres driven, increasing by more than 2.5 times
between 1970 and 2007.[95] This is the larger historic context that
Livingston's Ashbridges Bay initiation occurred within as past realities
were and are steadily subsumed within ever-growing colonial, industrial, technological, and urban processes.

In just over two centuries, this still expanding grid has participated in a worldwide urbanization that has become a significant
source of the greenhouse gases fuelling our climate of change. On a
yearly basis, the city averages 6,691 million litres in automotive fuel
sales and the international airport uses 1,830 million litres of jet
fuel.[96] While the north undergoes significant warming, the fossil-fuel
culture that Toronto participates in has brought a more modest local
rise in mean temperature of 0.6°C, recent records in summer heat,[97]
and smog increases that contribute "to about 1,700 premature
deaths and 6,000 hospitalizations" each year.[98] But Livingston suggests that the issue goes beyond these external changes, for his concern with "urban delusions" is that a powerful technological drive
is also transforming the way we mind relations. The colonizing
"garrison mentality" becomes a "technological imperative" that is

escalating our climatic crisis. Such a view calls into question long-assumed modern ways of living, from unrestrained transport to sprawling urban development to the speed of colonizing dynamics. Based on the Good Mind teachings of Thomas, Woodworth describes the persistent imperative of the colonizing mentality as "rolling heads." This way of minding is always on the move, with a driven pace that keeps it unconscious, or semi-conscious, of where it is situated on Turtle Island, of the ancestors in its midst, and of why Thanksgiving is vital to a good climate of mind.

The CN Tower that rises above the lighted grid I land in became a beacon for me to more closely attend the relations between our global climate of change, urban greenhouse-gas emissions, Toronto's colonial history, and this place's ancestral relations. This is also the central purpose of Woodworth's foundation, A Beacon to the Ancestors. Its goal is to raise awareness of Toronto's Indigenous and colonial history in its present global context. Though our starting points are differently situated, we share a concern with minding a severed past in light of current realities. In Woodworth's Haudenosaunee Good Mind, this means remembering and recovering "the practices that the ancestors of this place have already been given."[99] The CN Tower is not only a prayer stick calling people to this meeting place from around the world, but also a symbolic beacon that recalls the tall onerahtase'ko:wa and Good Mind duties performed on these shores. As with Livingston's painful paradox, the symbol of a "beacon" has two sides for Woodworth. It highlights "the impending shadow of danger in the darkness, and is a kind of protective light, warning device, as well as herald of an emergence from dark into light."[100] The darkness of a fallen dolon-do bears within it the hope that a peaceful meeting place can be renewed, that there is a reverential passage beyond an ever-colonizing rolling head and the climate of change it has initiated.

The two-sided nature of this beacon returned our conversation to the passage de taronto and a particular seventeenth-century moment of ancestral Haudenosaunee-French relations. On the ridge above where my class listens to Woodworth was a Seneca/Mohawk village of 5,000 people at its largest that was known as Teiaiagon, "the knife that cuts through the river."[101] Despite the earlier alliance of the French with the Wendat who had been displaced from southern Ontario, a tentative peace brought explorers like René Robert La Salle, *coureur de bois*, and missionaries to Teiaiagon.[102] Perhaps the earliest written record of this place was penned in December 1679 by Franciscan Récollet missionary Louis Hennepin while aboard a French ship in the same Humber River mouth that I often look upon from the Howard tomb. He wrote of trading for "Indian corn" with the village and being anchored in a freezing bay where they "were forc'd to cut the Ice with Axes and other Instruments" before the wind changed on 15 December, allowing them to sail to Niagara.[103] Throughout the decade or two of these relations, tension simmered and occasionally flared up at Teieiagon in singular conflicts like the early 1680s death of some coureurs de bois.[104] It boiled over in the 1687 French campaign led by Marquis de Denonville who, allied with the Anishinaabe, pushed the Haudenosaunee south of Lake Ontario. It seems that Teieiagon was abandoned or destroyed at this time.[105]

For someone of Canadien ancestry, the passage de taronto is a particularly good place for me to locate Livingston's initiatory pain in the actions of colonial ancestors. One who is particularly relevant is the coureur de bois I share a surname with, Antoine Leduc, who was killed by Haudenosaunee either in or while departing Teiaiagon. He was one of five Frenchmen with this surname who came to New France from Normandy in the latter half of the seventeenth century and found himself in this village at around forty years of age. There are two slightly different stories about the circumstances that led to his death. One comes from a 1682 memoir about the emerging war with the Haudenosaunee which suggests that the Seneca intended to escalate hostilities and thus began that "year by plundering three Frenchmen at Tcheyagon [Teiaiagon]," with Leduc being one of the three.[106] The other rendition, told by one of Antoine Leduc's descendants, tells of him coming to his end after participating in Denonville's campaign.[107] Drawing upon genealogical research and familial documentation, this story is that he, a couple of coureurs de bois, and two Wendat attempted to avoid the fallout of the military conflict by staying away from Lake Ontario and portaging up the Oak Ridges

Moraine toward Georgian Bay, Lake Huron, and then the French River where he was eventually killed. This incident is consistent with other records of a Haudenosaunee force "snatching up and torturing several straggling French soldiers they encountered along the north shore of Lake Ontario."[108] These are the ancestral relations that situated my dialogue with Woodworth as he described Denonville as conducting a scorched-earth campaign on both sides of Lake Ontario.

The French campaign of Denonville burned down villages, fields of the three sisters (maize, squash, and bean), and the extensive tracks of forests that included the great onerahtase'ko:wa.[109] Its excessive nature is one historic symbol for Woodworth of a colonizing rolling head that has little regard for where its severances are enacted. In much the same spirit as the garrison mentality, the teachings of Thomas guide Woodworth to questions about the seemingly never-ending movement and driving pace of the ship's colonizing mind. We are perpetually footloose, ready to find greener pastures without really getting a sense of where we are. As Thomas explains, elders "used to say it wasn't good to wander too far from 'home' – that is was dangerous" – and people do it only under exceptional circumstances.[110] A rolling-head mentality tends always to move quickly to other horizons beyond the presences in their midst, an approach I embodied in my original view of Toronto as a place for school or a job that, I hoped, I would pass through, not live in. The result is that we sever any significant relation with the Indigenous, ecological, and climatic past as they manifest themselves in our presence, a process that began in periods of war like Denonville's campaign and then, as is detailed in later chapters, became more comprehensive with the epic modernizing of the nineteenth and twentieth centuries.

From Leduc and Denonville to Livingston and Thomas, such stories ask us to find ways of ancestrally situating our lives in a Toronto, or wherever we live, that may no longer offer the same diversity of presences as once existed before its colonial, industrial, and urban transformations. As both Woodworth and Longboat explain from a Haudenosaunee Good Mind perspective, there is decreasing space to imagine relations with the ancestors because of both external changes to the land and internal transformations of how we live in this place. The imaginal presence of the land has "diminished in ratio to the decline of its biodiversity and so do cultures and languages rightfully allied to that biocultural diversity."[111] There is an interesting resonance between this critical understanding and what environmental education theorist Peter Kahn refers to as "Environmental Generational Amnesia." It is not only that our modern ecologies have been transformed over time, but our childhood experience of that progressively degraded ecology becomes a norm which is seen as the non-degraded condition.[112] My experience in the Humber ravines is of a degraded ecology when compared to Livingston's childhood, and his was likewise compromised in comparison to preceding colonial and Indigenous experiences. The severance that emerges from not belonging anywhere seems to make it more difficult to engage a bird's-eye minding of our relations, let alone also find a place to ground that view in an ancestrally rich Good Mind.

The diagnosis of a progressing amnesia helps to further highlight other dimensions of Livingston's two-sided paradox: the painful urban recession of ecological spaces that help us to remember that our present-day world has a colonial grounding. More than that, today's technological quickening is further abstracting us not only from our ecological and climatic dependence but also from the intergenerational support of the ancestors. Beyond being severed from an Ashbridges Bay, Taddle Creek, or Humber Plains, there is a progressive forgetting of what brought us to this place and time due to our immersion in the encompassing power of technologies like those vehicles that currently light up Simcoe's grid. There is an increasing recognition that our urban roadways "structure the experience of the environment and separate us physically and temporally from the world through which we pass."[113] The modern ship is like an individual who has decided to drive "a car by concentrating on its performance as indicated by the instruments on the dashboard as opposed to watching the road."[114] From such a dashboard experience, knowledge is dominantly conceived as information disconnected from culture, let alone wisdom, and is refined in what Livingston characterized as the ever-expanding technological imperative.[115] As we saw in the previous chapter, this tendency can be seen in the way climate researchers focus on the traditional ecological knowledge of Inuit and other Indigenous cultures because of the way it can be utilized in established models, all the while marginalizing culture and silatuniq. The earlier documented calculus of Toronto vehicle-energy efficiency, vehicle-kilometres driven, and fuel sales, to which we could add carbon markets, is informative while also indicative of this "dashboard tendency." The danger is not with management calculus but with how its prolific nature alters "our consciousness so as to make us almost unaware of our condition" and thus weakens critical insights that could help us escape a rolling-head pattern.[116]

Going one step further into the Good Mind of Woodworth and Thomas suggests that we need to renew our minding of not only such management calculus but also the broader Hekademos and modern tendencies of attempting simply to read and research our surroundings as texts. As was touched upon in the Introduction, the Haudenosaunee sense of ancestors engages the "objects" and "texts" of concern to the modern academic as presences that we need to be receptive to. For Woodworth, this duty rises up from the land and the ancestors it entombs, even in a deeply colonized city like Toronto. In fact, his "principal premise" is that this urban reality continues "to express the presence and even intentions of its place-specific Aboriginal Ancestors,"[117] a view that he holds in common with Anishinaabe and Wendat who continue to live here.[118] There are cultural differences across these Indigenous ways of minding relations, but there are also recurring commonalities. One is the sense of knowledge as something that "cannot be owned or discovered but is merely a set of relationships that may be given visible form."[119] Situated knowledge is ultimately grounded in relations that transcend what can be rationally known by a driven modern ship, and for this reason an Indigenous Good Mind dutifully engages the ancestral, ecological, and climatic presences – not objects or texts – in our midst.

Another rolling-head difficulty highlighted by both Woodworth and Longboat is that people are having regular experiences of the sacred in nature but have little sense of the protocols and duties that can help mediate those encounters beyond the textual reading, management calculus, and recreation of the modern ship. As ancestral ways of minding are severed first by colonial, then industrial, and now ecological and climate changes, another not unrelated issue presents itself in urban places like Toronto. Not only are many of us first- or second-generation migrants here for school, work, or some other reason, we have from an early age been aligned with a technological culture that makes place-based knowledge and ancestral stories seem irrelevant to our daily lives. The issue of being locally decontextualized becomes more stark for many urban dwellers who have no intergenerational connection to the place in which they live or any ecological sense of present and past cultural ecologies that for many Indigenous peoples are embedded in stories like those of the Peacemaker and the onerahtase'ko:wa. This colonizing amnesia is the ground out of which environmental amnesia emerges.

To follow Woodworth's beacon suggests another duty for a Canadian climate of mind. We must identify ways of remembering the place's past in relation to our ancestral roots, regardless of where we come from and what passage we took to arrive here. The ancestral further situates us in broader historic relations with city, nation, culture, ecology, climate, and, as we come to see in later chapters, cosmology. Though we have to heed these critical concerns about our forgetfulness, the Ashbridges Bay experiences of Woodworth and Livingston suggest that there are still places where presences can foster other imaginings and remembrances of where we live. Such experiences have a certain consistency with critiques of Frye's garrison mentality as being overstated. As present-day Toronto naturalist Wayne Grady writes, "Canadians are not afraid of nature" and Toronto is not a garrison, or at least is quite ineffectual in performing that role given that "there is as much of nature in the city as there is out of it."[120] This city was built over the wilderness and thus does not "always displace the plants and animals that live there"; rather, it continues to be part of that changing land.[121] This place's Indigenous presence also was not completely severed following the expansion of Simcoe's grid, though Bobiwash clarifies that Indigenous peoples did not "re-emerge again in any significant number until after the Second World War."[122] There are many historic stories that trouble any sense of the rolling-head mentality as being universal across the colonial past.

The first two centuries of contact in Canada are filled with examples of colonists like French coureurs de bois who engaged Indigenous ways and came to live in places across the nation as settlers, not as conquerors mentally trapped in a fearful garrison.[123] It is an issue that

John Ralston Saul addresses when he critically nuances Frye's garrison mentality as being largely only a reality "in the mind of colonial elites,"[124] like Simcoe and before him Denonville. Though many colonists recognized the land as being uncontrollable, this did not mean that it was only feared. Taking the rolling-head mentality a step closer to the Haudenosaunee Good Mind, many colonials who wrote of their experiences actively recognized a fearful unknown as a transitional "first stage in an imaginative ascent."[125] The land was "not a wilderness of horrors but a space of opportunity" at the edge of Western European power,[126] and its navigation often required close relations with Indigenous peoples. As is detailed over the next few chapters, alongside troubled relations during this early colonial period with Haudenosaunee that led to participation in official French war campaigns such as that of Denonville, the likes of Brûlé and Leduc had significant friendships with Wendat and Anishinaabe. These ties transformed the way in which the Canadien cultural traditions of coureurs de bois came to mind waterways like the passage de taronto and the forest ecologies lining them. The beacon-like tension between a rolling-head severance and deep reverence for the land was in many ways a formative, though often marginalized, experience of Canada.[127]

Both Livingston and Woodworth encountered the paradoxical nature of this two-sided beacon, though in a context of a quickening urban pace that makes such an imaginative bird's-eye minding increasingly difficult to ground anywhere. Lighting the beacon where the great onerahtase'ko:wa, dolon-do, fell requires for Woodworth a deep remembering of the land's Indigenous presence. Central among his ancestors is the Peacemaker, the one who "indicated that we would have to welcome peoples from the four directions of the four great white roots of the white pine tree to sit in consensus-building council," thus extending his original act of bringing together the Five and then Six Nations.[128] But it is not simply "good" ancestors who need remembrance, for we will see over the following pages that those of a more mixed nature like Brûlé, Leduc, Simcoe, and Denonville also have important stories to tell us on the passage to renewing relations. This embrace of ancestry is reminiscent of the critical genealogy outlined by Jungian psychologist James Hillman when he calls for a grounding of ourselves in a "family tree, including its twisted rotten branches."¹²⁹ The aim is not to reconcile but to engage the different dimensions of our ancestral participation in both the good and the bad, though with an intent of leaning toward the light of the former as a guide for future acts.

People from around the world are finding passage to the CN Tower Prayer Stick along Lake Ontario, and this towering beacon can symbolically assist us in remembering that there have been other ways of minding this place – ways that may be helpful in navigating the increasingly turbulent passages of our climate of change. While Woodworth's Toron:to vision is about recovering the ancient ceremonies of his Haudenosaunee ancestors, it is valuable because it does not discount Anishinaabe or Wendat presences in this city and southern Ontario. My intent is not to appropriate Indigenous understandings outside their cultural or regional contexts but rather to enter into a dialogue with ancestral ways of imagining relations beyond our modern rolling-head drive. In much the same spirit as the Haudenosaunee Good Mind, Hillman further writes that affirming our ancestry inevitably "ties you down with duties and customs" that require declaring your full attachment to our common waters on this planet.¹³⁰ There is a numinous feel to following such an approach of the ancestors that resonates with Woodworth's beacon, Livingston's paradox, and the mist surrounding Toronto's meaning along this old passage.

Walking a few feet north of the onerahtase'ko:wa grove on the Humber Plains, I come upon one of these trees laying across the bottom of a ravine where two creeks meet. It reminds me of Woodworth's dolon-do. The base of its trunk is about one-half metre wide and its length spans some twenty-five metres as it stretches out of the creek up the ravine and slightly above my head on this path. While not the sixty-metre tree of its pre-colonial ancestors and Woodworth's imagining, it still offers an inviting bough of cover that provides me with a sense of his impression of this meeting place. A great one of these evergreens fallen along the portage could have been an important marker of a spot to rest before Teiaiagon became a presence here, a shady place along this vital route that underlies our proliferating greenhouse-gas-emitting expressways. If I follow Woodworth's beacon below the hum of traffic some sixty metres to the east, I can further situate the Livingston-inspired climate of mind that began to emerge in chapter 1 on an older ancestral passage de taronto. To symbolically make that connection and help us respectfully remember the ancestral dimensions of the various places this book comes into relation with, I will from this point on use Indigenous names once they are introduced (see Glossary at end of book as a guide). Where we have started is in Toron:to on the shores of Lake Ontar:io, an ancestral meeting place that can be imagined as having sacred proportions and related duties for a good minding.

There is a symbolic richness to Woodworth's dolon-do that, as I listen to and talk with him, makes any desire to get the origins and meaning of the city's name factually correct seem superfluous. It makes me think of Indigenous storyteller Thomas King's affirmation that "we live stories that either give our lives meaning or negate it with meaninglessness."[131] Depending upon how invested we are with the stories we tell, they can change our lives or keep them stuck. The fishing-weir interpretation of taronto has a rich Indigenous background that is rooted in an important Wendat and Anishinaabe place. It also carries colonial references that remind me of the city's relation to a nation built on resources, from fish to fur to oil. That is a good story for us to contemplate because it allows us to reimagine critically where Canada has come from and what is blocking a sustainable way forward. In a profound way, our climate of change can be seen as an uncertain beacon from a phenomenal creation that is calling us to find ways of recognizing the violent mistakes and alternative passages in our ancestries. These old passages are means for renewing a spirit of peace after so much violent and painful severances. It is to this historic challenge that Woodworth's other story of Toron:to as a sacred meeting place around a fallen dolon-do speaks.

To practise the Haudenosaunee Good Mind is for Woodworth to help the Peacemaker, the Creator, his elders, and ancestors like Thomas to lift up the onerahtase'ko:wa wherever we are, even urban Toron:to. This is one of Thomas's central teachings, and it does not simply refer to physically lifting up the ecological species still found near the passage de taronto through conservation or restoration efforts. Rather, he is asking us to help lift the sacred tree skyward because when centred there, in our hearts and relations, peace can manifest itself. People are integral to enacting the duties of a Good Mind, otherwise the onerahtase'ko:wa falls just as the common ecologies, climates, and waters of the Two Row that sustain us are currently falling into disarray. All duties aligned with the Peacemaker aim to reconcile people with the ancestors and creation. This duty is consistent with the spirit that informs the role of a traditional peace chief or *royaner* like Thomas, whose position stretches back to the time when the Peacemaker chose from each nation "the wisest and kindest men to be the chiefs" and then set up a process whereby the elder women would choose subsequent leaders.[132] According to Woodworth, a royaner's central duty is never to take "sides on an issue for or against any group," for the focus needs to be on that which can spread peace in the face of severing violence.[133] It is clearly a difficult position to hold at the best of times, let alone in a context of a colonial nation's rolling-head expansion and now global climate of change.

Beyond recognizing our relation to today's crisis, the Good Mind teaches that we must consciously and peacefully find ways of raising the fallen dolon-do through duties that connect who we are to where we are and whom we are with. The ancestral depth of the Haudenosaunee tradition helped me recognize that I was engaging Livingston as an ecology of mind ancestor who, like me, was situated in the modern ship but critical of its severing tendencies. His passionate bird's-eye pleas to step outside Hekademos reaches up toward the onerahtase'ko:wa, but from a very different position that often seems to have little awareness of ancestors as vital. This environmental tradition lacks temporal depth for roots to take hold, as displayed in the fact that I became aware of it only while undertaking a graduate education. It seemingly had no connections to my family, culture, or ancestries. Something in Woodworth's Good Mind asked me to return to the mixed nature of my Canadien origins as a means of rooting me in those places and relations that made me who I am long before coming into Hekademos and Toron:to. It is to this ancestral passage that I now turn to contemplate not simply a sustainable but a good Canadian climate of mind.

3

Shapes of Violence

A shudder, a strange motion downhill into a vast confusion and a vaster sound, and one is in the pool which is the climax of the rapids of the Long Sault. So steep is the winding rush downslope into the pool and out of it along a furious curve that the rims of water close along the banks and stand higher than the tumult in the pit ... The forms of water rising and falling here, onrushing, bursting, and dissolving, have little kinship with the waves at sea, with those long bodies of the ocean's pulse. They are shapes of violence and the instancy of creation, towering pyramids crested with a splash of white, rising only to topple upstream as the downcurrent rushes at their base.

Henry Beston (1942)

Sun-drenched waves lap against the rock-lined north shore of the Saint Lawrence River, and over its waters blow a stern cold wind. A cormorant flies by the backdrop of the evergreen-dotted islands to the south, and as my eyes follow its path my awareness is carried to the columns of electric towers going southward through Mohawk territory toward New York State and northwest through the Ontar:io energy grid. Their source is the Moses-Saunders Power Dam that sits some 450 metres west of me, its turbines maintaining a constant background buzz. These waters have been a significant source of hydro electricity ever since 1958 when the energy from the rapids known as the *longue sault* was harnessed. If I closely attend the electric hum of the turbines, I can almost hear underneath it the sound of climaxing rapids "onrushing, bursting and dissolving" that for thousands of years prevailed here and informed very different ways of relating with water, rock, and energy. It is a good place for continuing

8 Power dam where the longue sault once ran (spring 2015). The power lines that stretch north and south from the Moses-Saunders Power Dam along the Saint Lawrence River in eastern Ontario harness the rapid energy of what was once the longue sault.

9 *Looking south at Lake Ontario*, by John G. Howard, 1870. This painting shows the industrial beginnings of the railroad that replaced the canoe as the primary means for getting resources to colonial markets, and it also offers another view of Lake Ontar:io before the driving severance seen today from the Howard tomb.

10 Bitumen sands development in Alberta, 2012. The current dominant resource
narrative in Canada that we discuss in this chapter has a long colonial history.

to weave Indigenous, colonial, and environmental understandings on
our passage toward a good Canadian climate of mind.

As I sit on a large rock at this place that was home for the first
twenty years of my life, it is not simply water that laps against my
being. Memories also arise of childhood fishing with my brothers and
father, camping and beaching on its shores, swimming its waters, and
driving the old No. 2 Highway along its winding route toward
Quebec. It was in following the guidance of Livingston and Woodworth
that I returned here and began searching below the river's waters for
not simply personal memories but stories ancestral to this place like
that of Henry Beston quoted above. From within the roaring sound of
a longue sault that was in its last years before being harnessed, he
wrote that one can be overwhelmed with the feeling of being engulfed
by "shapes of violence."[1] The four kilometres of large rocks, swirling
winds, and eastward-rushing water had for centuries called forth feelings of respect, fear, and, with the right skill, ecstasy. After traversing
the rapids in 1794 with her French guides, Elizabeth Simcoe wrote to
her husband: "Your sight must be terrified, tho' knowledge makes you

rest satisfied ... I should be in ecstasies if you were here to partake them."² Similarly, Beston referred to a calmness in the midst of its violence, for "enclosing the pool, in a strange contrast of mood, stands an almost sylvan scene, a country shore of grass and trees and a noontide restfulness of shade."³ This was a different ecological "shape of violence" than those industrial and colonial forms experienced by Livingston, Woodworth, and others, one that for centuries summoned feelings of ecstasy, peaceful calm, reverence, and humility.

The four-hour drive that takes me from Toron:to to the now drowned longue sault near Cornwall makes it difficult for a modern mind to imagine the winding water route which at one time brought coureurs de bois from the French centre of the fur trade in Quebec to the passage de taronto. Prior to the creation of Loyalist settlements like Cornwall, the route Étienne Brûlé, Antoine Leduc, and others took often headed northwest from the Saint Lawrence River up the Rivière-des-Outaouais near Montreal's western edge over to Lake Nipissing and the French River before turning southward on Georgian Bay. There were many reasons for making this journey rather than what seems to our driven minds the more direct Saint Lawrence River and Lake Ontar:io passage of our major highways. These include the location of Wendat villages north of the Oak Ridges Moraine until the mid-seventeenth century; the fear of Haudenosaunee attacks from south of these waters over much of that century; the relative insignificance of the passage de taronto in the colonial experience prior to the founding of Teieiagon, Fort Rouillé, and then Fort York; and the danger of portaging the longue sault. From the latter reality there emerged among my coureur de bois ancestors a braided way of relating to its shapes of violence. Such energetic meetings of rock and water signified a sacred power where Brûlé was taught by Indigenous guides the importance of offering "tobacco, which they throw into the water against the rock itself, they say to it: 'Here, take courage, and let us have a good journey.'"⁴ If I listen closely with knowledge and imagination, stories of human-energy relations can still be heard below the dam's turbines.

Beyond the dam and hydroline corridors that feed electricity to the grids of Ontar:io and New York State are drowned stories of painful severance that echo those of our first two chapters. The 1950s Saint

Lawrence Seaway project required moving fifty-three homes and six 130-year old villages before flooding their surrounding ecologies. One person who remembers this move to the newly created town of Long Sault – one of two towns created for the displaced – is Jane Craig, the past president of the Lost Villages Museum. A child at the time, she recalls coming home from school excited by the immanent move when her mom dampened the mood by leading her "out to the back steps overlooking the yard."[5] There she saw the three willow () trees her father had planted for each of their children, which were already making "the yard a shady bower." They were being cleared by Ontario Hydro workers with chainsaws in preparation for the flooding, and her father was on the steps crying. The Craig family were a few of the people who had lived on these shores since the early eighteenth century, not long after Simcoe's first excursions on the passage de taronto. Concluding her remembrance of the 1958 relocation, Craig asks us to "imagine the government flooding 20,000 acres in the heart of Ontario, displacing 6,500 people," for this is what happened just over half-a-century ago.[6]

Though the explosion of the cofferdam that drowned the villages and ecologies in the vast new reservoir for the power dam was a violent and relatively quick change, the previous chapters situate it in a history of colonial, industrial, and modern severances that is imaginable to many. In fact, the majority of those relocated before the flooding were not Loyalists but rather Mohawk of Akwesasne, which can be seen across the river. Such impositions are also not of the past, since they continue today in our unquenchable search for energy. This is epitomized in former prime minister Stephen Harper's description of digging Alberta's "bitumen out of the ground, squeezing out the oil and converting it into synthetic crude … [as] an enterprise of epic proportions, akin to the building of the pyramids or China's Great Wall. Only bigger."[7] While such a venture further aggravates climate change and disproportionately affects places like the north and island states, the extraction of oil from the bitumen simultaneously threatens ecologies that have long sustained Indigenous communities in that region. As Chief Allan Adam of the Athabasca Chipewyan First Nation states, "the survival of our communities, lands, animals, waterways and peoples lay in the hands of bureaucrats who are failing to uphold protection of Treaty rights and the rights of Mother Earth."[8] Just as with today's energetic developments, the watery relations of the longue sault were harnessed as part of modern Canada's ever-growing need for energy.

While the sudden severance that Livingston experienced in Toron:to is easily forgotten because of the seeming permanence of the urban form and drive of city living, here the harnessed hum of energy almost lulls me to a deep forgetting of where I am. It was not while growing up along these shores that I first became aware of stories like those of Jane Craig, even though I played hockey with her son. For much of my life, Long Sault was simply the village with our local arena. The hydro-power plant was a distinct reality, but it was disconnected from my experience of a Saint Lawrence River that was largely a backdrop for recreation. My working-class experience in industrial Cornwall unconsciously inherited what the previous chapters described as an intergenerational environmental amnesia. This common severance that connects a small town, Toron:to, and a myriad places across Canada raises for me questions about how to respond. Coming from unique ancestral traditions, Livingston and Woodworth suggest that stories reside in the land, water, and climate, awaiting us to remember with interdisciplinary flights and imaginative grounding. This chapter contemplates the energetic presence underlying the longue sault, a French term we will engage as a double entendre for the place where the book's recurring tension between severance and reverence is now situated, and as a symbol for the temporally long rapids we now need to mindfully traverse in the form of a climatically changing world.

John Mohawk tells an important story about taking the well-known Hopi elder Thomas Banyacya for a tour of Haudenosaunee country while he was north to deliver a talk at State University of New York where Mohawk taught. On the south shore and western edge of Lake Ontar:io, they "followed the Niagara Gorge to Niagara Falls, past the myriad chemical plants near the now famous Love Canal, and out toward the Robert E. Moses Power Project."[9] As they approached the Niagara Falls power dam and its "endless maze" of electric lines,

Banyacya asked to have the car pulled over so he could get out. "For a long time, in silent awe, he stood beside the road" before finally speaking: "In the Hopi teachings, we are told that toward the end of the world, Spider Woman will come back and she will weave her web across the landscape … I believe I have just seen her web."[10] Contemplating his words in relation to the faith modern societies continue to have in technology, Mohawk concludes that this Hopi vision has much wisdom to offer. In his words, the web of "power lines are a symbol of the end of this world as we know it."[11] His critique is not unlike Livingston's technological imperative, but it adds more Haudenosaunee dimensions to Woodworth's "rolling heads" and the colonial processes that fuel our current climate of change. What is the nature of this web that both gives us our energy and, for some, forebodes a dark future?

Before I was aware of Banyacya's vision, I was likewise struck by the electrified web radiating from the Saunders-Moses Power Dam some five hundred and seventy kilometres to the east of Niagara Falls. At the centre of the radiating power lines is a two- section dam that straddles the American-Canadian border and contains "sixteen turbines, with a combined annual power generation capacity of 2.2 million horse-power."[12] Its construction required "1,890,000 cubic yards of concrete, 116 million pounds of reinforced steel, and 15 million pounds of structural steel," as well as "the dredging and excavation of 360 million tons of materials costing in excess of one billion dollars."[13] This epic change to the place was in many ways the grand finale in Ontario Hydro's modernization of the province. Over the first half of the twentieth century, the public utility had harnessed the "thundering potential of Ontario's great rivers" into a web of transmission lines that radiated from fifty generating stations to link major cities and towns.[14] It was "an era of unquenchable confidence and expansiveness" that started in 1905 with the harnessing of Niagara Falls and ended with this world-scale hydro project on the Saint Lawrence River near the Long Sault and Cornwall where I grew up.[15] This harnessed electricity supported the intensification of Canada's industrial revolution in a province that became the nation's manufacturing heartland.[16]

Across the water, just to the east of the electrified web, is the Mohawk territory known as Akwesasne, the "land where the partridge drums."

A brief look at this place offers a sense of the Haudenosaunee context for John Mohawk's historic critique of modern technological progress. Akwesasne is situated on the islands and land between industrial developments on the Canadian and American sides of the river that, following the power dam's creation, included a Domtar pulp and paper mill, a General Motors foundry, and Alcoa aluminium works. There was consequently a significant rise in local pollution that biomagnified up the river into animals and fish. Not far from where I sit, "at the heart of a continent that traditional Iroquois think of as Great Turtle Island, a turtle was discovered to have 3,067 parts per million of PCB in its body fat – sixty times the amount needed to qualify as hazardous waste."[17] Elder Raterehratkense James W. Ransom says that this pollution has driven his people from the river and "everything that is associated with being Mohawk has been impacted."[18] The web of impacts include loss of elder knowledge that "is not being passed on to the next generation as fishing, hunting, trapping, farming, gathering and sustainable economies have been replaced ... [by] more uncertain means ... [of] existence."[19] As pollution drives people from this river known in Mohawk as *Kaniatarowanenneh* or the great waterway, ancestral values and ecological relationships are increasingly severed.[20] This is a different colonial history than that which led to Toron:to's current reality, though they clearly participate in common changing dynamics.

Prior to the 1784 arrival of Loyalists north of the Kaniatarowanenneh, the longue sault was just beyond French colonial settlements that halted near today's Quebec border with Ontar:io.[21] Few Europeans had gazed upon these lands before Sir John Johnson brought families of a royal regiment from New York to settle Cornwall, at first named Johnstown.[22] As with the Loyalist arrivals to Toron:to, the land seemed like "largely untamed wilderness" with no roads or communication routes to major centres like Montreal and thus western Europe.[23] Elizabeth Simcoe bypassed this place when shooting the longue sault in 1794 and again with her husband on 28 July 1796 upon their departure from York. During the latter trip, she wrote that the boats had such velocity they "appeared to fly" over the rapids, with whirlpools sometimes turning them round such that "the head of one and stern of another boat appeared buried under the wave."[24] They stopped after the longue sault on the river's south side, just beyond the Mohawk communities of Akwesasne and Saint Regis. The emerging British town they passed was in Haudenosaunee territory, and thus settlement had required negotiations between Johnson and Joseph Brant for the Mohawk. These men had fought the American revolutionaries together and Johnson's father had married the chief's sister, Molly Brant.[25] His familiarity with the Mohawk allowed Johnson to face "the Council of St. Regis chiefs and warriors over the question" of this land with a sense of fraternal relations.[26] While Brant negotiated five kilometres of land along the river's north shore for the Mohawk to continue trade with their Anishinaabe allies of La Petite Nation,[27] Johnson's Loyalists settled east of a longue sault that left Simcoe with "nothing to think of but the present danger."

The shift of Cornwall from a farming community to an industrial town largely began with the 1843 construction of the first British canal around the longue sault.[28] It was part of Upper Canada's canal-building projects that started in the 1820s as means to better navigate the waters with river steamboats.[29] The seventeen-kilometre canal and six locks that allowed safe passage along the longue sault made Cornwall an ideal place for entrepreneurs who wanted the lower costs of efficient transportation routes for their factories. Between 1868 and 1882, three Cornwall mills were established as part of an overall trend that led to late-nineteenth century Ontar:io becoming responsible for over half of the nation's industrial production, with Toron:to being a dominant player in the manufacturing belt along Lake Ontar:io's north shore.[30] French Canadians began arriving in Cornwall at this time owing to two interacting circumstances: rising labour demand at the industrial factories, and Quebec farmers running out of land to pass on to their male children.[31] These developments partook in a late-nineteenth century provincial trend of Ontarians leaving rural farming life for industrial cities.[32] By 1921, Cornwall almost equally embodied Canada's two colonial solitudes, with 2,542 inhabitants of French ancestry and 3,318 British.[33] On the eve of constructing the power dam and the Saint Lawrence Seaway that drowned the longue sault in 1951, the city's French-English population had expanded to 16,899.

While the energetic potential of such a project had been discussed over the early part of the twentieth century, it was the events of the Second World War that clarified the need for more easy access to the industrial production of the Lake Ontar:io region and the value of harnessing the longue sault's power.[34] National and provincial support was echoed in the city of Cornwall by business interests who "hinted at the untold wealth that could come from the untapped hydro-electric power" of the longue sault.[35] Dam construction began by diverting the rapid waters to a catch basin, leaving dead fish littering the rocky bottom. At the same time, agents began convincing people in the immediate area to move their towns and houses at the government's expense. Many like Craig's family were forced to relocate. After years of work, the project's completion was announced in 1958 by explosions that drowned the foundations of homes, villages, and ecologies.[36] Ceremonially opened on 26 June 1958 by Queen Elizabeth II representing Canada and President Dwight Eisenhower the United States, the Seaway heralded a new era in which large ships could find their way from Montreal to Toron:to.

For its first decades, Ontario Hydro's slogan was *Live Better Electrically*, and the longue sault's transformation was the culminating hydro project in this central myth of the modern ship.[37] Reflecting on the twentieth-century emergence of Ontario Hydro as one of the planet's most successful public utilities, the Canadian playwright and the utility's official historian, Merrill Denison, wrote that Hydro used "science and technology to unlock nature's bounty" from a rocky land that had long limited its development.[38] To this he would add that "the land provides the people with their economic sustenance, molds their character, determines their social and political outlook."[39] It was an epic vision consistent with the staples theory of Canada that was developed by Harold Innis over the first half of the twentieth century. As we saw in the previous chapter, Innis highlighted the way in which Canada's east-to-west development was historically grounded in the abundant waterways that were used to get resources out of the country to European and American markets. Resource extraction, from fish and fur to forests to minerals to the energy of hydro and increasingly oil, has been mythically connected to Canadian identity.[40]

How we harness the land for human needs would seem to have an influence in shaping our national character as much as the land itself. A focus on staples reflected, in Innis's analysis, "an economically weak country" whose wealth and character depended on providing resources to a more highly industrialized Europe and later the United States.[41] Canada's expansion followed an east-west line along the Kaniatarowanenneh, Great Lakes, and then western passages that eventually cycled their bounty back to the Atlantic Ocean and Europe.[42] These developments were often financed by government rather than private interests, as epitomized in the nation's role in replacing the waterways with the canals and then Canadian National Railways, thus fostering the "transfer of large areas tributary to the fur trade to the new industrialism."[43] Large hydro projects like the longue sault continued this tradition, though on a provincial scale, as public money "helped to iron out the ups and downs of the business cycle while allowing the government of the day to keep its friends in the gravel, cement, construction, and manufacturing industries happy."[44] After the longue sault, Ontario Hydro moved from rapid water to coal and nuclear power. From 1957 to 1974 the commercial sector's electricity consumption rose by over 600 per cent, while it increased by 200 per cent for mining, pulp and paper, automobile manufacturing, and other industries.[45] This is the era of *Living Better Electrically* that I was born into and is the context of my life, from the longue sault to Toronto and back.

The lines stretching before me are not simply external phenomena but on deeper reflection are also bound up with internal ways of minding relations. A world or nation of resources is not simply an idea or cultural myth but something that colours the quality of all our relations. Building upon the historic context outlined by Innis, the tradition of resource management has primarily conceived "the environment as an infinite supply of resources and a bottomless sink for wastes," with the most difficult problem being limited human means to utilize the land's abundant resources.[46] The emergence of industrial processes substantially increased the means for enacting these transformations. But with these developments something else also changed as the reality of natural limits emerged with conservation and then environmental awareness. That said, the resourcist logic has a strong hold. In the 1981 classic *The Fallacy of Wildlife Conservation*, Livingston argues that the recurring failures of conservation and environmentalism are ultimately due to the internalizing of a resource logic that continues to define ecology "as a human asset" in need of management for primarily human needs.[47] From an economizing logic to a technological imperative, Livingston's Ashbridges Bay initiation led him to highlight a powerful set of cultural forces that are affecting our minding of relations. It is a perspective that resonates with the Haudenosaunee views of Woodworth on "rolling heads" and with John Mohawk's story about Banyacya's comments on the radiating web of hydro lines.

From Mohawk's Haudenosaunee position, Western European cultures have tended to embrace technology in a way that has blinded them to the violent impacts of its ever-expanding severances. Beginning with the Greeks, the continual succession of civilizing powers suggests to him that the idea of "an unbroken story of cumulative improvements" leading to the present global culture is most fundamentally "an exercise in creative history."[48] He stretches this legacy back to Plato's Hekademos where the ideal of enlightened students who can "apply the principles and knowledge of the world to solve any of its problems" became a seductive mainstay of Western societies.[49] Its successes have been largely based on rationalizing activities that greatly harm others, a tendency that Mohawk sees as plaguing the colonial and then modern ship.[50] Being more firmly situated in the Haudenosaunee Good Mind tradition than academia gave Mohawk space from without to notice this deep historic pattern of minimizing failures in the face of progressive visions. In fact, he defines the goal of Indigenous studies as doing "what Western culture and anthropologists do to Indians: review it, point out what's wrong with it, explain where it went wrong, tell them it was wrong, and basically have a version and understanding of it."[51] Engaging technology as an external panacea is for Mohawk one of the prominent strands of the dark web observed by Banyacya that is blocking passage to a good sustainable future.

The epic transformation of the longue sault into the mythic ideal of *Living Better Electrically* is part of that historic idealization of technology that continues to fuel rising energy consumption on the road to an expanding modernization – from hydro to coal, bitumen, and fracking. For Mohawk, such globalizing ideals are flawed because their ever-growing scale inevitably manifests emerging problems that "the original visionaries did not anticipate,"[52] with one of the latest being global-scale climate change and its diverse local impacts.[53] But this numinous web of change that criss-crosses the continent also has internal dimensions. It is this dark hold that so concerns Livingston, Woodworth, and Mohawk because of the way it feeds a persistent expansion in the face of evidence demonstrating its destructive effects. As Mohawk observes, there continues to be an idealism in the modern mind that assumes that all humans are drawn to a lifestyle of middle-class consumption which brings about deforestation, destruction of animal habitats, mass animal and plant extinctions, and the waging of low-intensity war against regional economies.[54] What needs deeper contemplation is the way this persistent pattern binds humans and, more specifically, the modern ship to an energetic web that is fuelling our climate of change.

The westerly walk I take to this spot on the Kaniatarowanenneh from Cornwall's downtown passes a boarded-up fence that, when I peek through a crack, reveals another side of the web's dark "shapes of violence." The foundations of the old Domtar pulp and paper plant can be seen in the midst of what looks like a kind of post-apocalyptic land of piled rubble, roads going nowhere, ruined building foundations, and polluted soil along the riverbank. The sulphur smell of my childhood is gone and what is left are these hollowed-out remains of an industry that for a time manufactured one of Canada's staples. Until its closure in 2006, this mill was the economic foundation of the city. Though the Cornwall area was for most of the twentieth century largely stagnant in economic terms, the mill's unionized environment employed family and friends with good-paying work until that moment when it was no longer viable in a world of electronic media. In an odd way, the absent sulphur smell reminds me of earlier optimistic stories about the longue sault's transformation into a web of power and improved Seaway transportation that would "economically revitalize Cornwall."[55] The reality was that its completion coincided with departures of workers and collapse of the once central cotton industry, thus casting "a shadow over what had been a climate of optimism" that looked upon "the Seaway as an El Dorado."[56] As with many modern epics, they are full of promise until they move on and leave something like this behind board and fence.

Sometimes it feels as if I am looking through similar cracks in the modern veneer when news stories break like the one in the summer of 2013 about underground leaching from Alberta's bitumous sands. It was estimated that one million litres of oil had entered the northern wetlands, with general uncertainty about how to stop the leak.[57] Yet such cracks almost seem illusory, delusions, when watching a commercial like Cenovus's "Rising to the Challenge."[58] This hyped-up cartoon is concerned with getting the oil out of the bitumous sands in a sustainable way, what it refers to as the Cenovus Challenge. The voiceover describes abundant oil, but that oil is mixed with sand that is "hardened at times as hard as a hockey puck," "locked between multiple layers of thick rock," "sealed with a dense sprawling forest." There is a hockey-like atmosphere that plays on Canadian myths of frontier people using ingenuity, hard work, and strength to extract the wealth of its resources. The fast-paced stream of high-definition images does not on the surface look anything like these boards, but if I attend the cracks closely they almost seem to open up to reveal a hollowing out of land related to this epic attempt to extract oil.

Leaching was not the only event in 2013 to break open a different view on the Cenovus Challenge and Mohawk's dark energetic web. On 6 July a train disaster in Lac Mégantic, Quebec, killed forty-seven people and destroyed thirty buildings when the seventy-four cars carrying crude oil exploded in the town centre. It unleashed a fire thirty metres high that could be seen and felt two kilometres away, and also opened up a debate on the source of this tragedy. Various interacting factors were identified, such as the loosening of regulations around Canada's railroads, railway corporations focused on profits over safety, older tank-cars not up to current standards, and a significant increase of petroleum shipments across Canada "from 500 carloads in 2009 to a projected 130,000 in 2013."[59] For those focused on the industry of petroleum, this unfortunate disaster highlighted the need for more oil pipelines as means of reducing disasters and maintaining the increase in oil extraction – from Keystone XL to Northern Gateway to Energy East and the recently approved, though controversial, Line 9 that aims to take oil from Alberta's bitumous sands through Ontar:io, including Toron:to and Cornwall. And yet pipeline disasters are not unfamiliar news. It is this fear of a Northern Gateway pipeline spill that has spurred the "Save the Fraser Declaration" on Canada's west coast as Indigenous people, environmentalists, and Canadians fight to save the Fraser River and its delta on the Pacific Ocean. In the words of the declaration: "A threat to the Fraser and its headwaters is a threat to all who depend on its health."[60] For a resource-based nation, it seems logical that the only viable way forward, especially in times of economic uncertainty, is the epic processing of our energetic resources, despite the conflicting climate messages.

The seemingly addictive web of this Canadian dynamic is something I hinted at in *Climate, Culture, Change* when connecting industrial climate change to the nation's resourcist history and colonialism in the north.[61] My earlier experience as a social worker in a northern Indigenous community had brought me into contact with gas-inhalation addictions that are one common symptom of the cultural destabilization brought on by colonialism. Youth who sniff gas often become aggressive, suffer long-term memory loss, and display "feelings of indifference that have led some Indigenous to describe gas sniffers as those who 'cannot hear' because they do not comprehend the impacts of their actions."[62] These symptoms, particularly the deafness, seemed analogous to an industrial energy and economic system that, drawing upon Livingston, fosters pervasive "institutionalized delusions."[63] Modern cultures are, I proposed, "'without ears' for comprehending the indeterminate relation of the planet's external climate changes to an internal cultural mindset due to the West's historic proximity to powerfully disruptive fossil fuels."[64] There was more poetic flourish to this conclusion than substantive analysis, something I want to rectify here.

In following Livingston's bird's-eye view of technological, resourcist, and institutional impacts, I am reminded of Gregory Bateson's analysis of the addictive double bind that he saw as hindering the modern initiation into a sustainable ecology of mind. Double binds are described by him as situations where two or more conflicting messages present themselves in such a way that responding to one results in a failure to respond to the other, and in fact escalates the other message. There also seems to be no way out of the bind inasmuch as a response is required and yet the response is enmeshed in the issue, the mystery. These binds are tangles or web-like challenges arising from contexts that are responsive to habitual behaviours.[65] For example, an urban centre like Toron:to experiences increasingly high summer temperatures owing to a combination of warming trends and the urban heat-island effect, and these impacts of human behaviour encourage the burning of "more fossil fuels at power plants to air-condition overheated people – which in turn creates more GHG emissions and air pollution."[66] Such binds can also help us understand Livingston's initiatory paradox, for the response to human pressure on the water system intensified the ecological sever-ance by transforming Ashbridges Bay. All of this then informed his life's work that looked for ways out of the entanglement.

Binds and paradoxes are not necessarily the same, but in this case their web-like source in the modern ship puts those who are ecologic-ally minded in an intractable, paradoxical, position: to choose nature, even in times of turbulence, is seemingly to be against the human needs of their communities and national culture. The external forms of these tangles are for Livingston merely symptomatic of an internal double bind that affects various institutions, including environmental and conservation communities. There is within us, he wrote, a "fear of appearing to be emotional [that] requires us to fabricate our own rationalizations lest we invite the terrible censure of the technocratic age by committing its one unpardonable sin – irrationality."[67] So, while our climate of change highlights the limits to knowledge, deep historic inequities, and the need for diverse imaginings of our situa-tion, the modern concern with being irrational and potentially swamped by surrounding forces re-entrenches established patterns. A focus on more technology and an economizing logic as the only viable solution are acts that further bind us within the still modernizing walls of Hekademos even as we feel the web of changes escalating.

This tendency reflected for Bateson a kind of addictive pattern that can be seen playing out in Canada's approach to bitumen develop-ment, fracking, offshore extractions, oil pipelines, and climate change. At the core of this binding web is a feedback loop that keeps the modern ship from attentively minding relations, a resistance that Bateson attempted to clarify by considering alcohol addiction. The acts of an addict are, he writes, "guided by highly abstract principles of which they are either quite unconscious, or unaware."[68] Central is the belief that they are in control of their addiction and can resist drinking, a belief that is continually called into question by their predisposition to take risks that challenge their capacities in an escalating fashion. As the addiction takes hold, the alcoholic finds "it difficult to resist the social context in which he should match his friends in their drinking," and in time the escalating behaviour cannot be matched by friends and thus leads to solitary drinking.[69] There are differences across addictions, but what they tend to have in common is a pattern of increasingly risky behaviour and social isolation from what is seen as an external world that is becoming more critical of their actions and thus threatening. The isolation decreases social dissonance in the short term while maintaining addictive ways that are increasingly risky for their future.

This addictive isolation and pride-in-risk has certain symmetries with Stephen Harper's hubris in describing bitumen development in a time of climate change as "an enterprise of epic proportions." Discussing the reasons why national polls indicate that 42 per cent of Canadians are opposed to the bitumen epic, Thomas Homer-Dixon points not only to the popular concern about environmental damage to local ecologies and global climate, but also to the less obvious reason that "it is relentlessly twisting our society into" something that is exhibiting "the economic and political characteristics of a petro-state."[70] The bigger issue is the precarious nature of economically depending too heavily on one resource, with oil manifesting particular challenges. As he writes, nations "with huge reserves of valuable natural resources often suffer from economic imbalances and boom-bust cycles" that affect economic and social stability. Canada's move toward being a petrol state also partakes in a resource history of shifting political dependence on one-crop economies, from fur to forest to bitumen. Following the insights of Innis, there has been a tendency for resources to "monopolize the interests and behaviour of the state."[71] The addiction seemingly moves from one staple to another, with bitumen having an expanding influence on Canadian ways of minding relations and, through its greenhouse gases, global and local climates.

Just as the isolating nature of addiction fosters dependency on old patterns that lead to a deteriorating situation, Canada's response to economic and climate uncertainty has been increasingly marked by old resourcist patterns that escalate impacts in intensity and scale. The earlier research of Homer-Dixon analyzed the way in which arising environmental scarcities require technological, economic, political, and social ingenuity to manifest a sustainable response, with a particular focus on those isolating human systems that block such ingenuity.[72] Canada's energy politics participates in such a mal-adaptation by increasing not only potential economic instability but also social polarization and divisiveness. These effects are seen in the Harper government's denial of climate research and its portrayal of environmentalists who question the fossil-fuel industry as unpatriotic, even potential terrorists. Instead of "having an open conversation about the tar sands," Homer-Dixon writes that Canada "behaves like a gambler deep in the hole, repeatedly doubling down on our commitment to the industry."[73] Risky politics, social divisiveness, and increasing isolation from an interconnected climate, not to mention the international community, are some symptoms of fossil-fuel politics that resonate with Bateson's addictive bind.

The bitumen resource also requires extensive economic, social, and environmental investment for diminishing returns, as reflected in the endless debates and conflicts over cross-country and continental pipelines to get the resource to markets in the United States, China, India, or the European Union. On a deeper level, "bitumen is junk energy in that a joule, or unit of energy, invested in extracting and processing bitumen returns only four to six joules in the form of crude oil."[74] In contrast, the production of conventional oil returns about fifteen joules. This means that bitumen production requires more fossil fuels and thus emits significantly more greenhouse gases than the production of conventional oil. As with the escalating double bind, this epic calls for more economically rational acts that are simultaneously less efficient and heighten the surrounding climate messages. A similar fuel-intensive technological dynamic occurs in relation to more economically feasible climate-change responses. The oil industry and its supporters in government have often touted carbon capture and storage technology as providing a more managed response that requires less cultural change. But there is a significant "energy penalty" for this solution, since it requires burning "anywhere from 25 to 32 percent more fossil fuels" and developing a "new worldwide absorption-gathering/compression-transportation/storage industry."[75] In a sense, epic bitumen extraction, fracking, offshore development, and even carbon storage are analogous on a cultural scale to those dark corners where an addict tries to secretly maintain ways of living that are approaching, in this case, climatic freefall.

While Mohawk positions such tendencies in the West's ever-expanding technological web, Innis expressed similar concerns. In his words, Canadian culture and education has been "disciplined by the spread of machine industry," with the signs of a mechanical narrowing of knowledge reflected in Hekademos's skewed preference for facts, classification strategies, and professionalization[76] – to which we have added carbon calculus and other overly rationalized responses. The web stretches beyond those who advocate more conservative approaches to climate change and energy development. It is also internalized, as Livingston warned, by many of us on the ship who desire a more extensive and positive engagement of ecological science and policy options. As Mike Hulme points out, the "confident belief in the human ability to control Nature is a dominant, if often subliminal, attribute of the international diplomacy that engages climate change."[77] On a technological level, this goes beyond carbon storage to geo-engineering as the modern progressive myth continues to focus on climate mastery and control.[78] It is a binding set of tendencies that can hinder the interdisciplinary breadth and imaginative grounding of a good climate of mind.

Looking through the widening cracks of modernity's boarded-up epics we can today see natural disasters like Lac Mégantic, seeping oil sands, the Gulf of Mexico offshore fire, pipeline ruptures, and the world's changing climate. It appears that the era of *Living Better Electrically* has left us bound to a technological imperative that is increasing the climate crisis in intensity and scale while reducing time for navigating toward a more revolutionary response. This is what I merely hinted at in describing Canadian and, more broadly, modern climate responses as being "without ears." Technologies and fossil fuels are dulling our felt sense of what is moving toward us, of the presences in our midst that are signalling the need for imagining more significant change. Meanwhile, diverse human and ecological communities are being brought into the experience of what it means for a modern ship of global scale to "hit bottom," an experience Bateson describes as entering a state of panic reflective of someone "who thought he had control over a vehicle but suddenly finds that the vehicle can run away with him."[79] So what is the passage out of this web-like bind if not the power of technology for ever-more encompassing surveillance, management, and consumption?

A gusting cold wind brings my mind back from the old paper mill to the dark waters before me as they roughen with the approach of grey clouds and first trickles of rain. The elements remind me of a more vital current that can be observed in times of uncertainty and crisis. One such moment occurred across a great swath of eastern Canada and the northeastern United States in August 2003 when a cascading failure of the hydro system resulted in the largest blackout since the dawn of *Living Better Electrically*. The blackout was, Jane Bennett writes, "the end point of a cascade – of voltage collapses, self-protective withdrawals from the grid, and human decisions and omissions."[80] The electric grid is not simply a conglomeration of objects systematically connected but rather "an open-ended collective" of members each with its own vital force and agency to respond to messages.[81] This web-like assemblage of agents responded to each other as the dissonance of events escalated. From the resulting darkness came a momentary reminder of Bateson's words that the "unit of survival … is not the organism or the species but the largest system or 'power' within which the creature lives."[82] The energy humming through the wires before me is not simply a harnessed unity but a plural web of forces that is the energetic milieu of our lives. This is a social challenge for an isolated modern ship that is being asked to remember a deeper social ancestry than a Canada of resources.

On that fateful 2003 summer day, a complex series of interactions led to a crisis on two levels: the immediate power outage and the larger implications for modern ways of minding relations. This dual challenge began, Bennett explains, with several isolated withdrawals by generators in Ohio and Michigan that "caused the electron flow pattern to change over the transmission lines, which led, after a series of events including one brush fire that burnt a transmission line and then several wire-tree encounters, to a successive overloading of other lines and a vortex of disconnects."[83] One after another, generating plants then responded to the increasing pressure by separating from the web, though investigators remain unclear as to "why the cascade ever stopped itself, after affecting 50 million people over approximately twenty-four thousand square kilometres and shutting down over one hundred power plants, including twenty-two nuclear reactors."[84] With an immediate band-aid for the power outage initiated within a few days, though some places on the margins were without energy for more extended periods, the second more binding issue of how we renew energetic, climatic, and ecological relations returned to a state similar to amnesia.

What the blackout showed Bennett was a grid that is "a volatile mix of coal, sweat, electromagnetic fields, computer programs, electron streams, profit motives, heat, lifestyles, nuclear fuel, plastic, fantasies of mastery, static, legislation, water, economic theory, wire, and wood – to name just some of the actants."[85] Interacting with this energetic network is an increasingly numinous climate, which draws me back to the longue sault's almost forgotten wild "shapes of violence." It seems that there may be more agency in the web observed by Banyacya, Mohawk, and Bennett than is easily acceptable to the modern mind. This is perhaps the deeper reason why the internal dimensions of a challenging blackout are often boarded over and forgotten. But if we entertain the possible need for more profound internal change, then this vital world can bring to the surface premodern ways of minding places like this longue sault where rock and water once met prodigiously. Places such as this revealed to Indigenous peoples and then some Canadien colonials like Étienne Brûlé the human position in a greater power, and what this means for minding a web of forces that have internal and external dimensions.

The alternative colonial passage that Brûlé navigates us toward began at around the age of eighteen when Samuel de Champlain sent him in 1610 to live with the Wendat as the first French cultural interpreter whose task was to learn the language and gather knowledge.[86] Over the subsequent years, he learned much, including the value of

Thanksgiving when engaging the animate power of water and rock. Recalling one such ritual in the early 1620s for the Franciscan missionary Gabriel Sagard, he described the Wendats' belief that different spirits "rule over one place, and others over another, some over rivers, others over journeying," and observed that occasionally they offer "tobacco and make some kind of prayer and ritual observance to obtain from them what they desire."[87] He remembered one particularly big rock where they often made tobacco offerings, especially "when they are in doubt of a successful issue to their journey." One such journey, during which Brûlé participated in such an offering, "brought him more profit than any other he had ever made." Over the next two centuries, many coureurs de bois and then voyageurs followed his passage toward an appreciation of Indigenous ways for minding the vital energetic relations between rock and water.

The standard historic view is that Canadiens like Antoine Leduc were important cogs in the machinery of an emerging colonial nation based on staples, but such an objective analysis of general patterns misses the granularity of what was happening in the water. While traversing the Rivière-des-Outaouais in 1686, a French army captain, the Chevalier de Troyes, wrote of observing an interesting scene along a high steep rock whose mid-section was blackened: "Perhaps this comes from the fact that it is where the Indians make their sacrifices, shooting their arrows over it, which have small bits of tobacco tied to the ends. Our French have the custom of baptizing at this place those who have yet never before passed. This rock is named the bird by the savages and some of our people wishing not to lose the ancient custom throw water on themselves."[88] Indigenous sites that were considered sacred along the river passages became vital places for the coureurs de bois. Canoe journeys were marked with points of initiation as they more deeply entered Indigenous territories of the *pays d'en haut* and their French reality receded. These passages gradually led to a reimagination of their French Catholic tradition through braiding it with the ancestral Wendat, Anishinaabe, and, later, Haudenosaunee ways of minding the creation as a web of vital energetic connections. This pattern exemplifies what prompted John Ralston Saul to describe Canada as originally rooted in a weaving set of relations that attempted to find ways of traversing the Two Rows of canoe and ship.

The journey into the pays d'en haut started at the western edge of the Île de Montréal with a ceremony at Saint Anne's Church.[89] The mother of the Virgin Mary was patron saint of Brittany and then New France, drawing upon the tradition of sailors and fishers who prayed to her before setting out. It was this ancestral tradition that led Saint Anne to become patron saint of coureurs de bois, with her church just beyond the Lachine rapids being a place to acknowledge "the risk of travel in the pays d'en haut."[90] But this was merely the first place to pause and humbly recognize dependence on larger powers. The next stop was on the Rivière-des-Outaouais where the bedrock of the Canadian Shield presents itself about three hundred and twenty kilometres northwest of today's national capital, Ottawa. Here a ceremony of baptism was performed to initiate newcomers and remind others of their entry into the land's Indigenous reality, with its beauty and harshness. There were various baptism sites as one went deeper into the pays d'en haut. A coureur de bois described one such experience along the Grand Portage of Lake Superior, where his companions sprinkled water "in my face with a small cedar bow [bough]."[91] Features of land along the water, from significant rocks to cliffs to islands, became ritualized in ways that asked for support from Catholic saints and Indigenous spirits. Even the regular use of song to maintain canoeing rhythm and pipe smoking were like acts "of prayer, mimicking the common Aboriginal custom of offering tobacco to spirits."[92] As with Bennett's web of vital actants, there was an "instability and fluidity" to the coureurs de bois' engagement of Canada's energetic water that made rituals vitally important.[93]

If the fluid shapes of the world are ultimately just beyond control, then it seems necessary to foster a patterned way of navigating that vitality. This is what the coureurs de bois embodied, and it arose from particular cultural and land relations. Most of the ten thousand original French colonists who are the ancestors of today's six million French Canadians came from western France, primarily Normandy and Brittany.[94] This is the case with Antoine Leduc who arrived from Normandy in 1656 at the age of thirteen, in time became a coureur de bois, and then eventually met his end in the 1680s north of the passage de taronto. Many of the usually male and unattached colonists were on the brink of indigence back in France, thus mitigating what was seen by many French as a difficult move.[95] The view of Canada was largely non-paradisal, as reflected in one story of prospective colonists who passed through a small Norman town and "provoked a riot on the part of townspeople who, refusing to believe the travelers were leaving France voluntarily, insisted on 'rescuing' them from the colonial exile."[96] What the new arrivals found was a land less dangerous than imagined since it offered a slightly better standard of living.[97] In contrast to seventeenth-century France, the pays d'en haut also gave coureurs de bois a space beyond the social constraints of official French authority.[98]

The worldview that coureurs de bois brought into these vital spaces was often described by missionaries as a lax Catholicism. As with many in early modern Europe, they tended to understand the sacred in more fluid and magical ways than that which was preached by the colony's church and governing elite.[99] While watching a ritual performed by his Algonquin and Wendat friends that scattered tobacco to the spirit of a river, a fictionalized Leduc ruminates over the Norman Christian ritual where Saint Médard is "carried in procession to ask for rain and then dipped in the holy-water vessel to make quite clear to the Saint what was expected."[100] Both missionaries and later anthropologists noted similarities between the pagan European sensibilities of many colonists and Indigenous beliefs concerning "other-world contacts, spirit-helpers, the ability to talk with animals, shape-shifting and ecstasies."[101] The majority of coureurs de bois also came to align themselves with Indigenous customs, took on Indigenous female companions, adopted "items of native dress, accustomed themselves to use canoes and snowshoes, hunted alongside Indigenous men, and joined them in their ritual steam baths."[102] Such cross-cultural mixing was so prevalent that the Swedish naturalist Peter Kalm remarked how odd it was that, while many European nations imitated French customs, this was not the case among the Canadiens. He observed that they "follow the Indian way of waging war exactly; they mix the same things with tobacco; they make use of the Indian bark boats and row them in the Indian way; they wrap a square piece of cloth round their feet, instead of stockings, and have adopted many other Indian fashions."[103]

In much the same way that Ashbridges Bay opened Livingston and Woodworth to different ways of minding relations, it was in the unrestricted spaces of the pays d'en haut that a braiding with Indigenous ways became possible. Among Indigenous customs, egalitarianism was particularly appealing to the coureurs de bois but not to the colony's authorities, who saw it as "most threatening to Western social thinking."[104] There was a recurring concern expressed by governors and missionaries that the assimilation of Indigenous peoples was becoming less common than the pattern of uneducated and undisciplined colonists adapting Indigenous ways. The assumption had been that cross-cultural marriages between male colonists and Indigenous women would bring more conversions, and so the reality of Canadiens more often becoming "savage simply because they lived with them" was shocking.[105] Responding to this issue, the governor general in 1685 wrote: "I could not express sufficiently to you ... the attraction that this savage life of doing nothing, of being constrained by nothing, of following every whim, and being beyond correction has for the young men."[106] Jesuit missionaries issued similar protests and thus made concerted attempts to limit contact between French and Indigenous peoples, though that policy had limited success since the pays d'en haut was so accessible.

These issues for the colonial power structures were in many ways initiated by Brûlé. When he described his participation in the rock-water tobacco offering and its seeming benefits, the missionary Sagard wrote that "we rebuked him sharply."[107] It is interesting to note that Brûlé's initial journey partook in a cultural exchange that saw Savignon, the son of a Wendat headman, stay with Champlain for a shorter period of time. The two were roughly the same age when exchanged, but Savignon's trip to France did not appeal to the young Wendat man in the same way that Brûlé's experience did to him. Though he praised the good treatment he received, the spectre of public corporal punishment in France's capital was deeply disturbing and Savignon "never expressed any serious desire to return."[108] In contrast, Brûlé spent decades with the Wendat, despite being captured, tortured, and released by Seneca while living around lakes Huron and Ontar:io. By the time of his death in 1633 at the hands of some Wendat, he was seen as a traitor to France and would be historically stigmatized as a scoundrel. Though Brûlé started off as a cultural interpreter, he is considered by many to be the first coureur de bois because of the way he weaved French and Indigenous ways, an approach that was often rejected in the centres of colonial power where the garrison rolling heads of the previous chapter prevailed.[109] For this book, Brûlé is one mixed ancestor who carries stories about the potential and dangers of a braided way to minding relations in Canada, someone whose death reflects a mysterious mix of colonial relations that we consider in more detail in the next chapter.

There is much value in Innis's powerful staples history of Canada, but it also tends to impose a rationalist model on the past that marginalizes the existence of radically different imaginations. Coureurs de bois like Brûlé and Leduc open up more complex historic passages than a colonial nation bound by a resource consciousness from its earliest moments. Their impact on the environment and Indigenous cultures was minimal, and as such they "represent a time of possibility when the pattern of colonization was not inevitable or inexorable."[110] This is not to deny that coureurs de bois and Indigenous peoples actively participated in the fur trade, and in fact the former often found themselves in paternalistic work relations that they rarely challenged directly. If things got too bad, one option was desertion to the life of a freeman. This was for many the ideal because of its position outside official power structures.[111] Though Innis moved toward the exploration of oral culture as a place for resisting "the center's expansionary monopolies of knowledge with their dangerous lack of dialogue and reflexivity," he did not consider Indigenous peoples or coureurs de bois as part of this inquiry.[112] This was in spite of his recognition that "the Indian and his culture were fundamental to the growth of Canadian institutions,"[113] not to mention whole new cultural ways like those of the coureurs de bois whose Indigenous relations are the context for the emergence of Métis identities.[114] They participated in the fur trade as we do in the eras of *Living Better Electrically* and Harper's bitumen epic, but their lives entered liminal spaces of diverse vital forces that fostered vastly different and often resistant ways of minding relations.[115]

Attending the longue sault deep within the dam's hum, I am reminded of this French ancestry that looked upon places where rock and water energetically meet as more than simply a resource to be harnessed. These were places for remembering and giving thanks to vital, awe-inspiring, and challenging "shapes of violence" that were the context of life in Canada then and, we are learning, remain so now in the midst of our climate of change. Of course, this wild rapid violence is fed by a prodigious energy that goes beyond human control and minding, and thus prior to our modern time it often evoked some kind of recognition. The act of giving thanks is more than just a pause and offering, it is an attunement that slows us to "the rhythms of the land and the dependency we have" on presences beyond the human, thus instilling an awareness that any energetic harnessing of nature requires a duty of thanksgiving.[116] Such acts allowed coureurs de bois to live in liminal states whereby marked passages along rivers transformed their identity as they more deeply entered Indigenous lands beyond their Canadien homes along the Kaniatarowanenneh.[117] Their passages lead me to a recasting of our receding modern period not as an inevitable progressive reality that it has been long assumed will stretch inexorably into the future, but as a set of relations that is leading to an increasingly rocky and turbulent stretch of rapids we need to mindfully shoot and portage.

We are entering what I will describe from this point on as a *climatic longue sault* of fast changes that require us to navigate in quite different ways than the traditional speed and epic intentions of the modern ship. To return to Bennett's observation of the vital reality underlying the 2003 blackout, she writes that a lot of things were momentarily lit up by the blackout. These included "the shabby condition of the public-utilities infrastructure, the law-abidingness of New York City residents living in the dark, the disproportionate and accelerating consumption of energy by North Americans, and the element of unpredictability marking assemblages composed of intersecting and resonating elements."[118] The blackout emerged from various agencies: electron flows, generators, a fire, governmental faith in neo-liberalism and a self-regulated market, societal energy consumption, and so much more.[119] The uncertainty of such vital relations had begun to be experienced by Ontario Hydro over the last two decades of the twentieth century. This once global model of a successful public utility was hard-wired to various modern ideas, including ever-expanding electrical energy as a requirement for fuelling eternal economic growth, and nature as a reality that "could provide both limitless resources to do the job and a bottomless pit for the waste."[120] These are some of the energetic instabilities that moved us toward the blackout and beyond as we entered a period of debate and heightening conflict around bitumen development, oil pipelines, peak oil, and climate change. There are vital dimensions to our climatic longue sault that coureurs de bois and Indigenous peoples understood in their engagement of rocky rapids.

An energetic web of relations is calling us to a more diverse and nuanced minding of the presences that surround the modern ship. But, as we have seen with Bateson's addictive double bind, these rocky passages are not simply outside us. Our modern mentalities need to learn how to let go in a way not unlike the coureurs de bois' braided way of going with the watery flow. Considering the internal dimensions of this ecological challenge, Bateson suggested that the modern position is akin to the addict's experience of hitting bottom. Putting ourselves into increasingly risky situations, we want to respond by increasing an addictive dependence on technological and economic ingenuity.[121] An addict's panic is grounded in the increasing recognition that he or she is not in control, which is particularly devastating for ways of living based on control. Each act ratchets up the bind to a maladaptation that further isolates addicts from their social and ecological supports. This dynamic resonates with national energy politics, which has isolating effects in relation to the international community and a climatically responsive world. In Bateson's assessment, such a trajectory is suicidal: "It is all very well to test once whether the universe is on your side, but to do so again and again, with increasing stringency of proof, is to set out on a project which can only prove that the universe hates you."[122] We seem to be teetering on the edge of a whirling spiral within the climatic longue sault.

The humming turbines whisper stories of many epics to my ears, from Harper's bitumous sand vision to the dam before me to the approaching darkness of Spider Woman imagined by Mohawk when he concludes that Banyacya is right in seeing these power lines as "a symbol of the end of this world as we know it."[123] The sacred web that underlies our energy, climate, waters, and ecologies is initiating the modern mind into a more vital reality. Spider Woman is in the Hopi tradition the great weaver of this life that we find ourselves existing within, one where spirit interpenetrates the physical realities of our existence.[124] In the midst of such prodigious energy we need to find ways of re-engaging cosmologies like those of the Hopi, Haudenosaunee, and coureur de bois traditions because, as Mohawk states, they illuminate "a set of moralities that the west has generally found to be both confusing and inappropriate."[125] A similar insight led Bennett to conclude her reflections on the 2003 blackout with "a kind of Nicene Creed for would-be vital materialists": "I believe in one matter-energy, the maker of things seen and unseen. I believe that this pluriverse is traversed by heterogeneities that are continually doing things. I believe it is wrong to deny vitality to nonhuman bodies, forces, and forms, and that a careful course of anthropmorphization can help reveal that vitality, even though it resists full translation and exceeds my comprehensive grasp. I believe that encounters with lively matter can chasten my fantasies of human mastery, highlight the common materiality of all that is, expose a wider distribution of agency, and reshape the self and its interests."[126] Such a prayerful minding of relations has deeper ancestries in the human species than our darkening modern web, and this may be our saving grace. The braided coureur de bois approach is one ancestral way for me to return to such a spirited navigation of our climatic longue sault. This is what I am trying to renew in these pages by weaving Woodworth's Good Mind with Livingston's bird-inspired thought and the mixed strands of my French Catholic ancestry.

As I look at the tower of lines extending from the dam, the coureurs de bois symbolically inspire in me a Catholic view of the life energy being steadily crucified and yet rarely acknowledged, let alone given thanks in the Indigenous spirit of the Kaniatarowanenneh.

To return to the Haudenosaunee tradition passed down by Thomas, we saw in chapter 2 that our life energy is seen as emanating from the Sky World through a central onerahtase'ko:wa that we can affirm by performing the duties of the Peacemaker. The way in which this sacred energy of a Good Mind comes into the world reminds me of David Suzuki's environmental understanding that all the energy we receive comes ultimately as a gift from the sun. Solar energy "is constantly flooding our planet, providing high-quality energy to compensate for the steady decay of energy," and thus without energy from the sky above our lives and epics would end.[127] This humble acknowledgment of what fuels the longue sault's turbines and where humans are cosmologically situated in an energetic creation is something we return to in the book's last chapters because, we will see, it is the source of Thanksgiving.

In contrast to the dominant conception of climate change as an external phenomenon that does not explicitly symbolize its rough, rapid, and vital potentiality, the climatic longue sault further elucidated in the pages that follow is meant to highlight a turbulent, wild, and numinous presence of external and internal dimensions. It is one manifestation of a reality that is slowly teaching us, in Latour's words, that "we had never been modern."[128] Despite our burgeoning science, technology, and political capacities, the modern ship is becoming more "fragile, frail, threatened … back to normal" with each passing year of inaction. Of course, I like the warmth of my home and the comforts that came with the epic era of *Living Better Electrically*, but it is clear that a more mindful attunement of our relation with energy, oil, climate, ecology, and water is being called for today. I can see the modern resistance to the climatic longue sault's social reality playing out in my life. From a hometown that elected Harper's Conservatives to a home that depends on a Canada of staples and energy to a mind that is inexplicably moved by the Cenovus Challenge commercial, I hold on to a hope for emerging technological alternatives and a viable economic response that will allow life to continue as is. Such personal, national, and cultural binds are rarely just given up based on strong rational proof and arguments; instead, we need to navigate these obstacles by imagining alternatives that are simultaneously rooted in our being. If we cannot find the energy for such renewal, then our immersion in this rocky and turbulent bind will likely continue to be denied even as it worsens and the vital space for a good response recedes.

Our tendency toward individualized economic self-interest means that nothing is received as given but rather that everything is taken as a right to harness, own, and forget to acknowledge. This pattern is consistent with that of an addict who is hitting bottom and yet resists the detoxification that is essential for treatment, except that the scale is cultural rather than individual. Our aim is to reduce the energy epics that fuel our time of rapid changes. We are increasingly being confronted with a challenge that Bateson describes as being like "a spiritual experience,"[129] one that we need to attend if we are to mind our way through the climatic longue sault we have been drawn into. The difficulty is that our ship almost seems incapable of moving, on its own, toward a spirit of Thanksgiving and its lived implications. As with any addict, it seems to lack the ability to bring about healing. The modern pride which is also predisposed toward epic solutions is in need of transformation. As Mohawk concludes, the central modern idea, often touted on television and other technological mediums, that "supercivilization technology got us into this mess, and super-supercivilization technology will get us out" is the primary reason this darkening web is coming to its end. Extreme vulnerability and need for healing is a very different epic moment than what the modern mind still wants to rationalize.

There are different ways of engaging the uncertain end referred to by Mohawk, just as we are doing here with the mystery surrounding Toron:to's name and mixed ancestors like Brûlé and Leduc. It can be seen as representing the progression of colonial severances to their logical end and what that means for Indigenous ways that existed here prior to the arrival of rolling heads. Alternately, my Catholic-influenced mind can read it as a kind of global apocalypse that encompasses modern and Indigenous worlds. There is also the cyclical sense of endings found across Indigenous cultures which would see this outcome as representing a difficult passage beyond the modern era to something new, yet old.[130] In a similar spirit, Bateson writes that we must relearn how to surrender to a non-competitive or egalitarian relation with the world based on "the nature of 'service' rather than dominance."[131] Though all these mindings of the modern end are considered in the remaining chapters, there is something in Thomas's call for Woodworth to help him lift the fallen dolon-do, onerahtase'ko:wa, that inclines this book toward the third vision of where we are going. As dark as the modern ship's bind seems, our energetic relations are calling us to imagine ways of navigating the climatic longue sault to a new cycle. Renewing a braided approach may well be vital for our safe passage.

4

Protectress of Canada

In her, faith and culture enrich each other! May her example help us to
live where we are, loving Jesus without denying who we are. Saint Kateri,
Protectress of Canada and the first native American saint, we entrust
to you the renewal of the faith in the first nations and in all of North
America! May God bless the first nations!

Pope Benedict XVI (21 Oct. 2012)

On an overcast, wet, and cool spring afternoon, a brisk wind blows
over the rippling water of the Kaniatarowanenneh, penetrating my
bones. Yet it is not simply the elements that strike me to the core; I am
also acutely aware that yesterday, 17 April, was the last full day of my
father's life. Back in Cornwall for my first extended stay in over ten
years, I had come to this exact spot east of the longue sault before
being with my father on that day, realizing the end was drawing near
– though I did not know it was only a matter of hours. It was with
death's approach that I began to appreciate the importance of not
"denying who we are," for the last week with my father began clarifying the ancestral roots I needed to reconnect with. Each morning's
walk from my mother's house to his room at the Saint Joseph care
facility took me past his childhood church of Saint Columbans and
the Kaniatarowanenneh, known to most as the Saint Lawrence River.
In contrast to my Toron:to existence, the Catholic faith and French
language are ever present in this place, as they were for the first twenty
years of my life.

Two blue herons sail over the rough water close to land, their
powerful wings and curled necks guiding my mind east with the
river's current to the Île-aux-Hérons between Kahnawake and
Lachine on Montreal's western edge. With my father's last full day

11 *Saint Kateri Tekakwitha*, by Kevin Gordon. This painting
was commissioned by the National Museum of Catholic Art and
was featured on the altar of the shrine Our Lady of Martyrs during Saint Kateri Tekakwitha's 2012 canonization ceremony. Its
depiction of Haudenosaunee, Algonquin, and Catholic symbols
reflects the braid that is central to her being and a principal concern of this chapter.

falling on the date dedicated by the church to its first Indigenous
saint, the day Saint Kateri Tekakwitha died in 1680 along these same
waters, I am drawn to this isle where her romanticized story of
renewal began. It was while facing the Île-aux-Hérons in 1678 that
she and two other Indigenous women, in the words of her Jesuit
biographer Claude Chauchetière, "chose that place to be their home"
and spiritual hermitage, though the priest adds with a critical tone
that many "things entered into their deliberations for they had
no real understanding of what the religious life was about."[1] Yet,

12 Colum Cille's cell on the Isle of Iona (spring 2005). It was in an isolated cell represented by this ruin of stones still found on the Isle of Iona that a tradition of Christian spiritual practice engaged with natural places emerged. Colum Cille or Saint Columba is one of the ancestors considered in this chapter, along with other more popular figures like Saint Francis of Assisi. The paradoxes of this practice are central to my discussion, particularly as they meet the Canadian land, waters, and people in a period of colonial change.

between then and Tekakwitha's death two years later, she displayed a faith that convinced the Jesuit that a "sauvagesse" could be a saint. His hagiography or saint's story set the church on a journey that led to her canonization on 21 October 2012 by Pope Benedict as "Protectress of Canada." This saintly ancestor is for me a vital guide for contemplating the dangers and opportunities in braiding a good Canadian climate of mind during times of rapid change.

Beyond the "renewal" intimated in Pope Benedict's canonization speech is a Haudenosaunee reading that is not the interpretation he intended when, with a certain cultural conceit, he made Saint Kateri Tekakwitha the patron saint for renewing "the faith in first nations."[2] Her mother was an eastern Algonquin who had been captured by a Mohawk raiding party along the Kaniatarowanenneh and was eventually adopted into the community. During a period of much death caused by war and disease, capture began violently and for some ended peacefully with the adoption of a new name and role associated with a deceased clan member.[3] The Haudenosaunee ceremonies of "adoption, requickening, and initiation were solemn occasions to mark an important transition in someone's social identity,"[4] and it was through this that Tekakwitha, "She Who Puts Things in Order,"[5] was born of an adopted Algonquin mother and Mohawk man. It is her response to living in such a mixed colonial history that inspires many Indigenous and non-Indigenous devotees to see her as one who responded "creatively to what the Euro-American world had to offer without sacrificing or betraying her indigenous culture."[6] She embodies a spirit consistent with Saul's characterization of Canada as grounded in "a métis mindset,"[7] one that has the potential to engage the social complexity of Algonquin origins, Mohawk requickening, and Catholic conversion. What this chapter further reveals is that such a land-mediated braid of cultures, which in the past resulted in the creation of the Métis Nation, can today offer new alternatives to the recent violent history of a colonizing mindset.

The way in which "She Who Puts Things in Order" sits between cultures is a reminder of Canada's mixed nature, in its past, present, and future. In using the term "mixed," I am referring not just to the positive potential of braiding cultures as represented in the coureurs de bois of the previous chapter but also to the darker web of colonial binds that often sever such potential. The Catholic-Haudenosaunee dimensions of such mixed relations are carried by Saint Tekakwitha in her historically changing relations with Kahnawake, as well as with Akwesasne and Saint Regis, which were mostly created by inhabitants of Kahnawake in the mid-eighteenth century. For Haudenosaunee Longhouse traditionalists, she is often seen as a colonial victim who abandoned her community to join the French and then, after her death, became a powerful tool "used by the church to stamp out Iroquois culture and to promote a vision of assimilation that remains alive."[8] As we saw in the colonial severances of Toron:to and the longue sault, the religious missions she and Kahnawake experienced at the hands of the Jesuits continued into the modern era of *Living Better Electrically*, though in the different forms of the Indian Act, residential schools, and epic land changes. This continuing "conversionary" spirit is what John Mohawk observes in Banyacya's web when he talks about the need to remember the "devastation and horror brought by efforts to suppress or destroy those differences" that are the basis of a plural society, one that can accept "competing ideas about the ideal."[9] It is an awareness of this violent history that leads the Akwesasne Mohawk historian Darren Bonaparte, in *A Lily among Thorns*, to call for Tekakwitha's reintegration into traditional longhouse culture.[10] He calls for us to recognize her as a Mohawk woman rather than just a Catholic convert. In a similar spirit, this chapter attempts to engage Tekakwitha as someone who carries the painful conflict of a still ongoing colonial period focused on conversions of various kinds, as well as the potential of a braid based in nurturing our unique strands even as we approach such a mixed ancestor.

Though the coureur de bois way of the previous chapter resonates with me as one ancestral alternative to the ship's dark bind, it feels distant from my French Catholic experience growing up in late-twentieth-century eastern Ontar:io. Colonial severances were more the reality, though, unlike Toron:to's modernizing amnesia, an Indigenous presence was clear. With Akwesasne, Saint Regis, and, farther east, Kahnawake across the river, I grew up well aware of the Mohawk, from black-market cigarettes to the 1990 Oka conflict and the local casino. Relations always felt uneasy, with the ear often hearing slurs of various types that defined those who live on the other side of the Kaniatarowanenneh as different. The contrast between Cornwall and Toron:to reflects the way in which colonialism continues in smaller cities "where Native-white relations are immediately and visibly shaped by racial domination," while urban centres like Toron:to assume that such racism has been transcended.[11] There was in Cornwall more a climate of fear than the peace of the Haudenosaunee Good Mind, as shown in the fact that I became aware of this tradition only decades later in Toron:to. It was in the midst of my conversations with Woodworth and exposure to the teachings of Thomas that my father's death returned me to these waters and the Île-aux-Hérons for a different braided view.

For someone who grew up Catholic along the north side of the Kaniatarowanenneh, the Haudenosaunee valuing of ancestors recalls for me the tradition of Catholic saints. A saint is generally someone who has lived an extraordinary life dedicated to spirit and who, in their death, becomes an important mediator between the human community and the divine.[12] This common appreciation for the dead does not mean that the two traditions saw ancestors in the same way. As the following pages describe, there is a colonial pattern embodied in Tekakwitha's life and death wherein perceived commonalities were engaged and dissimilarities marginalized as means for conversion. In contrast, this chapter treats her as a saintly ancestor who offers insight on renewing or requickening Catholic and, more broadly, Canadian traditions that from the colonial period to her modern canonization have often displayed a conceit about who knows the true way. To symbolize this intention and honour her Haudenosaunee and Algonquin ancestries, I refer to our current guide from this point on as Saint Tekakwitha rather than the more common Saint Kateri. This name highlights her mixed ancestry and thus asks us to engage her as a saint who can help put our minding of relations "in order."

The herons that carry my mind to Tekakwitha's wild isle also lead this chapter to consider ways of renewing the modern ship's mixed relations with water, energy, ecology, and climate. What connects her to such an inquiry is that for many devotees she is not only the canonized Protectress of Canada but also patron "saint of ecology, sharing that billing with Francis of Assisi."[13] This view follows a Western tradition that conceives her conversion in the solitude of a wild isle opposite "the warring waters of the rapids" known as La Chine. Beyond the romanticizing of wilderness as a vital place of spirituality in both this Catholic tradition and the environmental movement, the reality of Tekakwitha's nature experiences were braided with Haudenosaunne approaches like "freezing baths and voluntary exposure to cold air" as ritual gestures for accessing the sacred.[14] We will see that there is something in her mixed approach to a sacred creation that can help us, following Bateson, mind ways of surrendering addictive modern binds so as to move toward a "noncompetitive relationship to the larger world."[15] As the turbulence of the climatic longue sault surrounds us, Saint Tekakwitha can help us imagine what it takes to renew a good pluralistic braid for minding relations in Canada.

The months of March and April on the Kaniatarowanenneh are marked by overcast skies, melting snow, and brisk winds blowing off dark and rough waters. Though the pine and cedar trees to my back offer a little solace, they are scattered across a grey-brown land of barren willows, matted grass, and mud. Despite a layering of sweaters, the surrounding wind pierces me, with my hands slowly going numb as I try to write on a large rock that reaches out into the river. There is something about being in the uncomfortable cold of this isle that helps me to imagine the climate Saint Tekakwitha spiritually engaged in her final spring of Lent penances over three centuries ago on these same waters. The romantic story is that her forty days of fasting "wore toward their climax in the death and resurrection of Our Lord," with Tekakwitha's strength wearing "away like the spring thaw that had set in over all the Grand River, freeing the ice, freeing the spirit."[16] The Lenten season leading to her final surrender on 17 April 1680, the day before Good Friday, saw her increase penances and privations in devotion to a Christian faith to which she had converted only four years earlier at the age of twenty. But her immersion in the river's elements was not simply Catholic, for it was mixed with Indigenous ancestries that traversed these waters for centuries.

For the romantic Western mind, the Île-aux-Hérons fills that old and still powerful trope of retreating from the civilized or urban into the wild for inspiration. Except for the herons, the isle where Tekakwitha deepened her conversion has been described as an uninhabited "place of supreme solitude" that acted as a kind of natural hermitage,[17] not unlike what the young Livingston experienced with the bittern and birds of Ashbridges Bay. The close-by Lachine rapids inspired Samuel de Champlain to write a half-century earlier that he "never saw any torrent of water pour over with such force."[18] It was from this energetic place of rock and water that coureurs de bois began their braided passages at Saint Anne's Church on the Kaniatarowanenneh's north shore. Across the river on its south side, Tekakwitha began contemplating a religious life after becoming curious about Catholic nuns she saw in Montreal and hearing the local Jesuit priest describe their vocation. Here with her friend Marie Thérèse, and an older woman, Marie Skarichions, who had lived with the Wendat Christians of Lorette, they talked about the nuns and described the isle across from them as a "place to which they could retire and lead a conventual life."[19]

The romantic imagery of Île-aux-Hérons tends to minimize the colonial position of these women between their village of Kahnawake and the French settlement of Lachine at the western edge of a Montreal that was then a fur-trade hub. Kahnawake was founded in 1669 by Jesuit priests as a Catholic mission community for diverse Indigenous converts. It was in its early days known among Canadiens as the place of "praying Indians." The village's diversity was embodied by Tekakwitha and her two friends, who between them had Mohawk, Algonquin, Oneida, and Wendat lineages. But Kahnawake also emerged amidst a northern movement of Haudenosaunee, with ten settlements created around this time in valleys along the Kaniatarowanenneh and on Lake Ontar:io's north shore, including the village of Teieiagon on the passage de taronto.[20] Though the Jesuits reported growing numbers of "adult baptisms at Kahnawake after 1675 to support their claimed influence," they also lamented the fact that most Haudenosaunee were little inclined to their Christian teachings and moved north primarily to renew disrupted familial relations and reconstitute their presence along this river.[21] In many ways, the universalizing aim of Catholic conversion was held in check by the arrival of Mohawk traditionalists and the village's diverse ancestral origins, thus giving Tekakwitha various ways of interpreting and responding to Christian teachings.

During the two years between Tekakwitha's decision to follow a nun-like vocation and her death, she, Thérèse, and Skarichions likely made many canoe passages to the Île-aux-Hérons. Such a practice resonates across cultures and time. In the Indigenous traditions of these women, it was recognized that spirit imbued creation and could be encountered in particularly strong forms away from the village. As we saw in chapter 3, some "spirits were associated with particular islands or large rocks," and since they could be friendly or violent it was important to propitiate them by offering tobacco in campfires, clefts of rock, or water.[22] Meanwhile, in the Catholic tradition, Saint Francis and his followers entered Italian forests around Assisi and were seen by the population as "more like Indigenous denizens of the

forest than fully human beings – *quasi silvestres hominess.*"[23] His
practice had precursors. The twelfth century saw Saint Bernard of
Clairvaux counsel those who were considering the emerging urban
universities that in the forests "trees and rocks will teach things that
the masters of science will never teach you."[24] His advice shared in
two old traditions: of monks founding hermitages in the still extensive forests of Europe's early medieval period; and, prior to that, of
Celtic Christian monks finding passage to North Atlantic isles.

The ecological and spiritual resonance between this latter tradition
that influenced Francis's nature mysticism[25] and the experience of
Tekakwitha on her isle hermitage makes it worth a short diversion. In
563 CE, Saint Columba, or Colum Cille by his Irish name, initiated
the Celtic Christian way by voluntarily exiling himself from Ireland
on the Isle of Iona off Scotland's southwest coast.[26] It was his legacy
that inspired my father's home parish of Saint Columbans, as well as
my pre-teen confirmation lessons at a retreat centre dedicated to
Iona. There was a mixed nature to Colum Cille's way as he was initiated in the traditions of Christianity and the Irish bards.[27] Unlike
much of Western Europe, Ireland did not experience the destruction
of earlier traditions in the course of its conversion to Christianity.[28]
Its monks did not reject pagan symbols, knowledge, and practices, as
was often the case elsewhere.[29] Vital to this Celtic tradition were
"thin-spaces," such as islands, hills, and forest clearings, where "the
human world could open onto the world of the gods and vice versa."[30]
While late-second-century Christian mystics of the Middle East
entered the desert to live in solitude,[31] Colum Cille initiated a wave
of monks who, on curraghs, sailed across to Iona and other isles on
what became known as the "sea-road of the saints."[32] A thousand
years later and an ocean apart, the canoe passage of Tekakwitha to
the Île-aux-Hérons likely had its own spiritual significance.

As I sit alongside this river in early spring, it is clear to me that there
is some validity to the romantic trope of associating the retreats of
Tekakwitha, Livingston, and Francis with abundant birds. The story
of Francis preaching to the birds is one of many that led Livingston to
follow others in describing him as a patron saint for ecologists, someone who was doing "essentially the same thing as the modern conservationist is doing – trying to create a perspective, in nature, for man

as a part of nature."[33] These words are brought to the fore of my mind by the swooping movements of kingfishers, the birdsong of robins, hovering terns and gulls, a myriad cormorant and ducks, a circling hawk high above, and the powerful flight of the great blue herons. When one is quiet in a place somewhat distant from human activity, like the watery edges of lakes, marshes, and rivers, the winged beings present a diversity of auditory and perceptual flourishes. The reason Livingston, Gibson, I, and many others have become bird lovers is that they carry our minds to flights of vision that are simultaneously grounded in the relations of a place. Such experiences led Livingston to write that there is "a very special, very personal, and at some times highly spiritual quality about the experience of the naturalist."[34] A reverential spirit also arises in me quite easily along this Kaniatarowanenneh.

Not unlike the romantic view of Tekakwitha on the Île-aux-Hérons, the Haudenosaunee Good Mind of the Peacemaker arose from a spiritual experience. As told by Rice, the Peacemaker's move toward the tree of peace was inspired by the song of a bird while he was alone in the woods. When his message of peace was first brought to a powerful Mohawk chief, he sang a song. It was so beautiful and peaceful that "everyone soon joined in," thus confirming the move toward peace that came to include the Onondaga, Seneca, Cayuga, and Oneida nations.[35] The song, it is said, was learned as the Peacemaker listened "to a bird during a moment of reflection, when he was contemplating what to do next while in the forest."[36] It is an experience resonant with Saint Francis's hearing of a heavenly melody while "alone in the midst of creation"[37] and with Livingston's description of being "transported by the spring songs of birds."[38] A Good Mind is based in creation, as John Mohawk states: "One learns from water as one learns from hawks or deer," and not primarily from the ideas and images one has of them.[39] Birdsong allowed the Peacemaker's mind to soar and imagine a different way of relating, one that would "plant a great tree of peace."[40] From the Peacemaker to Tekakwitha to Colum Cille, the common images are of "thin" spaces, somewhat remote from other humans, where elementary presences like birds can act as divine communicators.

The Franciscan appreciation of natural spaces found its way into seventeenth-century Canada, as can be seen in the writings of Gabriel Sagard. It was this missionary who met Étienne Brûlé in 1623–24 during his one-year stay with the Wendat, a journey that he began by walking barefoot and begging on the way from Paris to Dieppe where the ship departed for the Kaniatarowanenneh. The walk continued in this new world as he journeyed to and from Wendat country wearing a long grey habit and thick wood-soled sandals that were not well suited for "clambering over fallen trees on dim forest pathways, or wading in streams strewn with sharp rocks."[41] Sagard wrote one of the first Western views on southern Ontar:io in his book *Long Journey to the Country of the Huron*. While there are mistakes in his nature writing, he is clearly "impressed by the wonders of God's creation."[42] In the spirit of Francis, for whom creation is a mode of divine communication,[43] he concludes that God "filled the earth with various species of animals, as well for the use of man as for the embellishment and adornment of the universe."[44] This tradition of nature mysticism, however, was largely marginalized in the Catholic Church. As Livingston lamented, the Franciscan spirit never became "an effective presence in the struggle on behalf of non-human nature."[45]

What makes wild "thin" spaces a romantic trope in a negative way is the tendency to minimize darker issues, such as the austere ascetic practices and colonial context of Tekakwitha's experience. To begin with, asceticism is not easily amenable to romanticizing partially because it is unappealing to moderns focused on convenience, comfort, and the knowable rather than self-imposed sacrifice to connect with an uncertain beyond. Saint Francis fasted, mixed ashes with food, lived in a monastic cell lined with gravel and branches, and endured harsh environments.[46] He also promoted poverty and practices that aimed at remaining "in humble intimacy with the natural environment – not to control or impose on it systematically."[47] For Colum Cille, asceticism meant regular pilgrimages from Iona to the more remote Isle of Hinba where he employed a "flesh-subduing austerity" in the midst of the North Atlantic.[48] In the case of Tekakwitha, though her circle was culturally diverse, they collectively joined a "serious program of chastity and penitential asceticum" that was stronger in Kahnawake than in any other Christian Indigenous community.[49] One of her last penances was inspired by a missionary who talked about a saint lying on a bed of thorns. She went searching the woods for thorn bushes where, while feverish from her advancing illness, "she rolled on the prickles for three nights."[50] This is a far cry from romantic images that tend to portray her in a "generally attractive environment of birds and flowers."[51]

Île-aux-Hérons may have been a hermitage where the blue herons "rested on their long legs by the water's edge, as quiet as if they were carved out of alabaster,"[52] but it was also a place where Tekakwitha practised what is to the modern mind an unnecessarily painful asceticism. Even for late-seventeenth-century Jesuits who were accustomed to penances and treated the journey to wild Canada itself as a penitential practice, Tekakwitha's brand of asceticism was difficult to accept because it seemed to have little in common with Catholicism. Prior to the thorns she had placed burning coals from the longhouse fire on one of her feet while reciting an Ave Maria, an action that left a hole in her foot. This form of mortification was virtually unknown in the Christian tradition but had Haudenosaunee precedents.[53] Long before the arrival of the ship, Haudenosaunee engaged a host of ceremonial gestures aimed at connecting with the sacred that included "freezing baths and voluntary exposure to cold air."[54] Here we seem to be in the company of the Inuit shaman who, we saw in chapter 1, was initiated into the silatuniq of the tundra's "great loneliness," as well as Livingston, whose painful epiphany occurred at Ashbridges Bay – though the latter was involuntary.

The fostering of a wild asceticism by Tekakwitha distressed her Jesuit guide, who suggested European devices like sackcloth that were "susceptible to [more] limitations than some of the dangerous Iroquois techniques."[55] But she persisted until her death in employing rituals that wove together Catholic and Indigenous views of human affairs, with both traditions linked to "unseen forces that need to be cajoled, appeased, or thanked."[56] This was quite a different approach than that of the Jesuits, who saw Canada as a "menacing wilderness of dangerous savagery that haunted the imagination."[57] In contrast with the Franciscan engagement of creation, the seventeenth-century Jesuit view of wild nature was of something largely outside divine grace. The *Jesuit Relations* often depict Indigenous peoples as *sauvages* who are "like the uncultivated landscape," embodiments of the European myth of the Wild Man and Woman.[58] This myth suggested that sauvages had originated when sometime in the past they were separated from the protection of the Christian church and civilization.[47] Through reacquainting Indigenous peoples with the church, the Jesuits believed that this separation could be reversed. In relation to their missions, this meant that hunting was problematic because it took individuals back into the wilderness. As we saw in chapter 3, the Jesuits also displayed a persistent fear that the proximity of religiously naive French coureurs de bois to the sauvages and the wild "was a dangerous contagion, always threatening to undermine the health of civilization."[59] The pressure for conversion needed to be steady or a descent into wildness would readily occur, something the Jesuits critiqued in the figure of Brûlé and then extended to the coureurs de bois.

It is a concern that also plays out in their depictions of Tekakwitha. With the fields surrounding the Kahnawake mission seen as good and the surrounding forest "almost entirely evil," she was portrayed as communing with civilization while being "out of her element at the hunting camp."[60] To make a saintly ascent, Tekakwitha had to move out of the wild. Her story thus followed familiar saint plotlines in exaggerating the "persecution she suffered as a convert in her native village."[61] It simultaneously likened her asceticism to that of European mystic women saints like her namesake Catherine of Siena, with little to no acknowledgment that Indigenous traditions may have been involved. In the words of Chauchetière, "though her body was in the forest, her soul was wholly at the Sault."[62] There can be a tendency to interpret Tekakwitha from the perspective of a Catholic tradition that wrote this history and often saw a vital materiality as sinfully sauvage, and in need of conversion. This is the dark context for Saint Francis's limited influence on the ship as its ecological and then climatic impacts intensified.

It is possible to sense other elements of a Haudenosaunee Good Mind in Tekakwitha's spiritual acts that are quite different from Jesuit concerns with the wild and sauvage. Because this Indigenous tradition makes no radical distinction between body and spirit, "self-denying behaviour can be pursued both to strengthen the self and to gain spiritual benefits."[63] Unlike the Jesuit severance of spirit from body that fuels its desire to convert wild Canada, here ascetic practices have the potential to guide the Catholic tradition toward the land. In the words of historian Allan Greer, it is "reasonable to suppose that the ascetic women of Kahnawake were engaged in a hardening exercise designed both to enhance their physical courage and to help them achieve mystical ecstasy and spiritual power." It seems that Tekakwitha was practising a form of asceticism based on her mixed Indigenous ancestries, and it is this intent to link worldly experience with underlying spiritual presences that she demonstrated in the "thin" space of the Île-aux-Hérons. For the Jesuit and subsequent modern mind, such sauvage ways are in need of conversion. The impacts of this belief are at the root of Kahnawake's uncertain relation with Saint Tekakwitha, another rocky shoal of historic scale that we now need to address as part of our passage through the climatic longue sault.

When Pope Benedict entrusted the Protectress of Canada with "the renewal of the faith in the first nations," it is unlikely that he was envisioning a renewal of Catholic relations with the Haudenosaunee Good Mind. The conceit of a renewal conceived primarily in the direction of First Nations forgets the role of the Catholic faith in colonial severances, and it is this that needs to be looked at differently. From the perspective of Kahnawake traditionalists, Tekakwitha is a victim of colonization, a young woman who, like other survivors of the diseases and wars of the seventeenth century, was preyed upon by Jesuit missionaries. She survived smallpox at the age of six, though her parents died and she showed the disease's impact in her pock-marked skin, squinting, light-sensitive eyes, and poor health that persisted the rest of her life. These afflictions were experienced by many Haudenosaunee in Kahnawake and Wendat in Lorette, and indeed their presence and severity were often used by Jesuits to encourage conversions. This earliest of rolling-head severances, which later took the form of Canada's residential school system, is a significant dimension of Tekakwitha's role as Protectress of Canada, and remembering it is central to reweaving a good climate of mind.

The Haudenosaunee concern with Jesuit conversions has a long history that predates Tekakwitha's birth and is probably best epitomized in the 1649 ritual burning of the missionaries Jean de Brébeuf and Charles Lalement in Wendat territory north of the passage de taronto.[64] Prior to the 1634 arrival of the Jesuits, it is estimated that the Wendat Confederacy had "between eighteen and forty thousand people, distributed in eighteen to twenty-five villages."[65] As with the Haudenosaunee, they were an agriculture-based culture, and that very fact persuaded the Jesuits that Catholicism could take root among them since their descent into wildness had not progressed as far as it had among Tekakwitha's Algonquin relatives, who were still primarily hunters and gatherers.[66] Yet, even so, it was noted their corn fields were so large, unorganized, and wild that it was easier to get lost there "than in the meadows and forests."[67] In contrast to the agricultural discipline envisioned by Jesuits, the Wendat planted corn stalks, beans, and squash together in a way that took advantage of their potential ecological interactions. This practice merely exemplified a host of cultural differences, though with points of similarity that could be utilized to further conversions.

The Jesuit approach of making their missions accessible and attractive was rooted in universal theism, the theory that elements of Christian faith and virtue like agriculture can be found in cultures across the planet.[68] But it was not simply religious strategies that were utilized. The Jesuits – called Black Robes by the Wendat – also gave converts exclusive access to French celestial knowledge, metal technology, trade, and medicine.[69] Then there was the uncertainty of epidemics like smallpox that affected the Wendat, Haudenosaunee, and Anishinaabe in the latter half of the 1630s, with the Wendat population being reduced by about half at that time.[70] It was often those "who had lost or suffered most who were the most receptive to the evangelical message," thus the traditionalist critique of Tekakwitha's conversion.[71] By the 1640s, there was a Catholic Wendat village, Sainte-Marie, where new converts could ideally live without being corrupted by sauvage beliefs. This village was also given exclusive access to French trade. Wendat traditionalists referred to the Sainte-Marie mission as Ossossané, the "believing village." An unbearable fragmentation of the Wendat Confederacy was occurring as the religious values of the Jesuits "led them to place the salvation of individual souls ahead of the collective safety and well-being of the Huron people."[72] These colonial dynamics in southern Ontar:io have complex connections to Kahnawake's traditionalist critique of Tekakwitha.

In the Haudenosaunee stories told by Rice, the Wendat who aligned themselves with the Black Robes were becoming powerful at the expense of others, and thus some traditional Wendat began talking with Mohawk and Seneca about the divisive impact of disease and conversions.[73] Concerned with their survival, they "asked to join a war party to kill off all the Black Robes and Christian followers." The Haudenosaunee rarontaron, or war chiefs, replied: "We will help the traditional Wendat be rid of the Christians. Although they have done much harm to us, they are now one with us."[74] The result was a two-year war that used Wendat villages as a base for attacks against the French and their Wendat allies. Not all Haudenosaunee were in agreement with this war; Rice tells of one royaner in Thomas's Good Mind tradition who, "distraught" because he had just negotiated a peace with the Wendat, took his own life. The instability related to colonial diseases and trade struggles was leading to "the ascendancy of the war chiefs over the royaner" in the Haudenosaunee Confederacy, a shift that Rice heard of from Thomas and which is supported by historic documents.[75] According to Thomas, the pressures brought by the colonial arrivals resulted in a reversal of the Peacemaker's teachings as the Good Mind that did away "with the warring, fighting and killing" increasingly receded and people "went back to practicing the old ways."[76]

In 1649 Ossossané was destroyed, two Black Robes were martyred, including the later canonized Brébeuf, and many Christian Wendat were killed.[77] The Wendat were also dispersed from southern Ontar:io; some joined Anishinaabe allies, others followed the French to the Jesuit mission of Lorette on the Kaniatarowanenneh, and many were absorbed into the Mohawk or Seneca either willingly or as war captives like Tekakwitha's mother.[78] Distressed about the loss of their mission in the heart of Canada, the Jesuits depicted the Haudenosaunee attacks as destroying the Wendat.[79] Their writings have fostered the view – dominant in the historical literature – that there were economic motives in the Haudenosaunee offensive, but historic records and oral tradition suggest that the conflict "actually resembled the final phases of a 'civil war.'"[80] Returning to Thomas, he talks of the fur trade as being the initiator of bitter fights among Indigenous nations.[81] The Wendat, he explains, were the first to organize themselves into a Confederacy like the Haudenosaunee, and in fact the Peacemaker was born a Wendat along the north shore of Lake Ontar:io. With the arrival of the fur trade, the Wendat moved down a path that resulted in them "breaking up their own Confederacy."[82] It was the Tahonteanrhat of the Toron:to area who guided the Wendat challenge to the Black Robe conversions, which makes sense since they historically had closer relations with the Seneca.

The beginning of these internal divisions seems to have been a factor in Brûlé's death fifteen years earlier. After traversing the passage de taronto with what were most likely Tahonteanrhat guides, he was captured and then released by the Seneca. His release was based on the agreement that he would help the Haudenosaunee become "friends with the French and their enemies,"[83] an action that is perhaps reflective of an early royaner attempt to deal with emerging colonial challenges. This led to the 1620 peace embassy and then Brûlé's 1632 death at the hands of "Wendats opposed to his efforts to broker peace" because of trade implications, the same Wendat from which the Christian village emerged.[84] Just as the official Black Robe interpretation of the Haudenosaunee war on the Wendat radiated down through history, Brûlé came to be seen as a scoundrel and turncoat rather than failed mediator of peace. As for the Tahonteanrhat, their nation remained somewhat intact following this period of war in their own village of Gandougourae in Seneca country.[85] Within two decades the Seneca moved into Tahonteanrhat territory on the passage de taronto and Mohawk came to inhabit the mission of Kahnawake.

Long before Tekakwitha's move north, her mother participated in the Haudenosaunee renewal or requickening that many Wendat and Anishinaabe experienced during this period. She and close relations were captured by Mohawk on the Kaniatarowanenneh around present-day Trois-Rivières. Drawing on historic stories, Greer has us imagine an experience "of utter powerlessness and complete vulner-ability as she watched those closest to her injured, humiliated, and killed; her own physical pain and the anguish of knowing she might die at any minute."[86] This state of uncertainty persisted on the march back to the Mohawk community of Gandouagué in what is now upper New York State. It was decided that she would be adopted into a clan, whereupon the demeanour of her captors softened. Gandouagué was undergoing such massive change related to disease and requickening that, Rice relates, "it seemed like there were more adopted and captives living in the villages than there were original members."[87] Through the paradoxical forces of "war, brutality, and loving-kindness, Tekakwitha's mother became a Mohawk" whose new status was confirmed in the requickening ceremony.[88] In time she married a Mohawk man, gave birth in 1656 to Tekakwitha, and about six years later succumbed to another smallpox epidemic that almost took Tekakwitha as well.

Part of the Haudenosaunee requickening ceremony that Tekak-witha's mother experienced entails the adoption of a new name ancestrally related to their community. In this ceremony, which the Peacemaker initiated, she was given a new name that was in fact an old name of someone who had likely recently died and whose role in the community needed renewing. Though the distinction between those who were born Mohawk and adoptees was negligible once the latter had proven their commitment to the community, most of those who originally followed the Black Robes from Gandouagué to the mission that became Kahnawake had Algonquin and Wendat ances-tries.[89] In 1673 the composition of this mission began changing with the arrival of two hundred Mohawk who included the great chief Kryn. This seemed to upset the existing multi-ethnic and Christian order of the place, as reflected in "a great flare-up that resulted in a senior Huron chief leaving the settlement."[90] Considering these dynamics and Tekakwitha's decision to become Christian, she needs to be seen as someone who was neither an outsider to the Mohawk longhouse nor securely inside its walls.[91] The uncertainty of these rough colonial waters was central to her isle passage.

It was Easter of 1676, four years before Tekakwitha's death, when she was baptized Catherine, or Kateri in its Indigenous pronunciation, a name one Black Robe wrote as being "held in great veneration among the Indians."[92] There is an interesting resonance between Haudenosaunee requickening and Catholic conversions; the baptismal "rite involved an altered personal identity, a new name, and, because baptismal names always harkened back to a Christian saint, connection with a personality from the past."[93] Such ceremonial renaming would have been familiar to Mohawk in the seventeenth century because of the requickening tradition.[94] The revered ancestor for Tekakwitha was Saint Catherine of Siena.

While Tekakwitha had an isle hermitage, the stories of her adopted fourteenth-century Catholic Italian ancestor tell of someone who was initiated into a spiritual life after three years of hermit-like seclusion in a little room. Isolated from her family, the legend states, Catherine found in her house "the desert; and a solitude in the midst of people."[95] Tekakwitha would have heard of the saint's penances, which included "fasting, beating herself with chains, sleeping on a narrow board, and drinking the fetid water that had washed the oozing wounds of lepers."[96] The aim of these ascetic acts was "to break through the confines of mundane existence and partake of the divine." What concluded Catherine's hermit practice was a breakthrough awareness of the divine working in creation, manifested for her in a Mystic Marriage.[97] From this point on, a peasant girl who would die in her early thirties became known for a practical genius in politics and an immense power over ruling men that was inspired by visions of spirit working in the world.[98] There would have been for Tekakwitha many similarities between such stories and the duties of an Indigenous Good Mind that gives spirit the space needed to inform living relations,[99] even in a time of so much colonial turbulence.

The mixed position Tekakwitha holds on her isle hermitage between Black Robe conversions and Mohawk requickening resonates with aspects of Woodworth's passage. At our Toron:to presentation, he put up a picture of Joseph Brant and remembered his mother telling him that this man is us, people of Haudenosaunee and British ancestry. Brant's Haudenosaunee name, Thayendanegea, means "two arrows bound together" and symbolizes the way in which he was chosen as a Mohawk boy to undertake a British colonial education and become a fluently bilingual diplomat.[100] It was

from a position of Haudenosaunee and British allegiances that Brant
led some of the Iroquois Confederacy north of Lake Ontar:io following the American Revolution to resettle land their ancestors once
inhabited such as present-day Six Nations west of Toron:to. This was
also the ground from which he represented Akwesasne in negotiations
with Sir John Johnson concerning land for the arriving Loyalists east
of the longue sault. For Woodworth, the passage to identifying who he
is as an Indigenous person "is not as simple as it might first seem."[101]
Within him Mohawk and British blood has intermingled "since the
first arrival of my British precursors along the Atlantic Coast" twelve
generations ago. These ancestral relations are epitomized in Johnson's
father, Sir William Johnson, the British Indian agent much beloved by
the Mohawk, mentor of Joseph Brant, and husband of his sister Molly
Brant.[102] It is this intimacy that leads Woodworth to talk of a period
of love and fraternity in some early colonial relations, which came to
an end only in the past couple of centuries during the emergence of the
American and then Canadian nations.

The spirited life of Tekakwitha on the Kaniatarowanenneh similarly
straddles the divide of Indigenous canoe and colonial ship that intensified as Canada came into being. It was the Gradual Civilization Act of
1857 that began "a series of legislative acts designed to limit Native
participation in the economic, social and political life of the colony."[103]
Its goal was to restrict Indigenous participation in emerging urban
markets like Toron:to and limit "hunting and gathering activities which
supplemented farming," while also aiming to assimilate the population.[104] Though Haudenosaunee and Anishinaabe "rejected the Act as
an act of cultural war," it was followed by more extensive measures to
speed up the pace of civilizing, assimilating, and converting.[105] In contrast to their integral role over the first two centuries of colonial relations, Indigenous peoples were increasingly characterized as sauvages
who retarded economic development. Meanwhile, Métis were regarded
as half-breeds, neither Indigenous nor European. The Indian Act of
1876 had at its core a colonial urge toward social control that "was
predicated on legally identifying who was white, who was Indian, and
which children were legitimate progeny – citizens rather than subju-
gated Natives."[106] Its discrimination against Indigenous women had
long-lasting effects, accounting for "the majority of the twenty-five

thousand Indians who lost status and were externalized from their communities between 1876 and 1985," though Bonita Lawrence points out that when this number is projected into descendants the conservative estimate is of one to two million Canadians who were severed from their Indigenous ancestries.[107] With the fear invoked by the institution of Canada's residential-school system spreading through the twentieth century, this alienation became deeply internalized in different ways across Canada.

Colonial severances are for Tekakwitha and Woodworth not simply out in the world but intimately carried within their beings, as they are for many Canadians and Indigenous people. These actions of an emerging Canada led some Indigenous nations to adopt "a strategy of marginalization as a cultural defense."[108] Such isolation makes sense in the face of colonial processes that seemingly want to obliterate any ancestral grounding. But then, even within this isolation, new issues arise related to fears about political takeovers by community members of mixed ancestries. As Pamela Palmater explains, Indigenous communities are increasingly using traditions to exclude people in ways that often contradict the history of more fluid Indigenous identities.[109] Kahnawake is one such community that has recently been marked by membership disputes with the enactment of a residency by-law that "prohibits non-Mohawk members from residing on the reserve," leading to the eviction of non-Indigenous spouses of residents.[110] Membership has become defined by a "50 per cent blood quantum requirement, and a moratorium on mixed marriages." While these controls aimed at defining Mohawk status are seen as necessary in renewing the community's grounding in Haudenosaunee tradition, Palmater cautions that in putting the Indian Act behind it Kahnawake has "ended up incorporating the colonial ideologies about race and blood purity embedded in the *Act*."[111] It seems to recapitulate those dynamics that led to the Wendat dispersal from southern Ontar:io and the distrust of a canonized Indigenous saint. An internalization of colonial severances expands the gulf between the Kaniatarowanenneh's Two Rows, further isolating the braiding potential of Tekakwitha.

The pressure of such dynamics can, in Thomas's teachings, also distance people from a Good Mind. Prior to the ship's arrival, he teaches that the Haudenosaunee believed only in the Great Law of the Peacemaker.[112] It was with the turbulent colonial dynamics of Tekakwitha's time that the earlier tradition of warrior chiefs returned – something we saw with the mass requickenings. He adds that the latter half of the twentieth century has witnessed a resurgence of the Warrior Society as it "swept across the native territory in North America,"[113] a response to residential schools and other assimilative policies that was particularly strong among the Mohawk of Kahnawake and Akwesasne. Though the Warrior Society positions itself in Haudenosaunee tradition, Thomas firmly states that its members do not "follow the Great Law, and their own self-interest comes before the benefit of all nations."[114] Their actions intensified divisions "between Christians and Longhouse followers" and, more importantly, the Akwesasne longhouse became divided between supporters and opponents of the Warrior Society. If the Warrior Society is committed to the way of the Peacemaker, Thomas asks, why do "they turn their guns on their own people?"[115] He is referring to a 1989 conflict when two native people were killed at Akwesasne "during a crossfire attack between pro-gamblers and anti-gamblers," as well as death threats he received for confronting the Society. His response to being told to "watch himself" was: "I don't have to watch myself … The only one who will watch over me is the Creator." The lesson of Thomas's story is that colonial violence can get inside cultures and people, creating divisions that replace the onerahtase'ko:wa's roots of peace with "disunity, violence, and warring."[116] This is what occurs, he says, "when a nation no longer follows its traditional teaching," though change is possible if "we can practise a good mind once again."[117]

Finding passage through such historically mixed relations is a challenge Woodworth also embodies as he attempts to reconcile his mixed ancestries with a Good Mind. For his teacher, Thomas, the Two Row Wampum treaty is an ideal image for minding our way through a time of continuing colonial pressures. The current difficulty he observes is that many Haudenosaunee are struggling with the pressure to straddle two worlds, with "one foot in the canoe and the other in the great masted sail ship."[118] He asks Woodworth and others what they "will do if they are in a storm and have to separate."

Will they have a place to leap to or will they "fall fatefully onto the waters to be rescued or drowned?" As the shapes of violence increase around us, will we have a sustainable footing on which to mind our common waters? There was much in this Two Row teaching that made sense to Woodworth, and in one sense he follows it. Despite his English ancestry, he writes that "because I was born, nurtured and have lived my entire life on Turtle Island, the land of my Haudenosaunee ancestors, I consider myself primarily a Haudenosaunee man."[119] Similarly situating himself in Thomas's teaching, Longboat writes: "I work with my feet planted firmly within the canoe and study the sailing ship to see what may be useful, adaptable and beneficial to the continuance of our canoe and for our common future, the river that supports us all."[120] Practising a Good Mind is a choice that needs continual renewal, though Woodworth's mixed ancestry means that it also entails navigating between canoe and ship.

The tradition of requickening is for Woodworth vital to reminding our way forward in a good way.[121] His decision to follow Thomas's Good Mind led him to be requickened by his Deer Clan mother with the name Raweno:kwas, "He is Dipping the Words."[122] But affirming the value of requickening does not mean that we have to deny the struggles this Haudenosaunee tradition had with forced adoptions, such as during the time of Tekakwitha when war prevailed over peace.[123] Woodworth's mixed ancestries combined with other teachings to suggest an alternative sense of requickening, one that does not require a universalizing choice between canoe and ship. It is also possible, he writes, to "remember what a privilege it is to be able to carry two world views, since not all the people are able or willing to bear such a gift."[124] This third option of taking passage between the Two Rows is not for Woodworth about perpetually straddling our increasingly numinous waters, but actively participating in the creation of something new within the great river, the Kaniatarowanenneh, that is intertwined with the origins of Canada. Is there a braided alternative that we can foster as the climatic longue sault's shapes of violence quicken and "thin" spaces break in upon us? We are now guided toward a contemporary response to this question through the eco-mysticism of Saint Tekakwitha on her Île-aux-Hérons.

How could a sauvagesse be a saint? This question troubled the Jesuit priest Chauchetière as he considered writing an account of Tekakwitha's life. There was something in her asceticism and Lenten season death that led him to dream about her saintly calling. Yet, at the same time, his Jesuit education left him wondering whether the dreams and visions "were the tricks of the Devil."[125] What he was struggling with were the demons of a civilizing conceit that positioned the Canadian wild far from heaven, a conceit that continued to display itself in Pope Benedict's 2012 call for Tekakwitha's help in renewing the faith of First Nations while he remained silent on the church's own colonizing history and need for internal renewal. In Chauchetière's case, relations with the Haudenosaunee gradually opened him to a broader sense of human difference. His epiphany came in April 1680 as "he sat watching a young woman die," a woman who had written all over her face the "tragic Mohawk history of illness and death."[126] As Greer writes, the Jesuit's afflicted soul began to heal "through the agency of the saintly Mohawk."[127] He was healing from that conversionary spirit that is associated here with the experience of Tekakwitha and many Indigenous people. But this urge to convert has, we have seen, also fuelled the climatic longue sault that surrounds us today. In a sense, the Protectress of Canada now carries us to her island between the Kaniatarowanenneh's two shores. It is here that we can begin contemplating the depth of healing needed today by the ship for it to come back into good relations with the canoe and our common waters.

The modern ship's limitless appetite to penetrate, control, convert, and now sustainably manage all the river beads on the Two Row through an ever-developing technological imperative has disrupted our common river, on local and global scales. These changes have led some to re-engage the Two Row as a model for renewing relations between Indigenous and environmental knowledge, with the two retaining their unique strengths while "linked to one another by strands of truth, respect, and friendship."[128] The uncertainty of our situation is leading some on the ship to realize the need for a renewal of these old relations that for a couple centuries were largely the focus of conversionary efforts. A change is blowing over the waters that has the potential to shift our sense of knowledge. For John Ralston Saul, engaging Indigenous oral wisdom places us on "a line between myth and information" that is much needed if we are to respond to an ecological crisis that "calls out for a new, broad approach, one that rethinks how societies work."[129] The often rationalized interdisciplinarity of an ecology of mind needs to connect with the Indigenous Good Mind to offer different imaginings of the changes now needed.

But as with Mohawk, Livingston, and others, Saul also sees a dark web that is resistant to bold change, epitomized in the way that environmental and climate responses tend to be "stuck to the narrowest possible bookkeeper approach."[130] Looking at similar challenges in the context of our climate of change, William Ophuls considers the need for a braid of modern approaches with an Indigenous wisdom that the progressive era thought was no longer needed. Our responses, he argues, must primarily recognize what it means to become a "more experienced and wiser savage."[131] As with Livingston, part of this means we need to recover a "connection with nature lost when we ceased to be savages."[132] While the Black Robes were concerned with descents into wildness and subsequent generations have long been focused on progress, Ophuls characterizes the present period as the end of an evolutionary cycle that began with the human species unconsciously identifying with nature, moved during the modern period to an expanding severance of people from nature, and now requires us to consciously reidentify with nature based on experience and knowledge of our current crisis.[133] The severance of "primal participation and the antithesis of wilful scientific abstraction must eventually be resolved in a synthesis – a conscious and wise participation in nature that makes the human mind whole." The ship is being challenged to bring its ways back into accord with the common waters that have always surrounded it, even when its inhabitants forgot agreements like the Two Row.

It is a braided ecological vision that resonates with Saul's view of Canadian culture. In watery images, he writes that you can see nations as great rivers whose current is a kind of collective unconscious that "keep rising to the surface if you allow it."[134] The greater our awareness of "the current, the better we move down the river like an experienced canoeist, with purpose, taking full advantage of the eddy line." The coureurs de bois and Brûlé exemplify Canada's braiding tendencies, with about 40 per cent of French Canadians having "at least one Amerindian in their family trees."[135] Unlike those who went deep into the pays d'en haut and became the Métis Nation, those who stayed along the Kaniatarowanenneh had less space from colonial powers and populations to acquire a sense of themselves.[136] People of these ancestries differ from the Métis or particular Indigenous nations,[137] and yet they experienced a similar climate of fear. The last two centuries of conversion policies mythologized "the meaninglessness of a 'watered down' Indigenous heritage" and thus supported the active suppression of any Indigenous ancestry.[138] But even these conversionary acts mark Canada's potential to foster a braided way of being, only in a repressive fashion. The nation's greatest failure has been, in Saul's words, our inability to consciously internalize "the First Nations as the senior founding pillar of our civilization."[139] We need to stop viewing Indigenous as a sauvage problem to be fixed and relearn ways for commonly living on the Kaniatarowanenneh from a host of positions.

The canoe has travelled along this great river much longer and thus carries wisdom for those in the ship about good ways of traversing the eddy line, especially as the climatic longue sault intensifies. But, as the preceding pages suggest, it is not the sauvage mind that needs to become wiser; rather, the mind of the modernizing ship must return to the wisdom of the canoe. A *wiser modernity* is what we are in need of. That said, Ophuls's "wiser savage" terminology is meant to highlight our need to return to a wild sensibility, something colonials and moderns alike have generally denied as having value. Yet, even as we continue to hold onto our old beliefs, a realization that changes are needed is slowly arising at the edges of our ship as we come into contact with the turbulent climate and energy issues of our time.[140] Though these changes will increasingly have to reflect diverse climate, ecological, and cultural realities, not unlike those represented in the Two Row, Ophuls also envisions a kind of common river that flows with "the values of humility, moderation, and connection mandated by nature."[141] Instead of never-ending economic growth, resource development, urbanization, and technological "progress," the future "will of necessity be more simple, frugal, local, agricultural, diversified, and decentralized."[142] A wiser modern ship needs to traverse our common waters based in the Good Mind's egalitarian spirit of frugality and fraternity that so impressed many coureurs de bois.

Our current task is to make "a virtue of this necessity."[143] The difficulty is that the web-like era of *Living Better Electrically* has reached every corner of our lives, to the extent that our appetite for energy and emissions of greenhouse gases continually grows even as we tout the merits of frugal restraint and efficiency. Urban living in developed nations like Canada is built around the comfort and convenience of "central heating, air conditioners, climate controlled cars, underground malls and pathways, and weather forecasts,"[144] all of which predisposes us to the imperative of technologically managing climate change. While there has been a push for greater technological efficiencies in energy use, a noticeable rebound effect has also occurred, such that energy use continues to rise despite efficiency gains.[145] Since 1978, commercial energy use has increased over 65 per cent and home use by more than 30 per cent.[146] The major reason for the latter rise is the increase in average house size from 1,000 square feet in 1950 to 2,200 square feet a half-century later, with fewer people living in those dwellings.[147] Meanwhile, the energy-efficiency gains in home appliances have been cancelled out by the proliferation in the number of appliances, their increased size, functional innovations like computerized fridges, and new technologies epitomized in today's constant WiFi reality. We saw in chapter 2 similar automotive stories: increased fuel efficiency has occurred in a context of us driving further on more roads, with some research indicating that we drive on average four times the distance that we did in the 1950s.[148]

These facts highlight the point that sustainable frugality has to go far beyond efficiencies that largely aim to maintain the ways of the ship unchanged. The modern ship is also asked to reduce energy and resource consumption, something that is supported by the difficulties in shifting to renewable energy. Energy and environment scholar Vaclav Smil highlights three challenges to this transition.[149] First, the scale of fossil-fuel consumption is so large that it currently cannot be replaced by renewables. Secondly, there is an intermittent nature to wind and solar energy that requires maintaining back-up from existing power streams. Finally, the transition will be prolonged by existing infrastructure that is "worth at least $20 trillion across the world," an investment that is difficult for modern nations to walk away from.[150] Based on these factors, Smil concludes that lowering overall energy consumption is essential for speeding up a renewable transition. With reduced demand it will be more viable to retire old fossil-fuel sources and infrastructure and replace them with alternatives.[151] Efficiency combined with frugal reduction is essential to renewing the sauvage wisdom of a good climate of mind, practices that have a certain resonance with the asceticism of Saint Tekakwitha.

But just as the dark web observed by Mohawk and Banyacya binds the eco-efficiency approaches, it can also affect our approach to renewables. Considering the amount of land available for solar and wind production, it seems that current renewable energy technology could replace about one-tenth of present human energy consumption "without destroying the delivery of critical natural energy flows upon which all life now depends."[152] As a modernizing solution that attempts to maintain or even increase current energy-consumption levels, renewables can fall into a violent pattern not unlike the epic creation of the renewable longue sault hydro dam. That said, returning to sauvage wisdom does not mean a total rejection of modern technologies and ways. As Ophuls explains when he observes, "I am no Luddite," there does seem to exist "numerous possibilities for obtaining energy from non-fossil-fuel sources – thorium reactors, geothermal wells, solar furnaces, and many others."[153] Similarly, Suzuki writes that eco-efficiencies, electric cars, renewables, and other technologies can give us "time to wean ourselves from our current patterns of energy use" while constructing living spaces that further eliminate energy intensive consumption.[154] To take up alternative technologies in our move toward renewing ways of living will require immense energy and investment.

The potential for changing to a more frugal way of living is in our midst but we need the will and cultural energy to grasp it.[155] If we return to Bateson, perhaps it should be clarified that we are talking about "the will" to surrender addictive conversionary binds to a different way of relating with each other and our common river. This is for Bateson a multi-dimensional shift toward a renewed sense of our "relationship to others and to the universe or God" or however else we want to define the vitally common source of our lives.[156] Such surrender is implicit in calls not simply to shift toward renewables but to do so in a context of "reducing human pressure on global energy flows and lowering global energy consumption."[157] Perhaps we can conceive renewable energy as a kind of large-scale detoxification program for an addicted modern ship, with the transition inevitably accompanied by resistance and denial. We do not want to recognize how we have been drawn into a diversity of "thin" spaces that require very different ways of minding relations than what the modern ship deems efficient and rational.

Frugality is not simply a way to surrender for its own sake but rather is meant to open us to Ophuls's second dimension of how we need to approach our common Kaniatarowanenneh, that is, with a sense of fraternity. This term refers to our common brotherhood and sisterhood, a feeling of belonging and "social kinship that transcends biology but that nevertheless retains some of the force of" ancestry.[158] It is a sensibility that resonates with Thomas's teachings on colonialism and savagery. He says that most white people do not realize that it was they who "were the savages when they came here," for if the Indigenous were sauvage "then they would have killed every white person that set foot on Turtle Island."[159] In the Good Mind of his culture there was no belief that "we were better than the whites"; instead, there was a sense of shared humanity that was the basis for sharing the land. This egalitarian fraternity, which was appreciated by the coureurs de bois, was exploited by colonial powers to gain a strong foothold and in time enforce conversions.

Reclaiming a fraternal and frugal way needs to be justice-seeking because our climate of change partakes in the conversionary patterns of colonialism.[160] Global injustices, climate change, and wasteful energy consumption are "manifestations of the same set of problems," and thus we must distinguish between emissions that are necessary for survival and those that are mere luxuries, where the former are conceived as a basic right and the latter are not.[161] Such a perspective on our emissions leads to alternative vistas on the shadow price of the modern ship's freedom, for it depends on a kind of "energy slavery, which is the concrete manifestation of the slavery to appetite that impels us to enslave nature."[162] Stories about the epic conversion of the longue sault, Alberta's bitumous sands, and the place of the fallen dolon-do reveal to us the darkening shapes of violence while also illuminating the addictive binds that maintain unjust relations.

In bringing up the justice dimensions of a sauvage fraternity, I am drawn into another mixed sense of my French Catholic ancestors. While the Black Robe conversionary spirit must be critiqued, we need to do so while also recognizing the Jesuit move over the past half-century toward social and environmental liberation. In 1971 the Jesuit order pledged itself to "The Service of Faith and the Promotion of Justice," with the intent of working for justice from the ground of faith.[163] This pledge came together with Jesuit work among Indigenous people in poverty-stricken regions of Latin America to create a liberation theology that "is essentially transportable to any culture, particularly a culture that has a history of oppression as a determining feature of life."[164] Developed by the Indigenous Peruvian priest-theologian Gustavo Guttiérrez, it is concerned with liberating those who suffer from the oppressive injustice of sins that are at the "root of all disruption of friendship." [165] It has come to see the divine in those who have been oppressed, and over the past few decades it has extended this message from marginalized human communities to ecologies. Some descendants of both Jesuit and Franciscan missionaries now call for "ecological justice – respect for the otherness of beings and things and their rights to exist – and constant social justice, respect and concern for people."[166] There is an emerging sense that the need to return to human, ecological, and climatic fraternity is highlighted in the experiences of those who have been ill-treated by the ship.

This is the same spirit of renewing fraternity that we have begun connecting with Saint Tekakwitha. Just as with Woodworth's braided vision of Toron:to, she does not simply convert to the Catholic faith but rather can be imagined as adapting it to the lands of a Good Mind tradition that stretches back to the Peacemaker. As Greer points out, there is a tendency to emphasize the way in which Jesuits came "to terms with the strange ways of the Other, but parallel native efforts to bridge the gap of cultural difference and comprehend the European Other are just as noteworthy."[167] She attempted to be a bridge, the signs of which are woven in various aspects of her

life and practice. To say this does not mean denying internal severances that obviously arise in periods of any significant change, such
as is reflected in Thomas's teachings about the return of warrior
ways in Tekakwitha's period. Rather than "people following the
roots of the tree to its source, they were now being forcefully
absorbed as members."[168] That said, it is clearly a faulty view to see
Indigenous cultures as always "falling victim to a triumphant Euro-
American culture" when the more accurate image is that people on
both sides often chose to try and braid cultures.[169] Weaving this
potentiality seems more possible in "thin" times of turbulence like
Tekakwitha's period and our present moment, which is not to deny
the potential for epic conversion. Fraternal calls for justice and liberation is another common point of relation between the river's Two
Rows where a braid may be renewed.

Renewables, energy efficiencies, and reductions can offer us passage
to a kind of ascetic renewal of our fraternity with a myriad human,
ecological, climatic, and ancestral relations. In a sense, we are being
drawn into the presence of Saint Tekakwitha on her isle. According to
Ophuls, frugality "is not the same as stinginess or asceticism."[170]
While this is true, it is an assessment that is somewhat grounded in the
modern rejection of the emphasis in Christian asceticism on denial of
both the human body and nature. But Tekakwitha's approach in tying
her bodily being to a spirited energy embodies a different wisdom. It
avoids both the Black Robe denial of the wild and the modern mind's
denial of spirit. The frugality of a Good Mind is a means to affirm
relations, for duties like Thanksgiving renew fraternity with a spirited
creation. It is an approach resonant with that of Saint Francis, who, it
is said, applied "some of the most subtle canons of human friendship
to the creatures of nature, to this extent: that he could readily give up
his own sense of mastery, his own power to control."[171] Such friendship is committed to sustained relations, regular contact, giving
thanks, and other intimacies that nurture bonds. These are all features
of Livingston's bird relations, Woodworth's Toron:to, the coureurs de
bois' river passages, and Saint Tekakwitha's Île-aux-Hérons.

The eco-mystic fraternity of Saint Francis was in many ways renewed
during the closing decades of the twentieth century by the Catholic
priest Thomas Berry, who died in 2009. In his last book, *The Great*

Work, he explicitly draws upon the Haudenosaunee Thanksgiving address as a way of highlighting the radical move toward fraternity that is today required of the modern ship. This ceremony embodies for him an "intimacy with the natural world" and a formal recognition of human "existence as the gift of the various powers of the universe."[172] It also points to the kind of change needed, in science, politics, economics, and religion as they each come face to face with our climatic turbulence. To temporally contextualize the required shift, he sketches a brief history of the West that displays its movement from a dominant concern "with divine-human relations" to the current focus on the politics and economics of inter-human relations to a future where the ship has decisively developed a "capacity for intimacy in our human-Earth relations."[173] The drive to establish dominion over the planet and define all our relations in the economic terms of "use" value to humanity is now in need of change. This will, Berry writes, "be one of the most severe disciplines in the future, for the Western addiction to economic dominance is even more powerful than the drive toward political dominance."[174] Yet, as with Thomas's Good Mind optimism about the possibility for renewal, Berry says that we can move toward a future where "we understand the universe as composed of subjects to be communed with, not as objects to be exploited."[175]

Such an expansive view on our fraternal relations draws me to the Île-aux-Hérons and a saintly presence who has much to teach about how to renew ourselves as our climatic longue sault becomes increasingly turbulent. Just as nuns inspired Tekakwitha's braid, there is today a "green sister" movement that has been inspired by Berry's work and is renewing the Christian ascetic tradition in climatic, ecological, and energetic terms. Acts of fasting are used not to deny a sinful body but to encourage awareness of how our bodies and ecologies are "under siege from the hostile forces of environmental pollution."[176] These practices are concerned with detoxing environmental pollutants and shifting to more sustainable food relations, while also symbolizing the human duty to initiate a much needed planetary detox.[177] Beyond food, the green sisters are undertaking technological and carbon fasts as well. While affirming the body's relation with climate and ecology, such fasts offer "respite from the many stressors and the fast pace of modernity – a world that is all too much with us."[178] They are meant to help us enter into "thin" spaces where a radically different minding of our energetic relations can occur. Here we may suddenly find ourselves opening to a Good Mind Thanksgiving, to say with Tekakwitha and Francis: "Be praised, my Lord, for Brother Wind, And for Air, for Cloud, and Clear, and all weather, By which you give your creatures nourishment."[179]

To imagine myself on the Île-aux-Hérons with Saint Tekakwitha and Woodworth is to be guided not to a wildly romantic hermitage on the Kaniatarowanenneh but to a particularly "thin" space in a Canadian consciousness that needs to renew itself. In the Catholic tradition, a saint is often a patron of particular groups, places, or issues, an ancestral intercessor who can support the living as they deal with issues central to their being.[180] Saints carry something that is too heavy for the human community to bear, and the preceding pages suggest that what Saint Tekakwitha carries is twofold: those colonial severances that have deeply affected the canoe and repressed Canada's braided potentiality; and now the renewal of a potential alternative braid in this time of rapid change. Reclaiming our mixed past can, in Bonita Lawrence's words, "represent another stage of rebuilding the shattered hoops of different nations" and the possibility of demographically and symbolically healing Canada's conversionary dynamic.[181] A woman who began as a local figure of inspiration has been transformed by the Catholic Church into Protectress of Canada, a role that Tekakwitha takes far beyond the renewal of faith in First Nations. "She Who Puts Things in Order" offers to renew and requicken those colonial severances that the modern ship must mind to sustainably shoot and portage the climatic longue sault we are within.

With the cold wind and rain sinking in, this saint's mixed nature brings to mind a host of personal, familial, and genealogical memories. One is of a particularly frustrating drive through Montreal near the time of my father's death. In this island city that was built around the canoe, an accident or construction stalls the asphalt arteries that radiate from its bridges and tunnels. Taking a detour around one such blocked bottleneck, I was turned round on the non-grid-like roads that follow river contours until I was lost and then driving through Kahnawake. The experience of being lost raised spectres that filled me with dread about both the French of Montreal and Mohawk of Kahnawake. Familial remembrances at the funeral and afterwards led me to realize that my anxiety about being lost in Montreal and Kahnawake was deeper than the immediate moment. Though my father's early childhood was spent in Montreal, something happened that distanced him from the place. We rarely made the one-hour trip, and in Cornwall we were French Canadians who, except for my mother, did not speak French. In his last years he and his parents began to acknowledge that we were not simply anglicized French Canadians. Genealogy supports this as branches of our family tree connect us to the Mohawk of Saint Regis and the Wendat of Lorette in Quebec.[182] Are these Indigenous roots or French relations like Brûlé and Antoine Leduc who lived in close proximity to Indigenous peoples? Parts of our ancestry find some kind of refuge on Tekakwitha's isle, though the exact nature and quality of those relations is difficult to determine through the mists of time.

There is a pain underlying this familial forgetting that partakes in Canada's historic attempts at the sauvage conversions seen in this chapter, one that ancestrally situates me on a turbulent Kaniatarowanenneh in uncertain ways. A lack of cultural footing in the French row eventually led me westward toward the more comfortable English environment and grid-like roads of Toron:to, and yet this place surprisingly opened Indigenous passages for me to mind the mixed nature of Canada's ancestral severances. Though I am moved by the Haudenosaunee vistas on our common river that Woodworth and his teacher Thomas offer, I am not situated in their Indigenous canoe. My ancestral footing is clearly stronger in the French, Black Robe, and modern ship, and yet Livingston's bird-eye view made it clear that many of its conversionary ways are unsustainable. It was from this starting point that the coureur de bois and Indigenous passages of my ancestry began to emerge, passages that I traversed for much of my life without knowing it. Beyond some Haudenosaunee friends, my relation with communities like Saint Regis, Akwesasne, Kahnawake, and Six Nations is marginal, partly owing to the historic and familial fears of the Mohawk across the Kaniatarowanenneh that I inherited.

It was ultimately easier for me to go north as a social worker and then consider Inuit understandings of northern warming than engage those painful severances at the core of my personal, national, and ancestral being. The return to Toron:to seemed like another safe haven until the unexpected occurred as the Haudenosaunee Good Mind of Woodworth and Thomas called attention to these external and internal relations. As I struggle with my relation to Tekakwitha and my position on the Two Row, I am reminded of Lawrence's words. From the dominant modern perspective, for mixed-bloods who are grounded in the ship even to recognize Indigenous heritage feels "'inauthentic' and false – or indeed, even 'appropriative.'"[183] It is an internalized assessment that continues to cycle through their being as "a constant drain on their sense of self-worth." This denial of Indigenous roots and relations has many degrees and dimensions, and is a personal, cultural, and national issue that keeps us mired in uncertainty about where we have come from and where we need to go on our common waters. In my case I am clearly not Mohawk, Wendat, or Métis, and yet when I engage a teacher like Woodworth it is equally clear that there is much in the Haudenosaunee Good Mind that resonates at the core of my being. This is the common ground of the Protectress' sacred Île-aux-Hérons, which I need to approach from my position on the north shore of the French settlers.

It is along a severed Kaniatarowanenneh that my father and Saint Tekakwitha passed away during the Lenten season, a time of year that calls the Catholic part of my mind toward images, teachings, and practices focused on sacrifice, though exactly for what purpose was often unclear to me as a child. From her isle, it now seems that the green mysticism of a few can symbolize the need for and potential of renewing our ways of minding relations. But it ultimately needs to find resonance with functioning traditions like Thanksgiving and Lent or perhaps a more recent environmental ritual. The new Earth Hour occurs in the same spring season as Lent, and its approach to sacrifice is meant to raise people's awareness of their broader relations for at least an hour. Some Christian churches have promoted comparable carbon sacrifices during Lent through acts like biking and taking public transit, energy-retrofitting homes, and buying and living locally. But what do such seasonal eco-ascetic acts of energy reduction and consumer-based efficiency mean in a Canadian culture that largely supports an epic processing of oil and selling it to the highest bidder? Even Earth Hour at present seems largely bound to a resource-conservation logic rather than something more profound like the eco-fraternity of Thanksgiving. From such a stance, Earth Hour can feel as dry, uninspired, and thankless as my childhood Catholic rituals because they seem disconnected from powerful worldly realities.

On the "thin" space of Saint Tekakwitha's isle there is a less cynical and more optimistic way to watch these same lights go dark during the spring Earth Hour. That is as a seed that is moving us toward Thanksgiving and renewed relations with energy, ecology, climate, the land's ancestral elders, and our families. With the climatic longue sault calling us to slow down, do less, and be more frugal, we are being given a momentary pause to more fully mind our local and planetary fraternity with a host of relations. In such vital moments different questions can arise: Do I live in a place like Toron:to where it is possible not to own a car? If I live in a rural place like Long Sault, are there ways to rethink the type of car I should own and how much I use it? Is one airplane flight in the past five years too much? When I do fly, do I slow down long enough to recognize the implications of my action? Do I fast from luxury emissions and engage Earth Hour and spiritual seasons like Lent, Ramadan, and other cultural cere-monies as moments for renewing my ways of living throughout the year? What relations am I thankful for? These questions radically depart from the ways of the modern ship. They have the potential to renew our relation with the Protectress of Canada. With her herons flying farther eastward on the Kaniatarowanenneh, the next chapter is guided to a deeper cosmological sense of what it is she protects, and then to how stories can transform our sense of what every sacred Earth Hour asks of us.

5

One Mind, One Heart, One Mouth

Blessed Virgin, what joy we feel that, even before our birth, the town of
Chartres built for you a church … Oh, how happy are the Gentlemen of
Chartres, and how great are their merits for being your first servants! …
This is, Blessed Virgin, what we are doing today, in connecting ourselves
with the Gentlemen of Chartres, that we may have with them only one
mind, one heart, and one mouth, to praise you, to love you, to serve you.

Wendat of Lorette, Quebec (1678)

Making the old journey eastward along the Kaniatarowanenneh can
take us to a different Canada than a nation of epic resource development. It was on these shores near where they open out to the Atlantic
Ocean that in 1535 French explorer Jacques Cartier first heard the
term "Kanatha" from the Wendat ancestors of Georges Sioui. While
they were referring to their "big village" of Stadacona farther inland
near where Quebec City and the Wendat village of Lorette came to be
situated, this is not what the French understood.[1] Being immersed in
medieval feudal thought, the colonial ship applied Kanatha to an
"imaginary country of 'Canada,' reigned over by Donnacona, an
actual Headman."[2] As if symbolically foretelling the rolling-head history described in the preceding chapters, the nation's naming was
rooted in a misunderstanding that has environmental and cosmological implications. In Sioui's words, the planet's ecological and climate
crisis points to the failed "social philosophy that white invaders"
brought to Kanatha and elsewhere.[3] We are approaching another
"thin" space for minding the breadth of renewal that is needed as we
are tossed about by the climatic longue sault's quickening changes.

Following the herons from Saint Tekakwitha's isle toward Lorette
offers passage to an early Wendat-French discussion on the changing

13 Blue Virgin with Lake Ontar:io view (summer 2015). A replica of the Blue
Virgin stained-glass window from the cathedral at Notre-Dame de Chartres
is photographed here with Lake Ontar:io flowing behind her.

nature of the world we live in. In 1678, the same year that Tekakwitha,
Thérèse, and Skarichions looked upon the Île-aux-Hérons as their
spiritual hermitage, the Catholic Wendat who had come to Lorette
from southern Ontar:io sent a prayer with wampum belt to the great
cathedral of Notre-Dame de Chartres west of Paris in France.[4] The

14 *Turtle Island at Night*, 2012. An astronaut's photo from space of the web observed by Mohawk and Banyacya at Niagara Falls and which here is seen sprawling across North America or Turtle Island. The story of Sky Woman's view of this island as she fell toward Turtle's back adds cosmological dimensions to the discussion here.

letter that accompanied these gifts was translated into French by a Black Robe with the intent of expressing the Wendat's praise for the Virgin Mary and their desire to come to "one mind, one heart, and one mouth."[5] From the view of the ship, such a blessing is often conceived as arising simply from a colonized people who had been fully converted to Catholicism. But Sioui's position on the canoe offers a more nuanced sense of what is happening here. Consistent with how we imagined Saint Tekakwitha requickening Catholicism to the lands of Kanatha, he proposes that Wendat converts saw something in this religious tradition that they connected to their Indigenous ways of minding creation. Such resonances, as seen in the previous chapter, were exactly what the Black Robes utilized in their evangelicization work, and it makes sense that Indigenous converts made similar connections as they attempted to indigenize the ways of the ship.

In this chapter, I want to attempt to renew the ocean passage from the kanatha of Lorette to Chartres and back so as to imagine an exchange that could have occurred with the Good Mind but was stunted by the history of missionary conversions. The preceding pages have situated me in relation to ancestral presences like the Black Robes who are deeply implicated in these severances, more marginal ancestors such as Brûlé and coureurs de bois who have a mixed past, and a saint like Tekakwitha whose life suggests that a braid between the Two Rows may be possible. As we all do, I inherit a spectrum of potential stances in relation to Kanatha, originating for me with the diverse Europeans who came to these shores. From my mixed position a question emerges: Can we imagine coming here with an intent not to convert but to find ways of adapting our approaches to the Kaniatarowanenneh's waters? It is an act of renewal that, Sioui proposes, is already occurring. In his words, "it is the newcomers who eventually and naturally become assimilated into the basic civilization of the land they migrate to and that, therefore, in spite of superficial appearances, it is the non-Aboriginals who are in the process of becoming Indian, and not the opposite."[6] Can we make this change conscious, or will we wait for the climatic longue sault to violently ground our rolling-head ways? It is to these questions that the herons guide me as they approach the gift-bearing ship following its ocean passage.

One way of requickening people to the lands, waters, and climate of Kanatha so as to be able to praise with "one mind, one heart, and one mouth" is by finding connections between cosmologies. The Wendat of Lorette and earlier Stadacona had their own stories of a "Blessed Virgin" who bridged spirit and creation. She is referred to as Aataentsic, the ancient one, who, in Sioui's telling, fell from the "Upper World" into the unformed creation that we now inhabit.[7] It is a story that the Wendat share with the Haudenosaunee and Anishinaabe, though variations exist beginning with the name of the Sky Woman whom the Haudenosaunee call Otsi:tsia or Mature Flowers.[8] We will refer to her as Sky Woman. The common view is that she fell through the hole of an uprooted tree, with Woodworth writing that as she fell she clutched "the roots and earth around the edge, and took with her some seeds and cuttings from the tree."[9] As she hurtled toward the dark waters below, waterfowl and aquatic animals like those found in the marshes of Ashbridges Bay softened her fall so that the Big Turtle could, in Sioui's words, "receive her on his back."[10] This is why the lands of Kanatha are known as "Great Turtle Island,"[11] the place onto which Sky Woman fell and then birthed creation. Such a mythic vista of our place on the Kaniatarowanenneh takes us far beyond the fallen dolon-do, harnessed longue sault, ruined town centre of Lac Megantic, and epic bitumous sands.

In times of rapid change like ours, why contemplate a cosmological sense of what Kanatha sits upon and where life falls from? As with Woodworth's ancestral sense of Toron:to, Sioui writes that "as long as a person (or a people) has not found and understood the spirit of the land he or she inhabits, he or she is in a destructive stance in relation to that land."[12] Beyond the severance of ancestral, ecological, and climatic presences from our experience, the imbalance also has mythic dimensions that play out in relation to Sky Woman and Turtle Island. When recording the Wendat story, Jesuit priest Jean de Brébeuf wrote condescendingly about the miraculous birthing of creation from this sole woman, seemingly forgetting his own Virgin stories. In his words: "If you ask them how this could happen, you will make them very uncomfortable," adding that if pressed they will tell "she was pregnant" before the fall.[13] For Sioui, this disrespect is not an issue of the past; on the contrary, our environmental crisis is Turtle Island's response to a destructive modern mind that is marked by linear thinking and the conceit that the land "belongs to me."[14] It is a critique reminiscent of Woodworth's rolling heads and Livingston's technological imperative, and for Sioui it is the central feature of what "we have to change all around."[15] The climatic longue sault is renewing an awareness of the fact that we belong "to the sacred Circle of Life,"[16] a circular cosmology that is central to this North Atlantic Ocean exchange.

The letter from Lorette was destined for Notre-Dame de Chartres, one of the many cathedrals built in the twelfth century, all dedicated to the Virgin Mother. Considered an architectural jewel, this medieval building is closer to the cycling circle of Sioui's ancestors than the dominant linear tendencies of the ship's ancestry. Its central inspiration is a Neo-Platonic Christianity that expresses a cycling sacred reality which the rational mind can only partially attend and conceptualize.[17]

There are differences between the mythic models of Notre Dame and Sky Woman, but there are also interesting resonances that the Wendat may have intuited beyond the teachings of their Black Robe mediator. Reimagining this ocean dialogue is not simply an act of historic analysis. For these immersive mythic models also offer us a contrast with the computer-generated climate models we predominantly look upon today, and thus highlight another modern bind. Our proliferating flat-screen experiences tend to reinforce technologically driven thought which severs us from Turtle Island. From Kanatha to Notre Dame and back to our present moment, this chapter examines some mythic models for renewing our sense of climatic cosmologies that we can only partially know and control. By undertaking this ocean passage, we will begin to approach a clearer minding of what Saint Tekakwitha protects on our common Kaniatarowanennenh.

In the early spring of 1536, a sailor on Jacques Cartier's second voyage to the New World wrote that the men, after placing an icon of the Virgin Mary "against a tree about a bow-shot distant from our fort across the snow and ice," were ordered to "go in procession."[18] This party spent the winter where the Laurentian Wendat village of Hochelaga had been situated the previous year, near present-day Montreal. It was an ill-advised trip since by April the cold winter had coupled with scurvy to kill twenty-five. These losses could have been circumvented if the ship had listened to the warnings of the Wendat the previous fall at the kanata of Stadacona. Over three days in September 1535, Chief Donnaconna attempted "to impress on the French that they should not navigate towards Hochelaga."[19] On the third day, Cartier's two Stadaconian interpreters came "out of the woods and, after the Catholic way they had observed, walked around towards the French, their hands joined as if in prayer."[20] With eyes raised toward the sky, they offered "the words 'Jesus, Maria, Jacques Cartier.'" The level of admiration displayed leads Sioui to interpret the ritual as an attempt to ask "for protection for Cartier and his men." The French responded with laughter, with one crewman stating: "If you just believe in Jesus, he will keep you from the cold."[21] The French departed the following day, whereupon the Catholic faith began its initiation into the cold realities of Sky Woman's wild Turtle Island, the mythic presences to whom their prayers, perhaps, should have been directed.

The letter and gifts of wampum that departed from the kanatha of Lorette a century and a half later offers us passage to a different sense of Cartier's Blessed Virgin. For the Wendat and Haudenosaunee, these lands were known as Turtle Island, the place where Sky Woman fell and began birthing creation. Though the Franciscan Gabriel Sagard refers to this story in the early 1620s, its first full French transcription is recorded in the 1636 *Jesuit Relations* by Father Brébeuf while he was in the Wendat territory of southern Ontar:io. He begins by stating how astonishing it is "to see so much blindness in regard to the things of Heaven, in a people who do not lack judgment and knowledge in reference to those of earth."[22] These disparaging words echo Cartier's earlier rejection of Stadaconian pleas which were based on a message from their Great Spirit, Cudouagny, who "announced that there would be so much ice and snow that they [the French] would all die."[23] The captain's response: "Go tell your messengers that your god Cudouagny is a fool."[24] The tone of the Black Robes also tended to be "contemptuous, scoffing, or indifferent," which Sioui contrasts with the attitude of Sagard and the Franciscans, who were often more willing to exchange knowledge and appreciate the humanity of their hosts.[25] He writes that, long before Jean-Jacques Rousseau's romantic elucidation of the noble savage, Sagard "understood a new and transcendent moral quality in the Amerindian social genius, the product of life in nature."[26]

Though Brébeuf exemplifies the way that many missionaries, including Sagard, disregarded Indigenous beliefs as delusions, he was faithful to what the Wendat voices stated, "faint and garbled though they may be."[27] His and Indigenous versions commonly tell of a Sky Woman who, it is said, "descended from a celestial world" onto Turtle Island.[28] Where did the sky opening come from? What Brébeuf heard was that "the husband of Aataentsic, being very sick, dreamed that it was necessary to cut down a certain tree from which the people who dwelt in Heaven obtained their food," and when the tree was removed a hole emerged.[29] The Haudenosaunee stories in Thomas's tradition also tell of Sky Woman's husband, Rotea:he, who was ill and had the Sycamore tree of lights uprooted. It was while looking through the hole that Rotea:he threw "a blanket over her head and thrust her over the edge."[30] This series of events arose, Woodworth explains, because Sky Woman's husband had descended into a dark confusion after learning that she had embraced a powerful spirit and was now pregnant.[31] Offering other details, Rice describes the illness as part of a larger Sky World imbalance related to declining light energy or *orenta* – a term popularized in Joseph Boyden's 2013 fiction *The Orenda*.

While the uprooted Sycamore provided a point of passage to the dark watery world below that we now inhabit, the great celestial onerahtase'ko:wa was, as Rice explains, the axis from which life energy radiated from the purer light worlds above.[32] This tree's vibration was slowing, and the Sky World's orenta was giving way to a "freezing dark energy known as otkon" that was the root of the illness and fall.[33] With the sky people's fear rising as their protective light dimmed, the ill Rotea:he had the Sycamore uprooted to create a womb-like opening to the world below.[34] As Sky Woman was pushed into the dark abyss, she unsuccessfully grasped at the hole's edge to try and stop the fall. What she did grab hold of and brought down with her were seeds and tree cuttings from the world above.

Gravity doing its work, the world below gradually came into clearer view. It was a dark watery reality with, Woodworth states, "many creatures, some like and some unlike those we know today."[35] In his telling, it was the ducks who first saw her approach, while Rice and others claim that it was a character known as Loon. [36] But because Loon "had the habit of always having his head looking down into the water," he mistook her reflection and thought that she was arising from the depths below.[37] Following Loon's cry that a female is "coming up from the water depths," a small heron known to us as Bittern responded: "She is not, in fact, coming up from below; she is, in fact … falling from above."[38] According to this version, it was the ancestor of Tekakwitha's herons and Livingston's bittern who first saw her heavenly fall, as is fitting for a brown feathered bird who easily camouflages itself among marsh reeds and grasses, often escaping "detection by standing upright with bill pointed upward."[39] The few times I have seen bittern in Toron:to marshes has been in such a concealed stance of raised beak among reeds. With eyes looking upward, the Haudenosaunee call this being Kentsiokwas, "He Who Keeps Looking at the Sky."[40]

Now that it was discerned that a woman was falling from the sky, the waterfowl and aquatic animals like Beaver, Otter, and Muskrat of this marsh-like world agreed on the need to find her a place on which to land in the midst of the water. Across the stories, it is said that a great sea Turtle was called by the animals "to the surface to provide her a place to land."[41] In Brébeuf's Wendat version, Turtle raised "its head above water and perceived her" and then "called together the other aquatic animals to get their advice."[42] Such an island cosmology resonates with the name Wendat, which means islanders. The name more directly refers to the ecological realities of their land in southern Ontar:io known as Wendake, a region surrounded by the waters of Georgian Bay, Lake Huron, the other Great Lakes, lac de taronto, and many rivers.[43] As Sioui writes, the "most likely meanings for the word wendake are 'the island apart,' 'the separate country,' 'the peninsula country.'"[44] This watery place was merely a part of the larger, mythic Turtle Island that, in both Wendat and Haudenosaunee stories, Sky Woman landed upon.

The shell of Turtle was a good place but it needed some earth. Accordingly, a council of animals decided that someone "would dive to the depths of the primordial ocean and bring back the magic, sacred mud" that was needed to support her life.[45] After a series of animals unsuccessfully attempted to grab the mud, Rice's Haudenosaunee tradition tells of Muskrat asking to make an attempt. He was laughed at "because he was supposed to be the weakest of them all."[46] Yet, as Rice writes, when his dead body floated to the surface, Beaver "noticed that his paws were holding some dirt." In the Wendat version told by Sioui, it is Toad who brings back the mud because, as he similarly explains, Toad is "the humblest of these animals."[47] The determination of Muskrat and Toad teaches people not to underestimate anyone's abilities, that the "weakest are sometimes the most capable."[48] More than that, Rice explains that the sacrifice Muskrat made so that Sky Woman and later people have "a place to live and grow" is a continual reminder that human beings should "always be thankful to the animals for

sacrificing their bodies so that their children could continue to live."[49]
The duty of Thanksgiving is grounded in a humble recognition that
our life is based on the gift of other creative presences.

Beyond this general lesson, Muskrat also carries a specific reminder
for those ancestrally situated in the colonial and now modern ship.
The laughter of the other animals at his offer to help is reminiscent of
the self-assured disregard of Cartier's crew when the Wendat pleaded
with him to not go down the Kaniatarowanenneh to Hochelaga, an
attitude that continued with the Black Robe missions and, as we have
seen, in the epic web of converting the wild into useful resources (natural and human) and energy. Muskrat and Toad teach the ship about
minding relations in ways that include a whole cosmology of presences, and the humility needed to recount those lived relations in story
form. This creative humility is mirrored in an Indigenous Good Mind
that allows variations on the Sky Woman story to sit comfortably
with each other. It is something that Rice learnt from Thomas who
explained "that each person has his or her way of telling a story" and
that traditionally people got together "to compare stories and learn
from one another" since no one owned the story.[50] So, while there are
common features across these diverging stories, there can be different
emphases. For example, Rice's reference to the onerahtase'ko:wa in
the Sky World uniquely offers other wisdom on the cosmic flow of
orenta into Turtle Island from the Sky World and light realms beyond.
Such a humble and flexible capacity to storytelling is vital to the modelling of a good climate of mind.

The popular telling of Sky Woman's fall by Thomas King, a
Cherokee storyteller and scholar, can help us conceive what this modest flexibility means in relation to modelling our position on Turtle
Island. Here Sky Woman, named Charm, has more of an accidental
fall. After digging at the base of "the oldest tree in the forest" until she
could see "right through to the other side of the world," she stuck her
head through "and sure enough, she fell."[51] While the birds "broke
her fall," Turtle gave Sky Woman a back to land on, though with a
joking spirit added, "If anyone else falls out of the sky, she's on her
own."[52] Here it is Otter rather than Muskrat or Toad who floats to
the surface dead, with the others opening her paws and discovering
the dark, gooey sacred mud.[53] What makes King's variation

particularly striking is his reference not simply to "a fall" but rather to "the Fall." As with Woodworth, his Iroquoian roots mix with a Christian heritage that familiarizes him with a resonant story of the modern ship's ancestry, one that fosters a less flexible stance. The history of colonizing dynamics like the Black Robe concern with the Kanathian wild leads King to propose there was something in the biblical "story of being made in God's image, of living in paradise, of naming the animals" that fostered the ship's arrogance.[54]

The resonance of Sky Woman's fall with the Fall spoken of in Genesis did not go unnoticed by the Black Robe Brébeuf. In the spirit of the universal theism, he highlighted this similarity as an important sign "that in the past they [the Wendat] had some knowledge of the true God that was more than merely natural," and thus was hopeful about the prospects of conversion.[55] With a familiar tone, Brébeuf goes on to state that, among the various versions, one "seems to have something to do with Adam, though the story is predominantly falsehood."[56] Such disregard is also displayed by the Franciscan Sagard, whose inquiries into inconsistencies in the Wendat's cosmologies and spirited practices led him to conclude that "they do not really know and adore any divinity or God, of whom they could give some account and whom we could recognize."[57] It is a marginalizing tendency that King sees as grounded in a cosmology where "all creative power is vested in a single deity who is omnipotent, omniscient, and omnipresent."[58] In contrast, Sky Woman highlights a common Indigenous approach to storytelling deities as "figures of limited power and persuasion," falling into our creation because of an accident or illness.

The biblical Eden begins as a perfect world that falls, after the grasping after knowledge of Adam and Eve, into a state of sin and disharmony. In the words of King, their actions lead to a "chaotic world of harsh landscapes and dangerous shadows," a wild world that is at war with itself and, in the modern ship's economic language, is increasingly informed by competitive self-interest.[59] The juxtaposition of these two stories clarifies the way in which the communities that surround Indigenous storytelling tend to create more openings for a proliferation of perspectives.[60] This is exemplified by King's practice of quoting the Bible verbatim, a linear passage that contrasts with Sky Woman's orally diverging and yet interconnecting stories.

On Turtle Island, "the pivotal concern is not with the ascendancy of good over evil but with the issue of balance."[61] Different beings, from Sky Woman to Heron to Muskrat to Turtle, co-create a viable land that has its challenges, as seen with Muskrat's deathly offering. But ultimately the Indigenous Good Mind "imagines creation as a far richer possibility than any one story can evoke by itself."[62] To mind a cosmology that is co-created by diverse presences, it seems that more than one story is needed.

This mythic contrast returns me to the words of Toron:to's Howard tomb, "It was thou who caused the severance," and Livingston's diagnosis of the planetary crisis as rooted in Christianity's belief in original sin. His early 1970s examination of this original severance drew upon Lynn White, Jr's seminal article on the Christian ethos that underlies the ship's ecologically destructive marriage of science and technology.[63] In White's analysis, the Christian victory over an antique paganism that viewed "every tree, every spring, every hill" as having its own spiritual guardian is described as a great psychological revolution.[64] It replaced a sacred world of mysterious proportions with one that was simply wild. This revolution "not only established a dualism of man and nature but also insisted that it is God's will that man exploit nature." All of this was integral to the emergence of the Enlightenment, the scientific revolution, and today's technological imperative. Many have since clarified that White's analysis may have been too broad and lacked nuance, though its assessment has some degree of truth.[65] With fallen humanity situated between God above and nature below, there is, Livingston writes, "no conceivable reason for the existence of the blue planet apart from the needs of God and man."[66] Trapped in this perspective, he adds, our modern secularized selves are implicated "in blood crimes of such appalling enormity that none of us is fit for the 'service and fellowship' of God."[67]

Perhaps a way out of the ship's enclosing models can be found by minding the proliferating stories of the Indigenous canoe and the humility they teach about a world of divine proportions. With the dirt from Muskrat, Otter, and Toad now spread upon the back of Turtle, the waterfowl who had flown up to ease Sky Woman's fall then placed her "upon the Earth on Great Turtle's back" and flew away satisfied.[68] The contrast between this supportive landing and

Eden's harsh fall leads King to ask some poignant questions: "What if the creation story in Genesis had featured a flawed deity who was understanding and sympathetic rather than autocratic and rigid? ... What if Adam and Eve had simply been admonished for their foolishness? I love you, God could have said, but I'm not happy with your behaviour. Let's talk this over. Try to do better next time."[69]

Recognizing this same challenge leads Livingston and White to doubt that a sustainable response can simply entail "applying to our problems more science and more technology,"[70] and thus they hold up Saint Francis as our ecological guide to a different way. On these Kanathian passages, we are led by Saint Tekakwitha and her Herons to Lorette and the blessed story of a falling Sky Woman who initiates a dynamically balanced creation on Turtle's back when, Woodworth writes, "the seeds and earth from the Sky World fell from her open hand and took root."[71] We will return to this part of the story later. For now, we need to pause so as to follow the Wendat letter across the ocean to Notre-Dame de Chartres and a Christian model whose symbolic abundance has potential for weaving with Sky Woman rather than converting.

———

The words and wampum from Lorette went eastward down the Kaniatarowanenneh as it opened to the Atlantic Ocean, with the river coming to feel quite different from that which encompasses Saint Tekakwitha's Île-aux-Hérons and my origins along the longue sault. Its growing width takes in the ocean tide, with a rocky shoreline appearing and then disappearing daily. At land's edge where the gulf releases its waters into a dark blue-grey expanse, one can almost imagine the original abyss Sky Woman fell into, especially on an overcast day. The "long bodies of the ocean's pulse" have a tenor that Henry Beston contrasted with the longue sault's "shapes of violence," and it was into this expanse that the ship carried the Wendat letter to another heavenly woman, Notre Dame. In a spirit similar to Sky Woman's descent into water, the etymology of the Catholic Mary connects her with the Latin *mare*, sea, and an ancient lineage of pagan goddesses that are "the primordial womb of life from which all created forms emerge."[72] The Wendat would similarly exalt the Virgin Mary as "one who not only has given birth, or is giving birth, but who will always give birth until Jesus is perfectly formed in us all."[73] These gifts are being brought into another sacred model, a kind of ancestral womb where the modern ship can potentially renew relations with a symbolically rich cosmology. It is a "thin" space that offers us more flexibility for relating Mary to Sky Woman and, as we consider later, our flat-screen models of the climatic longue sault.

Walking toward Notre-Dame de Chartres, one is awed into still-ness by the abundance of stone-carved ancestors and angels that grace each of its three portals. Around the North Portal are the elders who prepare the way for Christ and his crucifixion. The South Portal depicts Christ's triumphant resurrection and the saints who followed in his footsteps, signs of the world's eventual renewal in the Last Days. Finally, to the west is the Royal Portal where most pilgrims enter to be spiritually reborn. Centred above its three doorways is the Mother Mary with child amidst many carved icons. Some are of an explicitly religious nature, like one portraying Christ's ascension, while others refer to scholarly knowledge. Pre-dating French univer-sities, Notre Dame was a cathedral school where students learned the seven liberal arts represented on the portal: grammar, rhetoric, dialectic, music, arithmetic, geometry, and astronomy. Beyond this interdisciplinary scholarship of its day was an experiential working knowledge symbolized in zodiacal signs paired with peasant occupa-tions. All this knowledge was to be crowned with divine wisdom.[74] While its symbolic abundance bewilders the modern mind, Notre Dame's cosmology and rituals made sense in the twelfth century because they intersected with symbols of nature, knowledge, and the sacred. All the symbols "remained irreducible to one another even when knotted together in a Christian structure,"[75] thus intimating the potential this space has for renewing a braid with Sky Woman.

Here one enters "the earthly reproduction of a transcendent model" that continually resanctifies the world and our knowledge by containing it.[76] Within this cool stone model one is bathed in the light blue radiance of stained-glass windows that depict more ances-tors and stories, though the atmosphere is inspired by one saint in particular. The model for all the Gothic cathedrals dedicated to Notre

Dame was Abbot Suger's Abbey Church of Saint Denis, situated about eleven kilometres north of Paris. Inspired by the Eastern Christian mystic theology of Saint Dionysius and its description of the One as the "initial, uncreated, creative light" that emanates through "every living creature, every natural object,"[77] the abbot attempted to give this theology a concrete, material form through architectural renovations of Saint Denis. These included increasing the church's height, enlarging its windows, employing the new art of stained glass, and reducing obstructions to light.[78] The interest in Dionysius was due to the association of the Christian Neo-Platonic writings of this eastern contemplative with Saint Denis, the third-century missionary bishop martyred in this region of Gaul. The "patron saint of Paris,"[79] he became symbolically woven with Dionysius whose reverence for emanating light was instituted in high-reaching stone and stained glass. It is a Neo-Platonic Christian model that is best evoked by the cathedral at Chartres.[80]

The feeling within is of a peaceful shore, particularly when one stands in front of the stained-glass window set high in the southeast wall. Dating from the mid-twelfth century, this window is known as *Notre Dame de la Belle Verrière* or Blue Virgin. Her shades of blue and white garments flow like water and air within a red surround that resonates with the Sun's warming rays. Upon her lap sits the Christ child with a green halo, while on either side are three angels ordered hierarchically one on top of the other. Above her the Holy Spirit in the form of a Dove descends, bringing in "the rays of divine inspiration."[81] In both symbolic form and positioning, the Blue Virgin affirms Dionysius's mixing of Neo-Platonic and Christian thought in various ways, but especially as it relates to light. While in the Christian view "God is light, the light of the world, the light of truth which illuminates" hearts, the Neo-Platonic added the sense of visible sunlight as the outward reflection of an invisible energy that radiates from the divine source of life.[82] It is a view resonant with Woodworth's architectural connecting of light to sacred building, from Haudenosaunee longhouses to Christian temples to even modern homes. Quoting Frank Lloyd Wright as a central inspiration, Woodworth states that "a higher order of the spirit has dawned for modern life in this interior concept of lived-in space playing with light."[83] From Notre Dame to Turtle Island to our homes, sunlight enters our creations as a reflection of the primal energetic source that is much more than just fuel for the modern ship.

There is a potential resonance here with the Good Mind's onerahtase'ko:wa *axis mundi* that radiates peaceful light from the Sky World to, following Sky Woman's fall, Turtle Island. During the first beautiful spring days "when the air shifts, the snow is gone and the warmth is comforting," Woodworth writes that a faith keeper announces "the early morning time of the Sun Ceremony."[84] In a spirit similar to the Blue Virgin's connecting of visible sunlight with heavenly light, the Sky Woman creation story conceives the Sun as "the brother of Sky woman, placed in the Sky world to give brightness so that the people can see to walk on the Mother Earth." The Sun Ceremony begins "with gratitude for the past winter when the prayer for return of these days was given." The men "recite in unison the honouring to their brother the Sun" as the women rhythmically clap to keep time. Describing similar ceremonies, Brébeuf writes that the Wendat call upon the Sun and Sky for "all their needs, and respect the great bodies in it above all creatures, and remark in it in particular something divine."[85] He respects this view since "nothing … represents divinity to us so clearly," radiating as it does goodness to all creatures.[86] Such words call forth Saint Francis's *Canticle of Father Sun*: "Be praised, my Lord, with all your creatures, Especially Sir Brother Sun, Who brings the day, and you give light to us through him … Of you, Most High, he bears the likeness."[87] The Haudenosaunee singing "is meant to amuse" Brother Sun, which is then followed by a Thanksgiving address with burning tobacco hanging in the air,[88] just as incense fills the nose in Notre Dame. In this atmosphere, the ocean between Lorette and Chartres, Indigenous and Catholic, does not seem so impassable.

Sunlight is central to the positioning of the Blue Virgin in the southeast where she takes in the rays of the Sun's daily rise and passage across the sky, thus bringing forth the life represented by the Christ child. Long before May became the official month of Mary in the eighteenth century, "she had been celebrated as the epitome of the delight of spring on May morning, as she is still, with flowers and dancing and processions in which her statue is carried around the town."[89] The baby Jesus's green halo with three golden rays suggests spring's natural profusion or what Saint Hildegard referred to as *viriditas*, a fresh green that "represents the principle of life, growth, and fertility flowing from the life-creating power of God."[90] Writing in the twelfth century's Neo-Platonic atmosphere, she praises this viriditas in her *Song to the Virgin*: "So the skies rained dew on the grass, and the whole earth exulted, for her womb brought forth wheat, for the birds of heaven made their nests in it."[91] Using natural images while making no direct reference to Mary, Hildegard associates her with the green Earth, the dew with the Holy Spirit, and the Christ child with a divine energy that sustains creation.[92] In the Good Mind tradition, the viriditas of the Creator spreads by first creating the Sunflower as a sign of the Sun's warm light, and the greening of Turtle Island continues with the red willow that is an important medicine, strawberry as the first berry of spring, and so on in an unfolding that Thanksgiving follows.[93] It is in this spirit that the words from Lorette seem to join in exalting the Virgin's viriditas as one "who will always give birth until Jesus is perfectly formed in us all."[94]

In the midst of the child's green halo and Virgin's blue countenance, my imagination wanders to memories of familiar Kanathian waters, from the Kaniatarowanenneh to the Great Lakes and a particular stretch along Lake Ontar:io near Ashbridges Bay. From late spring through summer, blue skies with streaks of white clouds often intermingle with sparkling dark blue waters, and jutting out into them is the green growth on Turtle's back. Looking at the Blue Virgin reminds me of an energetic presence that underlies our world. This is an old view inherited by the Romantic tradition, as seen in the story of Saint Tekakwitha. With the La Chine rapids churning as she contemplated adopting Catholicism, the story goes that the sky was "as blue as the robe of Our Lady."[95] The coureurs de bois also gave thanks to the Virgin Mary on their river passages, according her the consoling title of Notre Dame du Portage; one individual related that, when being transported over a sault, "I saw nothing but a noble Lady in white who hovered over the canoes and showed us the way!"[96] Such views arise again in the nineteenth century among Canadian transcendentalist poets who wrote of Nature as a Temple where one can "cool the world's mad fever."[97] Reverence can become a salve for the ship's driving severance when we slow down, or are slowed down, long enough to look out on a Turtle Island shoreline clothed in the coloured light of the Blue Virgin. The blue-robed lady of Saint Tekakwitha's isle would have also been coloured by Sky Woman stories that told of her pregnant fall to Turtle Island.

Informing the Blue Virgin's viriditas is the Dionysian sense that the divine energy of God or, in the Neo-Platonic tradition, the One emanates into a receptive creation through a hierarchy of angels highlighted on either side of the window. These angels ring bells, wave incense, hold candles, and bring food, for all sensory realities are intelligible transmissions of the One "from one level of the celestial hierarchy to another."[98] There are three groups of angels. Closest to the One are the angelic thrones, cherubim, and seraphim, which are described by Saint Dionysius as having many eyes and wings that allow them to receive "more directly the first enlightenments."[99] In the second group are authorities, dominions, and powers associated with the evolving history of places, cultures, and nations. The lowest consists of angels, archangels, and principalities that more actively guide individual lives. This celestial hierarchy emanates in a way that "the greater intensity of the superior being exerts an attractive force upon the one next below it and draws it upward."[100] There is a flow of orenta entering the creation that is metaphorically epitomized by the Sun, though it has spiritual dimensions in both the Christian Neo-Platonic and Good Mind traditions.

In the ninth century, the influence of Saint Dionysius is evident in the work of Jon Scotus Eriugena, who translated Dionysius's Greek writings for the Carolingian emperor.[101] Taught in the Celtic Christian tradition of Colum Cille on Iona, Eriugena brought this knowledge to Gaul in the wake of the Viking attacks that were occurring on those Atlantic shores.[102] He had an understanding of Eastern Christian mysticism and traditional Celtic thought, resulting in a hybrid philosophy that resisted any distinction between natural and supernatural. As Eriugena proclaims: "Nature is the Cause and Creator of all things, is wise, and is alive."[103] The Eastern Christian view of Nature as revealing a "divine mentality" was another tradition that Lynn White, Jr saw as having the potential to temper the rigid dualistic impulse of severing nature and divinity that came to inform the ship.[104] Eriugena's emanating light connects everything in an "ineffable friendship and insoluble unity," a divine fraternity that guides him to quote Saint Dionysius: "Let us understand love, whether we are speaking of the divine, angelic, intellectual, spiritual, or natural kind, as a unifying and blending power which moves higher things to forethought of the lower, joins equals in a reciprocal bond of communion, and turns the lowest and subordinate toward their betters, placed above them."[105] His translation of Dionysius marked the arrival of a Christian Neo-Platonic philosophy influenced by metaphors "of higher realities giving off an effulgence towards the lower," all sensible through the eyes, ear, nose, mouth, and touch.[106] Turtle Island is to be engaged as divine, not simply converted because of a perceived original sin.

The angels of this window are much less anthropomorphic than our current way of minding such presences. In ancient and medieval traditions, these winged beings often passed "by in a breeze or a ruffling of water, or in the heat and light of Sun and stars – in short, in any elementary fluxes that make up our Earth."[107] As Dionysius writes in relation to his highest angelic order, "they are also named 'winds' as a sign of the virtually instant speed with which they operate everywhere."[108] In the thought of Eriugena, this divine energy emanates into the Blue Virgin with an elemental feel that would have resonated with the Celtic Christian practices of Colum Cille. As with the Neo-Platonic focus on light, it is said this saint experienced "heavenly light" of an "immeasurable brightness" that illuminated the night.[109] The light is associated with the divine gifts of angels, for Colum Cille is often portrayed "as one around whom the supernatural world broke into this world" through angelic dialogues.[110] Though his legends do not talk of a hierarchical spiritual order, he is said to have had experiences of angelic presence while he remained "awake on winter nights or as he prayed in isolated places."[111] This Celtic tradition of engaging spirit in "thin" spaces pre-dates Christianity, something that can also be seen in the original placing of Notre-Dame de Chartres.

While the first Christian church on the site of this French cathedral dates to the third and fourth century, before this there was a Gallo-Roman temple built over what was at one time a Celtic *nemeton*. It was first described by Caesar as "a vast clearing surrounded by impenetrable forests" where "Druidic worship was celebrated."[112] As with islands for the Irish Celtic Christians, such forest clearings were "thin" spaces where "the human world could open onto the world of the gods and vice versa."[113] There are many symbols and built elements of the cathedral that harken back to this pagan sensibility, including the crypt and sacred well found below the structure. Even the Blue Virgin's symbols overlay earlier Celtic understandings and spring ceremonies, thus at least partially situating the Black Robes' use of divine aids for conversion in an old ancestry. The descending white Dove symbolized more than the Catholic Holy Spirit; it also referred to preceding pagan depictions of "ancient mother goddesses" with supportive birds, and, in the earliest traditions, these goddesses were "unified in one being called the Bird Goddess."[114] In many ways, the popular energy behind the flurry of cathedral building dedicated to Notre Dame stems from these deep connections, epitomized in the building of Chartres on an ancient sacred site. It is in such "thin" spaces that many on both sides of the North Atlantic imagined broader cosmological relations long before the arrival of Christianity.

Making these pagan, Indigenous associations with Notre-Dame de Chartres does not mean that everything that it symbolizes is good. During the period in which these cathedrals emerged, "more stone was cut in France alone than at any period in the entire history of Egypt,"[115] and the decimation of forests for lumber used to erect the Gothic stone structure and fire the stained glass was prolific.[116] Beyond explicit ecological impacts, there is also the mixed history of Saint Dionysius's celestial hierarchy owing to its historic relation to Christianity's often violent conversionary spirit. Since the idea of hierarchy comes into the Western Christian tradition with Dionysius's translation, the eco-feminist theologian Anne Primavesi critiques it as historically "validating violence in the name of God" and thus is the deep religious ground of today's environmental and climate injustices.[117] Though the Dionysian sense of the hierarchy is of "the overflowing of God's love" through the mediating role of angelic beings, she argues that it has fostered the reverse.[118] A divine hierarchical order has often resulted in the positioning of particular people as knowledgeable representatives of that order who thus gain power over others and the creation – with a proliferation of injustices as evidenced in the Black Robe missions to Turtle Island. Furthermore, the interconnected responsiveness of the planetary climate highlights the way in which "a God-concept which does not allow for reciprocal evolving relationships between God and the world is not only seriously, but dangerously, flawed."[119]

In much the same spirit as the Indigenous view that one story on its own lacks humility, the danger of a uni-directional hierarchy is that it promotes violence through excluding, degrading, and denying "value to the earthly substantiality of certain classes of being" and people. It is another dimension of that which severs the ship from good sustainable relations with wild Turtle Island. This important insight offered by Primavesi can be partially dealt with by returning to Eriugena's focus on a celestial theophany, rather than hierarchy. His theophany is "understood not as a hierarchy of self-determining and therefore internally multiplicative principles but as the theophany of a single self-determining God who is multiplied through his own act of creation."[120] In Eriugena's words, "creation subsists in God, and God is created in creation in a remarkable and ineffable way."[121] This is more consistent with Primavesi's call for a response to hierarchical abuses that grounds our minding in a "non-hierarchical, coevolutionary organizing principle."[122] It resonates with Sioui's Wendat sense of a "circular-thinking" that interconnects humans and Turtle Island in "a great chain of relationships linking an infinity of beings."[123] Each being's unique role in a dynamically circling creation is seen to express aspects of a common life energy that radiates into the creation, including humanity.[124] Rather than completely negating a Neo-Platonic hierarchy or Plato's Hekademos, we can follow Primavesi, Eriugena, and Sioui and other Indigenous thinkers to conceive divine and ancestral presences in a less perfect and mixed way so as to not forget the limits of our situated positions in a responsive cosmology. This less-perfect perspective of divinity calls my mind back to Sky Woman's fall from a Sky World afflicted with illness, with a dimming light that needed something from the dark waters below.

An emergent model of co-evolutionary relations in the midst of the Blue Virgin has recently been supported by the Catholic thought of Thomas Berry. As was touched upon in chapter 4, he proposed that the preceding eras of hierarchical violence for divine and then human motives are now giving way to what he referred to as an Ecozoic era when humans again "become a mutually beneficial presence on the Earth."[125] The role of humanity in relation to divinity and Turtle Island is continually evolving, and despite the hierarchical severances of the past few centuries, he writes that humanity can be seen as neither "an addendum nor an intrusion into the universe." Rather, our humanity must be seen as "quintessentially integral with the universe," as the way in which "the universe is revealed to itself as we are revealed to the universe."[126] It is an idea that resonates with the mid-twentieth-century writings of Jesuit priest and geologist Pierre Teilhard de Chardin, who similarly envisioned an evolving human consciousness as integral to an emerging planetary being of divine scale. Just as the universe has been spatially expanding in a physical sense, on Earth there has been a biological expansion from the simple to increasingly complex that is the ground from which consciousness emerged. Humanity plays a vital role not only in planetary evolution but also in fostering a particular quality of conscious relation with divine energies that are the source of life.

A colonizing era of extensive social and ecological violence is revealing intricate levels of interconnectivity on both external and internal scales of our experience within Notre Dame. As we have seen, our minding of relations is becoming more complex as the planet responds to our abuses, thus requiring us to increase the interdisciplinary scope and intercultural breadth of our knowledge. But there is more going on here for someone like Teilhard, who conceives a planetary noosphere that is becoming more self-conscious in its involution of emerging human knowledge as it responds to the current changes. This process is a further extension of the way in which life's evolving biological complexity has been, in his words, "bound up with a correlative increase in interiorization, that is to say in the psyche or consciousness" of individual beings as integral with the planet itself.[127] Because humanity partakes in the spirited mind of Turtle Island, we are continually asked to transform our ways of thinking and being. This viriditas is symbolized by the virgin birth of the green-haloed Christ child, as well as by the light-giving orenta of the Peacemaker. In the Haudenosaunee story, his grandmother was given "a message from the Sky World" that he should be named Tekana:wita, which means "like two streams flowing together he is born of both the Sky World and the Earth World."[128] There is in this emergent model a co-evolutionary play of energies that descends into the creation with Sky Woman and is continually transformed by Turtle Island relations, with humans now having the duty to play a vital Good Mind role in renewing those relations.

The cosmological commonalities between Chartres and Lorette are not, then, divine aids for hierarchical, linear, and epic conversions, but the overlapping of limited cultural views that when braided may provide canoe and ship with a better sense of our co-evolving place and time on the Kaniatarowanenneh. Within this model that traverses an oceanic gulf, the Blue Virgin's angelic theophany is an emanation of energy that vertically connects heavenly light to the viriditas of Turtle Island. These angelic presences resonate with the winged beings of Sky Woman's fall who help to initiate the creation. It is told that Heron and the other marsh birds were the first beings her ill partner threw down before she fell, with Rice describing them as "unique in that they could live on land, air or water."[129] In crossing elemental worlds, they are "the first order of creation in the new world and the ones with the greatest abilities." They brought Sky Woman down to Turtle Island just as the angels emanate light through a Blue Virgin who in earlier pagan understandings was also connected with birds like the Dove above her. This symbolism clarifies that "Mary has partly taken the place of Christ at Chartres," a spiritual presence in the field of time who is "first among all creatures."[130] As Hildegard writes, the One "is honored first, then the Virgin Mary, highest of created beings," after which comes "the angels and the various ranks of saints."[131] There are deep commonalities here for weaving Blue Virgin with Sky Woman's many names, a potentiality symbolically modelled from now on in the braided name Notre Dame de Turtle Island – our sacred window on the energetic source of not only the resources that the ship unsustainably harnesses but also of the climatic longue sault that is arising in response.

Over a decade ago I made pilgrimage to Notre-Dame de Chartres with the intent of finding a viable way of cosmologically modelling climate relations based on my predominant French ancestry. In entering the Royal Portal I came into a space that felt familiar, and yet the surround was ecologically, climatically, and culturally unfamiliar, strange. For more than three and a half centuries, my ancestors have lived along the Kaniatarowanenneh between the original kanatha and the longue sault. Many things came into relief at Chartres: I am a French Canadian who cannot speak French; I am a relatively good English-speaking Canadian following years of childhood speech therapy to deal with a phantom French lisp and extensive schooling including university; and I have branches of uncertain Mohawk and Wendat relations that until recently I would follow family in rarely acknowledging. While I could sense at Notre Dame an alternative model to the Catholic conversionary spirit of the Black Robes that came to Kanatha, it became clear that there was no stable ground for me until I renewed a dialogue with the land I am from and its Indigenous elders. This is the passage that the Lorette letter and wampum opened up. In response to this gesture of fraternity, the cathedral sent back a prayer, a replica of the Virgin's tunic found in the cathedral, and precious remains of departed saints. It is with these gifts that we weave our way back to Kanatha with the question of how to navigate our time of increasing turbulence.

Such ocean journeys can be filled with uncertainties that evoke a less rational minding of relations, something the former North Atlantic fisherman, student of Livingston, and environmental thinker Ray Rogers knows from experience. When he began fishing, there were two large tasks: "getting over the dread of venturing out into the North Atlantic night alone"; and "learning about the ocean bottom and the places where the fish might be."[132] The first was somewhat tempered with the familiarity that comes from spending time on these waters, though a storm or rough patch could raise dreaded spectres. As for the second, Rogers says he began by poring over navigation charts and maps, and then going to the wharf to "seek out Arthur Swansburg, a seventy-year-old fisherman who had spent his life on the waters."[133] The elder Swansburg described each place as he waved "his hand in a vague way over the chart" that Rogers was engrossed in. After some time Rogers began to realize that they were "having two different conversations." As an ocean newcomer, his charts and maps were necessary, while Swansburg left the chart behind to look out at the sea with eyes closed and "feeling the wind in his face."[134] A model of the uncertain ocean was inside him, imaginable. Perhaps we need to relearn to model relations as a familiar feel that can take the ship beyond navigational maps, computer climate models, and a sacred cathedral, which does not mean these have no value in renewing our relations with Notre Dame de Turtle Island.

Just as Rogers tried to get a bearing through charts, the proliferation of ever more detailed computer climate models reflects the ship's attempt to understand the complex interactions of multi-scalar phenomena. The goal of objectivizing climate through standardized measurements and quantification has, Mike Hulme writes, "led us moderns to see climate change as a physical transformation to be predicted and managed, if not mastered."[135] Even though scientists understand their embedded limitations, we seem institutionally and politically bound to managing external phenomena. This tendency minimizes the way in which limits to knowledge persists, such as the recognition of how some variables of all models are hypothesized based on assumptions. For example, modellers tend to concentrate on simulating the most likely climate outcomes based on "a subjective judgment about risk tolerance" and the potential of system variability.[136] Climate models have until recently tended to assume smooth gradual changes, despite paleoclimate evidence of interacting feedbacks that have the potential to result in abrupt changes – a projection that each succeeding report of the Intergovernmental Panel on Climate Change (IPCC) has found increasingly probable and is the focus of the next chapter. A related issue is that we are talking about neither smooth linear warming from year to year nor changes that are even across the planet. These non-linear, regional, and participatory qualities are about cycling relations, and yet many people, particularly non-scientists, tend to mistakenly mind climate change as an object or set of objects rather than as processes.[137] For a modern mind that wants objective maps before acting, mysterious relations that draw people into the centre of a process creates modelling difficulties. In contrast, this is what the stories and models of this chapter are concerned with representing.

One reason the human position in climate models has been minimized is because of the way in which powerful political interests have attempted to utilize uncertainties in the science to question its usefulness and deny the need for human change.[138] Because of this resistance, some have proposed that it is better for climate science to actively highlight the social limits of these models ahead of time rather than respond after the fact.[139] This aim of demystifying scientific knowledge by highlighting its social construction does not, contrary to the skeptics, deny knowledge of a real and troubling situation but rather raises important questions about how we understand and live with this mystery. Alleviating public doubt by attempts to increase climate research needs to, in Hulme's view, be better balanced with initiatives that increase "public understanding of and therefore trust in the social process through which those facts are scientifically determined."[140] There is a seeming need for a popular social turn that can build upon the social plurality of a Good Mind. It is a proposal consistent with Rogers's insight that "human social relationships have remained viable only because they have been underwritten by the sociality of natural communities."[141] Our minding of climate has to include shifting North Atlantic water circulation, wind patterns, the atmosphere, the Sun's seasonal cycles, and the many beings of Turtle Island, all of which commonly encompass the ship and canoe.

Whereas the stories of Sky Woman's fall and Notre Dame's stained glass carry knowledge about historic, ecological, and cosmological realities that are meant to meaningfully connect with particular people, Hulme makes the point that in increasing our climate knowledge and potential responses we have "lost – or chosen to ignore – what climate means to us."[142] The meaning he is referring to lies beyond science, economics, politics, and computer-mediated experience, and it needs to be re-engaged if people are to be drawn in and moved by our detailed climate models. Following Bruno Latour, he writes that "the purification of knowledge" central to modern science needs to "recognize that understanding climate change and responding to it demands a re-engagement with the deeper and more intimate meanings of climate that have been lost."[143] The source of this issue is another historic severance identified by Latour. He writes about how the flowing potential of science and religion have been turned "into an opposition between knowledge and belief, an opposition that we then deem necessary, either to overcome – to politely resolve – or to widen violently."[144] He adds that "*belief is a caricature of religion exactly as knowledge is a caricature of science.*"[145] There is a need for us to get beyond the static caricatures of pre-modern stories as immovable beliefs or modern scientific models as based on concrete knowledge.

Renewing this strand of the braid is quite difficult because of the way in which science and religion have seemingly been severed following the West's Enlightenment. The modern ship has vacated the world of irrational stories and symbols that suggest there is something beyond human power. Referring to the Promethean myth of an ancestor who steals fire from the gods, Latour writes that the critical progressive thought of moderns now conceive the fire, life energy or orenta as coming "from humans, and from humans alone."[146] His words recall Livingston's statement that "we have travelled so far in our cultural self-deceit that we actually believe we have no need of sensory stimulation or nutrition beyond that provided by ourselves."[147] This hubris has tended to grow with the ship's epic conversions as moderns become increasingly severed from bird's-eye flights of mind that can clarify the way in which orenta is not a human creation, the Blue Virgin is not simply a resource for human use, and knowledge can come from participating with other realities.

Though the abstract nature of climate computer models does not lead the science to explicitly espouse such modern beliefs in concrete objective knowledge, there is something in the way the technology places humans in a position of looking at a screen that implicitly supports centring the ship's passengers on Notre Dame de Turtle Island. This issue partakes in a global web of computer-mediated virtual relations between people and the world. We live in what Latour refers to as a "double-click communication" reality that situates moderns quite differently than the cosmological stories of the preceding pages. The double clicking of computers, tablets, and smart phones leads "us to believe that it is feasible to transport, without any deformation whatsoever, some accurate information about states of affairs that are not present to us."[148] Anything and everything is searchable with an answer at the ready, and thus we look for ever more distant connections in a virtual mental space. But this global minding lacks depth, familiarity, and appreciation of mystery, all of which are actively fostered when one enters storied models of Notre Dame de Turtle Island.

Brains conditioned by extensive use of the Internet tend to activate parts of the prefrontal cortext that inhibits the consolidation into long-term memory of short-term information taken from the web.[149] Reviewing this research, Nicholas Carr writes that the technologies and economics of Internet powers like Google "promote the speedy, superficial skimming of information and discourage deep, prolonged engagement with a single argument, idea, or narrative."[150] The title of his Pulitzer Prize-nominated book *The Shallows* refers to the way in which the networked life is turning us "into 'pancake people – spread wide and thin as we connect with that vast network of information.'"[151] He conceives these developments as an extension of various historic technologizing developments, and draws on many scholars including Harold Innis. What interested Innis was the way in which the great mobility of light-weight communication mediums supported empire building and the power hierarchies critiqued by Primavesi.[152] While the Internet has often been described as a more open space, it also has a structural hierarchy that is "woven into the hyperlinks that make up the web; it is economic, in the dominance of companies like Google, Yahoo! and Microsoft; and it is social, in the small group of white, highly educated, male professionals who are vastly overrepresented in online opinion."[153] What Carr and Latour commonly recognize is that this virtual web also shapes and shallows our minding of relations.

These networked models have in a sense extended the darkening web observed by Mohawk and Banyayca, which they conceived as coming from Spider Woman. She is another mythic female who Hopi see as a vital force behind the weaving of creation and its unravelling. Speaking on a similar cosmological scale in relation to today's virtual web, Latour writes that when the Virgin Mary "hears the angel Gabriel's salutation" and becomes pregnant, there is a kind of communication occurring but it "is not a case of double-click communication!"[154] The queries of a networked mind that wants to check "whether or not she was really a Virgin," whether the angel's rays of light were spermatic, or "whether Gabriel is male or female" are

caught in double-click minding.[155] Attention is directed away from
the story's meaning and toward questions that are a categorical mistake and "so irrelevant that no one has even bothered answering
them."[156] But we keep clicking and tweeting in search of needed
information or relations while being distracted from the real social,
ecological, climate, and cosmological contexts of our lives. As screen
time has expanded, people spend less time in nature, as evidenced in
dynamics like declining visits to national parks and the increasing
pressure for these parks to be WiFi-networked.[157] The shallows of a
double-click consciousness distract us from minding the more fundamental changes that are required for living in a relational world.[158] It
continues the web-like technological severance observed by Livingston
and Mohawk, one that simultaneously fuels the climatic longue sault
and informs a shallower approach to our modelling.

Following in the spirit of Indigenous storytelling, Latour advises
that the only way to understand cosmological stories is to repeat
them with an intent of producing "in the listener the same effect"
manifested in the telling. Notre Dame "impregnates with the gift of
renewed presence," of a transformation in consciousness that cannot
be rationally modelled or double clicked.[159] There is an energetic viriditas to this approach that draws me back to Roger's way of minding
what his elder Swansburg was sensing. He is guided by Livingston's
last book *Rogue Primate*, wherein he proposes there are three scales
of consciousness beyond the individualized form that currently pre-
dominates on the ship. The more encompassing fields are groups (i.e.,
from localized species groupings to species as a whole), communities
of regional ecological relations, and global interrelations that have
come into our awareness with the ecological and climate crisis. These
scales are consistent with models of nested ecological hierarchies that
highlight physical and energetic relations between various scales, so
that more encompassing eco-systems move over longer time frames
and broader geographies that are the context for smaller scale interactions of shorter time frames and more localized spaces. Such a view
also resonates with Leopold's land ethic, which "enlarges the bound-
aries of the community to include soils, waters, plants, and animals,
or collectively: the land."[160] But what Livingston adds is a consciousness dimension which suggests that healthy ecological functioning of

biological beings, including humans, requires times when consciousness is only secondarily focused on the individual and primarily
engaged in a participatory way with broader scales of being.[161]

Rogers observed something akin to this in Swansburg as his mind
left the wharf, carried by the winds and currents of the Atlantic
Ocean. This imaginative flight is difficult for moderns to make
because of the way in which economic self-interest, scientific objectivity, and the technological imperative enclose its inhabitants within
an individualized sense of self, even during its romantic turns.[162] In
the case of Livingston, he was jolted out of this mindset by an
Ashbridges Bay experience that led him, after a life as a birder, naturalist, and environmental thinker, to conclude that all beings have the
potential to experience the presence "of their own and of neighbouring species not as others but as simultaneous co-existences or co-
expressions of that place, perhaps as extensions of themselves."[163] He
observed such minding in many wild beings that have seasonal or
innate tendencies for identifying with the broader envelopes of self.
For example, flocks of "shorebirds and schooling fishes are highly
visible manifestations of group self-consciousness."[164] Also of interest were migratory birds that "are not fixed to a specific place, but
participate as a group consciousness that takes its cues from global
occurrences."[165] The Herons of Tekakwitha's isle and Ashbridges Bay
do not simply have knowledge of particular waters and seasonal fishing holes they move through, but co-participate in the familiar patterns of group, community, and global scale relations. Recognizing
this participatory way of minding is, for Rogers, helpful because it
"contradicts that of late-modernity in which all participants are
objects."[166] It goes in the opposite direction by asking us to imagine
ourselves in relation to a rough ocean or turbulent climatic longue
sault. The difficulty is that a lack of familiarity with such waters can
lock us in an individualized mind that is terrified at being engulfed,
thus leading us to become even more focused on external models.

To be moved by blowing North Atlantic winds, a telling of the Sky
Woman story, or the radiance of the Blue Virgin's stained glass is not
a function of believing in some storied model. As Latour writes, moderns only "believe in belief in order to understand others," and thus
find it difficult to realize that non-moderns and pre-moderns "do not

believe in belief either to understand others or to understand themselves."[167] Rather, ecological, climatic, ancestral, and cosmological presences are engaged in participatory ways that guide to deeper reflections on the kind of world we live in and how to mind relations here. From winged marsh birds to angelic emanations to scales of consciousness, all share features of a creation we are in the midst of. Reflecting further on the relevance of divinities to the present, Latour explains that they "are not substances – no more so, in fact, than lactic acid ferment" or, I would add, the climatic longue sault.[168] What these mysterious winged presences are is "all action." Another French scholar, Michel Serres, describes angels as helpful to mind during times of "turbulence or dissipation of energies" because these messengers are "unsystematic, troublemakers, boisterous, always transmitting, not easily classifiable, since they fluctuate."[169] For Latour, the dynamic boundary crossing nature of angels and, we could add, Sky Woman's marsh birds tells us about the limits of any knowledge, belief, or factish, including those arising from our pervasive double-clicking way.

The modern belief in the power of knowledge and the fallibility of mythic belief is inhibiting a more immersive modelling of what all the active movement underlying the climatic longue sault symbolizes. The ship has not, according to Latour, properly understood the second biblical commandment concerning graven images. We were not asked by God "not to make images (what else do we have to produce objectivity, to generate piety?)." Instead, "he told us not to freeze-frame, that is, not to isolate an image out of the flows that only provide them with their real – their constantly re-realized, re-represented – meaning."[170] The charts Rogers had on the wharf could for Swansburg be consumed in his ocean relations, which is exactly the kind of capacity that Livingston proposes the modern individual needs to renew. The increasingly turbulent shapes of violence that surround us can only be partially represented in the knowledge of our computer models and the information we double click, and thus we are being returned to the diverse storied insights of an Indigenous Sky Woman or the French Blue Virgin.

This can seem like a potentially dreadful ocean journey for those comforted by the rational objectivity of the modern ship. Knowing something about this experience, Rogers sees fear surrounding us as we create ever more intricate computerized models in the hope of navigating to a safe shore. Latour, also sensing such an increasingly precarious situation, asks whether we can recover pre-modern wild ways "of thinking for our own use?"[171] The models for navigating the ship toward some sustainable place need to find ways of transforming the ever-present computerized screens that we look upon into something we enter as subjects. Interestingly, it is our climatic longue sault that is initiating these quickening changes toward mindfulness. This marks a dynamic shift in where the ship has perceived the locus of agency, for that surround which not long ago was predominantly perceived as simply an objective backdrop for progressive human futures is now displaying a responsiveness as mysteriously complex as the human mind, perhaps more. Such a transformation is something Livingston appreciated when conceiving global environmental issues as a biospheric consciousness that arises at critical times like those of our present moment, and recedes "when recourse to other forms of self is more appropriate."[172] A social world requires ways of minding relations that aim for the same degree, quality, and scale of sociality. There is a need, building on Latour, to stop freeze-framing factishes by renewing ways of entering models of great interdisciplinary and imaginative scope.

Renewing the braid begun by Lorette and Chartres can offer us a more varied and nuanced sense of our potential to consciously participate with the presences of Notre Dame de Turtle Island. This is consistent with the spirit of Teilhard's insights on humanity's evolving relation with the noosphere. Considering the scientific tradition he was educated in as a geologist, he wrote that "as science outgrows the analytic investigations which constitute its lower and preliminary stages, and passes on to synthesis – synthesis which naturally culminates in the realization of some superior state of humanity – it is at once led to foresee and place its stakes on the future and on the all. And with that it out-distances itself and emerges in terms of option and adoration."[173] A good climate of mind is moved toward a spirit of Thanksgiving because of the emerging awareness that the ship has never been separate. We need the many beings of Notre Dame and Notre Dame has a role for us if we choose to partake. The participatory quality of our relations is of fundamental importance; this is what Teilhard's Omega Point, Livingston's biospheric consciousness, Chartres cathedral, and the Sky Woman story hold in common. A climatic theophany of angelic and ancestral presences is mysteriously drawing us toward a fundamental change in how we mind relations. We need to relearn how to appreciate "a reality which reason could not attain and which reason, even afterwards, could not conceptualize" or model.[174]

Re-entering the gulf and Kaniatarowanenneh brings us back to the shores of Kanatha. It was in one of these bays that Cartier, on his second journey to the New World, arrived on 10 August 1535, the feast day of Saint Lawrence in the Catholic calendar. He venerated this ancestor by giving his name to the bay, and by the early 1600s colonials extended it to the whole gulf and river. Carved with other martyrs around Notre Dame, this third-century saint, martyred at the hands of the prefect of Rome who had demanded that Lawrence turn over the riches of the church that he guarded as archdeacon, is said to carry gifts of bodily and spiritual mercy.[175] Legend has it that he tended to "the poor, the crippled, the blind and the suffering" and told the prefect: "The Church is truly rich, far richer than your emperor."[176] For the still medieval mind of Cartier's time, the naming of a bay was not simply a colonial appropriation. The uncertain ocean passage that could evoke dread in Rogers had reached a safe shore fifty days after departure thanks not only to human ingenuity but to cosmological presences. Even if some arrived with a view of Kanatha as wild and in need of conversion, all could find comfort in the mercy to be had in returning to the great river and stable land of Notre Dame de Turtle Island.

Before our modern era of flight, most entered Kanatha by the Kaniatarowanenneh, its rocky shores slowly narrowing as the ship was surrounded by treed lands of cedar, pine, maple, birch, and oak. As we saw in chapter 2, Northrop Frye described the experience of the settler entering this river as akin to "a tiny Jonah entering an inconceivably large whale."[177] Whereas in the United States the journey ended with the ocean meeting shoreline, here the watery passage continued into the interior of a land carved and traversed by waterways. But Frye's colonial emphasis was on the fear of being consumed by such wildness, whereas our return journey asks us to reconceive this experience of being immersed in something larger. How do we facilitate this passage? In a spirit similar to Sioui, Woodworth often states that, though "we are all Indigenous" to some place, we need to bring that awareness into relation with the ancestral elders of the land we live in. It is a move that makes increasing sense in today's carbon-constrained reality, for a requirement of sustainable models is that they be engaged where we live. This necessarily entails re-engaging local Indigenous stories for the knowledge and wisdom they offer, remembering progressive colonial injustices that have led us to actively forget our elders, and reconnecting with those historic dynamics that brought our families to these lands. It is a passage I have tried to follow here through renewing a dialogue between relations that are ancestral to my being.

When the French ship docked at the port it contained the gifts of the Virgin's tunic and saints' relics from "the Gentlemen of Chartres" to the Wendat of Lorette. This exchange participates in Notre Dame's viriditas and the merciful gifts she brings to all creatures, including humans. In the braided model that is beginning to emerge here, it is a sharing way that must be enacted even in "thin" times of turbulent change like Tekakwitha's life and our present moment. A good response to the present crisis has to be based on, in Primavesi's words, "a conviction that we would not, and cannot, exist without [Gaia's] gifts."[178] That which is given is in fact the basis of Kanatha's resource web, from beaver fur to bitumous-sand developments, but the ship's rolling-head ways offer little space for considering the gift of life and our spirited grounding. Yet, though on the one hand we seem bound, the climatic longue sault is simultaneously breaking open fissures from which we can potentially transform our relations with Notre Dame de Turtle Island from being primarily focused on extracting resources to honouring all that surrounds us as emanating gifts – even the rising "shapes of violence."

Beyond Lorette we approach the Island of Montreal and are given another historic reminder that underlying the nation's largest urban centres are colonial and Indigenous cosmologies that, if we slow down our drive, can tell us much about our relation with ecology,

water, and climate. It was almost a century after Cartier's fateful 1535
trip to Hochelaga that the French attempted to settle this island, with
the place at first being dedicated to the Blue Virgin in a kind of remembrance of Cartier's 1536 spring procession. On 8 May 1642 a group
of pioneers began building Ville-Marie as a place to further the
Catholic conversions of "Amerindians in what was then the remote
hinterland."[179] Its failure did not take long since the island position
amidst river passages entering the pays d'en haut made it an ideal
place for coureurs de bois, the fur trade, and, in a couple of decades,
another mission across the rapids, the Kahnawake of Saint Tekakwitha.
Ville-Marie was renamed Montreal in 1705. From within its present
maze of roads, bridges, and tunnels, it is difficult to imagine earlier
times when a wilder Kaniatarowanenneh evoked such deeply felt ritual acts and naming by colonists, not simply Indigenous people like
Saint Tekakwitha or missionaries like Brébeuf. That mythic place is
still here situated on the back of Turtle Island, though our drive leaves
us unconscious or dismissive of such an old storied sensibility.

    While the transcendent model of Chartres cathedral and Sky
Woman stories sacralizes all creation, in a carbon-constrained future
it is necessary that we find symbolic routes in our cities and kanathas
to inspire knowledge wherever we are. The search for models does not
need to be restricted to sacred spaces which are already imbued with
meaning.[180] Because our built environments are so intertwined with
the changes heading toward the modern ship, it is necessary for us to
contemplate the cosmological dimensions of our seemingly secular,
purified, and concretized urban places. Recognizing the sacred wherever we live is a way to be more mindful of our relations and to learn
about where we have come from, who we are, and what we can be –
the good, bad, and mixed. This is what Woodworth does when he
reimagines Toron:to as a fallen onerahtase'ko:wa and its architectural
core as a Skyscraper Medicine Wheel and CN Tower Prayer Stick. But
his Good Mind adds a vital dimension to this cosmological renewing
of the urban, for it centres the land's Indigenous nations and understandings as beacons for a sustainable way home. This is the light that
guided these pilgrim-like passages along the passage de taronto,
longue sault, Kaniatarowanenneh, Atlantic Ocean, and return to
Lorette, Ville-Marie, and Saint Tekakwitha's Île-aux-Hérons.

In contrast to Saint Lawrence and Sky Woman, both of whom evoke mercy, Saint Tekakwitha protects, fosters, and renews a braided way of minding Kanathian relations in our time of rapid change. Beyond the Blue Virgin viriditas is a dark cycle that the Protectress of Kanatha was also well acquainted with in both Indigenous cosmology and the colonial form of mass epidemics that took her parents, destroyed her own health, and brought many cultures into collision. This is the ground from which her asceticism came to nurture a frugal restraint whose goal is to affirm our sacred fraternity with each other and creation. The eco-asceticism inspired by her can now be joined to a moderating influence, asking us to turn off our flat screens some days and perhaps for extended periods during significant times of year like Sun Ceremony, Lent, or whatever comparable rituals we inherit. Guided by Saint Tekakwitha, Saint Dionysius, the Peacemaker, and other ancestors, we will now move in the last two chapters toward a darker mythic cycle. This cycle will deepen our sense of why it is vital "not to freeze-frame" our often digitized images, stories, and models of a climatic longue sault we are within – to remember again that, though the scholarly knowledge of Hekademos can increase in a seemingly unending way, Notre Dame de Turtle Island is sacred and thus "science and mysticism ought naturally to complement one another."[181] We are guided back to a sacred isle where it is possible to braid not only ship and canoe but also knowledge and wisdom.

The ocean-traversing gifts of this chapter teach that what is central to weaving our passage down the Kaniatarowanenneh is a common etiquette of Thanksgiving and praise for that which sustains us all. Returning to Sioui, if moderns can truly feel "that they belong to this land, instead of this land belongs to them, we'll be okay."[182] Just as this return journey goes westward against the river's current, we need to go against the current of the modern mind by bringing our overly abstract flat-screen models for living into a much more mythic cosmology – one that is on this side of the ocean closer to the spirit of Kanatha than Chartres. On a personal scale, this ocean passage guides me toward a similar embrace of uncertainty as it highlights my lack of firm ancestral footing. Not only must I agree with Latour that I was never modern, but I now realize that I am also not Wendat, Mohawk, or Métis. Moreover, my Canadien ancestry is a linguistic ocean away, and the Loyalist British culture of Toron:to has largely been a place of refuge for someone in between – just as this diverse "meeting place" has been for many immigrants and refugees over the past century and a half. I am in the dark shadows of what it means to be a Canadian who is Kanathian, one who belongs to Turtle Island.

It is clear that Saint Tekakwitha protects something of a more primordial scale than the Kanathian nation-state. She guides a good climate of mind beyond our individual selves to a renewed cosmological sense of what it means to speak with "one mind, one heart and one mouth." As with the varied meanings of Toron:to discussed in chapter 2, this land of kanathas that she protects on her isle does not require one story or model, though we do need to recognize their diversity through a common spirit of praise for fraternal relations that make life good on our Kaniatarowanenneh. What the Sky Woman and Blue Virgin add is the insight that good stories "attract other stories, and therein lies faster leverage for change."[183] Such models breathe stories out while offering space for others to be inspired before breathing back the gift. They are in contrast to the halting breath of conversionary stories that tend to be non-relational and discouraging of movement across models. A severed mind is bound to one objective story and ancestry until that moment when something bigger throws off darker cycles that, we see in the next chapter, are also well represented in the passages of Notre Dame de Turtle Island. The numinous shadows that make me feel uncertain about what it is to be Kanathian surround this isle. With that insight, the Protectress's Herons fly me back to the familiar feel of a watery wind blowing against my face as we return to the passage de taronto and Lake Ontar:io where we started.

6

Darkness Will Cover the Earth

If we stop giving thanks to the Creator for the things he has given us, the clouds will be still. The spirit of the waters of the springs and brooks will no longer look after them for our benefit. The sun will go dim and then darkness will cover the earth. A great smoke will arise in the air and cover all the earth. Then the poisonous monsters created by Sawiskera will appear ...

Skanyiatar:io (in Rice 2013, 309–10)

Back in Toron:to, my mind is cycled from Notre Dame de Turtle Island's viriditas to an increasingly numinous sense of the climatic longue sault we are within. Mediating this shift is a labyrinth situated at the centre of High Park, east of the passage de taronto. It is a replica of the one found at Chartres just beyond the Royal Portal entrance. Labyrinths are, according to the park's sign, "a universal, ancient symbol that dates back 3,000 years," a model we enter to "feel attentive and attuned to the sights and sounds around you." Surrounding it are black oak savannah grasses, a fenced-off conservation plot, recreation areas for baseball and soccer to the north, a small pine grove and hill to the east, and to the south a rich patch of plants indigenous to Ontar:io's threatened savannahs. And, while the ear is drawn to birdsong interspersed with the park's slower car traffic, the sign's sense of the labyrinth as a space for attuning "to the sights and sounds around you" is clearly bound to the materialistic factishes of the modern ship. Its maze-like passages were also traditionally meant to raise our mind to the cosmological mysteries of life and death, to renew "life through contact with the dead ancestors."[1] It is a good place for returning to Woodworth's ancestral vision of Toron:to and further contemplate what it means to help Thomas lift the fallen onerahtase'ko:wa in a time of mythic changes.

We go to "thin" spaces to remember that spirit energizes our lives and is to be found everywhere, even within the modern ship's urban hull. Cycling from the Blue Virgin's southeast position to the northwest where these labyrinthine passages begin, we are in the presence of a Black Madonna statue who symbolically resonates with darkening skies, winter, and the phenomenal terror of a cycling life. A feature of Chartres is its representation of the One's emanations through Nature in "regular and determinate sequences" such as seasonal shifts and celestial motions, not unlike Sioui's circular Good Mind.[2] The Blue Virgin's viriditas inevitably gives way to a darker potential symbolized in the Black Madonna. Today, science likewise attests to the ever-cycling nature of Turtle Island's climates and ecologies, and the challenges this poses to human life and thought. These changes have been mythically marked in stories found around the world and, we will see, in a Toron:to ravine geology that engulfs us within signs of Indigenous and planetary presences related to the last major global climate transition from Ice Age to Holocene. As we enter these winding paths, the stained glass, statues, stone walls, and vaulted ceilings of a distant cathedral become transparent to the immediate Kanathian ecology, climate, and ancestral cosmologies like those of Sky Woman, the Peacemaker, Saint Tekakwitha, and Skanyiatar:io, whose words open this chapter.

While the cataclysmic biblical myth of Noah reflects through stained glass on the same northwest side as the Black Madonna,[3] the Haudenosaunee story of Skanyiatar:io, or Handsome Lake in English, tells of a coming time when the Sun will go dim as great smoke emissions release Sawiskera's "poisonous monsters."[4] His vision came in his sixties while he was suffering from an illness related to alcohol addiction that came over the ocean with the ship.[5] In the story told by Thomas, this Seneca royaner, who died in 1815, had been "a heavy drinker, one of alcohol's many victims, and his people saw their leader's suffering."[6] He was bedridden "for four years, during which time he began to consider who we was, and what his place was in Creation."[7] There was a smoke hole in the roof above where he lay, and through it he could see "the top of the trees and the stars in the sky."[8] Many times he looked at this sight, and Thomas says he began to ask himself questions: "Where did all this come from? Why is this

15 Labyrinth east of passage de taronto (summer 2015). This photo is of a Notre-
Dame de Chartres replica labyrinth found in High Park north of the Howard
tomb, a place that allows us to bring the dialogue started by the Wendat of Lorette
back into some of their traditional lands along Lake Ontar:io.

all here? What am I doing here? What does this all mean?" Skanyiatar:io
also began to remember who he was as a royaner and his duty to the
Creator, known as Sapling, and so he resumed the act of giving thanks
each morning for having seen another day.

    With death imminent, Skanyiatar:io experienced "a great sunlight,
and the feeling of profound well-being," then a man approached who
called him outside. Surprisingly, he could now walk and came upon
three men with beautifully painted faces who were sent to Turtle
Island by Sapling "to look for a certain strongminded man who was
doing wrong."[9] It is from them that Skanyiatar:io received a vision of
how he and the Haudenosaunee had "once again transgressed from
the things" that Sapling taught and the Peacemaker renewed with the
onerahtase'ko:wa.[10] They also taught him how to find a way back to
peace in the midst of the colonial darkness brought by the ship. As
with the Peacemaker centuries earlier, his was an era of cultural

16 Staghorn sumacs in summer (summer 2015). Staghorn-sumac groves and their connection to staghorn deer have a symbolic and educational importance in Haudenosaunee and Wendat teachings, and it is one of the teaching places engaged over the next two chapters.

decline owing to the intensifying colonial severance carried by Saint Tekakwitha, though related at this time to the emergence of the United States and eventually Canada. The vision he received highlighted duties that needed renewing, including: abstaining from alcohol; staying with the canoe's way so as to keep it strong and distinct from the ship; showing tenderness to orphans like Tekakwitha in a context of so much loss; revering Turtle Island through recognition that it "is not merchandise to be sold" but rather a gift to be passed on; and continuing the duty of Thanksgiving for all that is given.[11] What he saw concluded with prophetic images of a "dark time when the earth will finally be consumed in a great fire" as Sawiskera's poisonous monsters reappear, thus clarifying that things have gone "out of order and that the end is near."[12]

In Notre Dame's labyrinth, this Kanathian story resonates with Noah's time when "all flesh had corrupted their ways on earth" and he was given the duty to make an ark to traverse the coming flood and thus renew life.[13] Across these stories of different times, cosmological cycles commonly respond to the quality of human behaviour, though the interpretive differences outlined in the previous chapter continue. In the Haudenosaunee Good Mind, such a dark cycle brings forth myths of the younger twin grandson of Sky Woman who manifests death and disorder. While the elder twin was given the name Teharonhia:wako Sapling, "He Who Grasps the Sky," by their grandmother to reflect his awareness of being one with the Sky World's emanating orenta, she called the other "Sawiskera, Flint Crystal Ice ... because you only think of yourself."[14] It all started with the way that Flint came into the world through the armpit of his mother, thus killing her. His deathly potential continued in the climatic construction of northern ice bridges that allowed "deadly monsters from the invisible farther shore to come across and wipe out" humanity.[15] In many Turtle Island stories "the north is generally associated with danger, demons and disease," and Flint's icy associations connect him to a difficult glacial period when ice stretched south of Lake Ontar:io.[16] We will see that he takes on other dark forms like the poisonous monsters seen by Skanyiatar:io and the violence responded to by the Peacemaker, to which can be connected a flood of stories that followed his receding glaciers.

The dark cycles of Flint call forth the creative renewal of Sapling's sky-reaching viriditas, and this is related to Thomas's call for help in lifting the fallen dolon-do. There is a dialectic between Flint and Sapling that Woodworth explains does not mimic "the Christianized world of good and evil" but are like the cyclical rise and fall of a breathing breast that "sharpens distinctions, and draws out the character of all things."[17] As noted in chapter 1, the more you try to solve a mystery the more its web "pulls you in," engulfing you in ways that increasingly take on shadowy hues. The experience is not new for humanity; in fact, the current prevalence of apocalyptic stories in relation to our climatic longue sault can be seen as harkening back to the cycling "experience of disruptions whose uncertainty" led people to mythologize them.[18] Recurring cycles of climatic uncertainty stretch into the deep past and are returning today, leading William Calvin to portray our challenge in this way: "Use our civilization's achievements to prevent the next whiplash, or lose much of civilization's gains as our warm period suddenly ends with a population crash."[19] We are being rapidly pulled toward a numinous awareness that not only has the potential to weave diverse perspectives but can also draw us into common "human instincts for nostalgia, fear, pride and justice."[20] These ancestral instincts are at the centre of this chapter's labyrinthine passages.

We are in a time of significant change that is asking us to reconsider some of our basic beliefs and factishes about the kind of world we live in. While the global scale and rising intensity of this challenge can seem unprecedented, we all at some point go through pivotal experiences that call us to reassess ways of relating with familiar presences. It can be the death of a loved one, such as when my father's departure brought with it questions about who I am in our changing climate. For Livingston, the severance of Ashbridges Bay initiated a stomach-wrenching pain that came to inform his life and work. Others may have an illness, injury, or even a near-death experience that brings forth a change in direction or recovery of something seemingly lost or forgotten. Then there are those who experience such calls as part of a broader cultural milieu that has been violently affected from without. This is what Skanyiatar:io embodies, a near-death experience and call for renewal of ancestral wisdom that can help his people find a way out of accruing colonial impacts. All cultures and families have dark stories of some form or other that tell not simply of "thin" times when apocalyptic change is unavoidable but the quality of human behaviour that is needed to find a good way to the other side. If we cannot instinctively find ways of minding the numinous dimensions of Notre Dame de Turtle Island, then the weaving paths of this chapter suggest that a dreaded darkness can engulf ship and canoe on our Kaniatarowanenneh.

Such mythic and yet personal binds are the inspiration of labyrinths, with many being connected to a spider web that entangles one in deathly patterns as easily as acting as a guide into a transformed state.[21] Most European labyrinths, including Notre Dame's, were inspired by the Cretan myth of Theseus and the Ariadne thread that allowed him to enter, confront the Minotaur, and find a way out. From Ariadne to the Hopi Spider Woman to Notre Dame, these finely knitted passages have often been connected to feminine forces that the individual and collective has been unconscious of but that now return in questions of "life or death, deliverance or entanglement, metamorphosis or fixation."[22] There are resonances with our situation. On the one hand, we are bound to the technological web observed by Banyayca, Mohawk, and Livingston. At the same time, energizing this binding web are sacred emanations from the source of life. The climatic longue sault challenges us to transform and untie our knot by walking the labyrinth of this "thin" time. This passage is in accord with the Good Mind, for to forget the human duty of giving thanks is to call forth Flint's poisonous monsters and thus intensify our immersion in a darkening surround. It is with this Kanathian story that we begin before returning to the mythic challenges of past and current floods of climatic change.

The labyrinth starts with a northwesterly path through matted yellow savannah grasses and snow patches of late winter. Edging into the ravine bush below, I am surrounded by the bare limbs of sassafras that slowly give way to a staghorn-sumac grove. Here the evening's encroaching darkness meets the snowy ground with an almost blue-grey haze that these trees reach into some six metres or so. Though sumacs do not seem to have the Sky World reach of the onerahtase'ko:wa, Woodworth learned from Thomas that such trees can foster a good minding of Sapling's spirit, "He Who Grasps the Sky." In Woodworth's words, the sumac's "deer horn like yearnings of the upward growing reaches of the branches will yield long slender green of leaves, and later, flame like torches of robust deep red seed pods," thus giving such groves a feel that make them a good place for telling storied teachings.[23] Just as a stag deer's antlers reach skyward, especially when they sense danger and stand attentively,[24] the antler-like sumacs are understood to be particularly receptive to spirit. At this time of year, the wind blows easily through the slight trunks and overhanging branches of this grove that is bare of all but a few yellowed red seed pods.

With dusk approaching and shadows growing in the ravines below, this is a good place to stop and renew relations with darker climatic stories of Notre Dame de Turtle Island. Back at Chartres the Black Madonna sits in this northwest position, directly opposite the Blue Virgin's viriditas. She holds the spot of the setting Sun and where the ocean's cold and wet winds originate. Softly lighting her dark presence is a red glow of candles that shines off her magnificent golden robe, crown, and sceptre, all of it evoking the swiftly approaching night sky "as a mother who gives birth ... to the moon and stars and each morning to the sun."[25] On a more primal level, she is associated with "the realm of darkness, of grief, of death, the realm of eternal mansions of the dead, who await the resurrection of the flesh."[26] All of these associations symbolically connect her, as with the Blue Virgin, to preceding Roman and Celtic pagan goddesses like Isis, Artemis, and Rhea. We are in the presence of the *materia prima*, the unorganized potentiality out of which everything emerges and to which it returns.[27] These dark associations bring us back to Sky Woman, who also moves closer to the numinous hues of Flint as the story of Sapling's creative viriditas unfolds. It is a story that offers us a Kanathian way of re-minding Notre Dame's Black Madonna expression, and thus we return to a still pregnant Sky Woman on the back of a growing Turtle Island.

With Sky Woman safe on the back of a shell covered with the sacred mud dredged from below, Sioui tells of an interesting series of events that has symbolic relevance to both these sumacs and the Wendat ancestry in Toron:to. It seems that the animals that helped her land did not stay long on the growing island. At a council the animals learned from Turtle how "Deer had climbed to the sky on the Rainbow's magnificent path of colours," and all of them except Turtle "decided to follow the same path and live in the sky."[28] While Deer's sacred name in Wendat, Dehenyanteh, means "path of many colours made for the Deer by the Rainbow," there is a comparable Haudenosaunee understanding. As Rice explains, buck deer are considered wise animals who are "messengers to the Creator, as their name means the first to the spirit world."[29] The Deer's many coloured paths would be renewed here by Sapling's birth and eventual creation of all the animals, including those who found their way back to the Sky World with Deer. What connects the Wendat to this part of the story is that the name of the Toron:to nation, the Tahontaenrhat, is thought to have meant Deer Nation or "white-eared people,"[30] perhaps related to their once prevalent nature in the savannah ecologies just to the east of these sumacs. This grove reminds me of a rainbow path to the Sky World that also embraces the emerging creation, which is why they are a good place for storied teachings.

To the west, the shadowy ravines below seem to call forth the Tahontaenrhat ancestors who may well have told, in spaces like this, what followed Deer's rainbow departure and Sky Woman's planting of the roots and seeds she carried. In Thomas's Haudenosaunee tradition, it is said that she gave birth to a daughter known as "Ohshe:wa, Budding Flowers."[31] While the name of Sky Woman is Otsi:tsia, Mature Flowers, which connects her with the fertile "potential of plant life that came with the rain," her daughter, Budding Flowers, is "the female essence of plant life" from which springs forth viriditas.[32] It is she who, while fetching some wood, met the Great Turtle one day. He said: "I am searching for a mate, and I would like to know if you would marry me."[33] They do, and Budding Flowers becomes pregnant, giving birth to plant life and the twin brothers. First came the one with a skyward tree-like reach, and then the other who broke into the world through her armpit, killing Budding Flowers. The twins and the two primordial women embody a complementarity of presence, not unlike the Blue Virgin opposite the Black Madonna.[34]

The different ways in which the two brothers came into the world gave their grandmother, Sky Woman, an indication of what to name them. Teharonhia:wako Sapling was given his name because he was aware of his Sky World relations, while the second was called "Sawiskera, Flint Crystal Ice … because you only think of yourself."[35] In contrast to Sapling, who extends the creative work of Budding Flowers, Flint was born with a dark power that his grandmother recognized when he showed her an arrow and said, "I can stop anything from growing or living and bring darkness with this weapon."[36] A flint arrow is what gave him the power to kill Budding Flowers, and not simply during his birth. As Rice relates, Sapling became aware that his brother further destroyed their mother by freezing her body, and confronted him: "It was she who first established herself in this world, and it is now she who will be the first to depart from it."[37] He eventually puts his mother up in the sky as the Moon whose monthly cycle is associated with the seasonal gestation of plants and animals as well as influencing human biology, psychology, and stories. Every destructive act of Flint is complemented with a creative response by Sapling, in this case raising her up as the Moon. The story's repeating cycles add layers to Turtle Island, thus representing "a continual state of change and transformation brought about by balanced forces interacting with one another."[38] As Woodworth writes, everything reflects "the ongoing interplay of Taharonhiawá:kon/Sapling and Tawiskáron/Flint," dynamic cosmological relations to which people must "offer 'thanksgiving' at the beginning of each day that is returned to us."[39]

What is particularly relevant in this labyrinth is that the icy darkness of Flint comes to live with Sky Woman in the west where the Black Madonna is situated, presences that seemingly crowd into these ravines on a late-winter night. Following Budding Flowers's death, Sapling's grandmother continued to live on Turtle Island and became closer to Flint's dark powers than Sapling's skyward reach. Both Wendat and Haudenosaunee stories say that she and Flint live together attempting to undo the good works of Sapling,[40] with her being the one who was in "charge of the souls of the dead, caused men to die, and bred epidemics."[41] It needs to be remembered that what Sky Woman's name, Mature Flowers, represents is the plant potential brought with rain. But rain does not come across all seasons, at times is crystallized in ice and snow, and sometimes is too heavy, and so she has a dark potential that is at times close to that of Flint.

The dynamic relations of the twins are like opposed seasonal gods, with Flint the congealing breath of winter that converts everything to a stone-like state, and Sapling the restorative warmth of spring that renews life in the viriditas spirit of skyward-reaching saplings.[42] Though their cycling movements reflect a kind of balance, it is also clear that there are succeeding periods of imbalance when Sapling and, later, people need to find ways of rising above Flint's insatiable urge to spread an icy darkness.[43] In Sioui's Wendat telling, it is said that the war between the twins led to Sapling, Tsestah in his tradition, driving Flint from the eastern home of the rising Sun.[44] To do this he received help. According to Rice and Woodworth, his father, the Great Turtle, told Sapling how he could limit his brother's powerful weapon and continual urge to control the creation.[45] It seems that Flint's dark power could be limited, even destroyed, by two things:

"flint stone and the horns of a deer."[46] If the coming people who Sapling would create used "their minds like the deer," then they could use hard antlers that reach skyward to chip the flint. That which was lifeless could be used to create arrows for hunting and to make a fire to warm the longhouse. It is for this reason that royaner like Thomas and Skanyiatar:io are chosen by clan matrons to wear Deer horns as a symbol of those who can "raise their antlers and watch over the people," and who can sense danger and embody a Good Mind that can transform hardness into life and darkness into light.[47]

Even in the hard core of Toron:to's contemporary urban asphalt and fragmented ecologies, I have come face to face with two deer that made their way into these sassafras and sumac. A deer this deep into the urban environment has to cross many driving severances that often lead to media reports of highway traffic mayhem, police tranquilizers, relocation, and sometimes death. So it is here a mixed blessing to look up into the eyes of a deer that is ten to twenty metres away and has already noticed your presence. With a cautious awareness, the deer's eyes calmly survey me in the midst of foliage and a hum of background traffic that does not pull its attention away, though the more immediate pounding feet of a jogger cause it to subtly shift its ears and head in a way that does not draw attention. Across many cultures, this animal has come to symbolize a sensitivity to realities beyond what humans immediately sense, often related to their antlers.[48] It is a gift to sit for as much as twenty minutes with such a presence and be given a phenomenal education into what Deer symbolizes in the Good Mind.

With Flint's power limited by the attentive Deer horns to periods of seasonal winter and daily night so that he could not destroy all creation again, Sapling now had more space to creatively green Turtle Island.[49] He brought forth plants, trees, and a host of animals to Turtle Island. But as it is the way of Flint to attempt to undo his brother's viriditas, it is told that he fashioned a kind of ice bridge from which his dark, death-bringing monsters could again make life difficult. His hardness is of stone flint and cold ice like those glaciers that were integral to the formation of the ravines surrounding this sandy savannah I stand upon. The last planetary glaciation began around 75,000 years ago with the ice eventually reaching its maximum in

Ohio south of the Great Lakes. When the shift out of this Ice Age began approximately 16,000 years ago, a two- to three-kilometre-thick Laurentide glacier encased southern Ontar:io and Toron:to in ice that dwarfs today's CN Tower.[50] In scientific models, such glacial expansion and retreat is seen as related to the planet's orbital cycling around the Sun, a regular process that will eventually bring another glacial advance – one whose power will scrape Toron:to off Turtle Island, dumping "the twisted and mangled remains far to the south."[51] These past events and future potentialities are told by Haudenosaunee and Wendat with the different imagery of the twin brothers.

It was in the midst of creating the expanding glacial bridge that Flint heard the call of a Bluebird, "Kwe, kwe, kwe, kwe, kwe!" Fearful, he fled and "as fast as he ran the bridge behind him disappeared."[52] Because Bluebird is one of the first migratory birds to return in spring as the Blue Virgin's viriditas bursts forth, its call signifies the breaking of winter's icy power.[53] The melting glaciers began to recede and about 13,000 years ago the shoreline of Lake Iroquois formed about nine kilometres north of today's Lake Ontar:io.[54] The bluffs around the city demarcate this more expansive lake, with the former ridge-winding path becoming Davenport Road and the Anishinaabe term Ishpaadina, "rise in the land," coming to grace the busy Spadina Avenue.[55] The continued retreat of Flint opened the Kaniatarowanenneh passage to the Atlantic Ocean around 12,000 years ago, draining Lake Iroquois to just below Lake Ontar:io's current extent, with these shifting waters further carving the sandy soils of Toron:to's rolling ravines, streams, and ponds.[56] Farther east, the receding glacier also released the waters that had formed into the Champlain Sea, thus creating the conditions for the great river's longue sault and La Chine rapids near Saint Tekakwitha's isle.

In this sumac grove, Flint's icy bridge and recession can be sensed underneath the whole labyrinth, from the descending ravines to the sandy ground I stand on. After these shifts Bluebird's call continues to rise with a chorus of other birds as southern Ontar:io is invaded by "plants and animals that had survived the glacial period in differ-ent areas."[57] A succession of arboreal changes occurred as tundra vegetation gave way to a boreal forest first of mostly spruce and then pine. Within a thousand years Sapling's growing forests had pine, hemlock, and beech trees, which were then joined around 7,500 years ago by warmth-seeking deciduous trees like oak, maple, and hickory, thus bringing the mixed forests that came to adorn this place right into the colonial period.[58] In the midst of this shifting balance of power, the first peoples are said to have moved north of what became Lake Ontar:io. Starting around 12,000 years ago, small groups tracked and hunted "herds of caribou, muskox, mammoths and mastodons" on tundra at the edge of the Laurentide glacier.[59] As temperatures warmed and tundra gave way to boreal forest, wide-ranging caribou-hunting bands were succeeded by groups that followed specific river valleys and surrounding regions.[60] These are the deep post-glacial roots of human relations with waterways like the passage de taronto and Kaniatarowanenneh that were carried down to the present by Anishinaabe and then the Iroquoian-speaking Wendat and Haudenosaunee.[61]

While Flint's ice and subsequent melting came to etch the climate and ecology of this sumac grove and ravine below, what can be more difficult to see is the role these changes had on the Indigenous minding of Turtle Island. As we saw in chapter 1, research is revealing the evolutionary relation of minds and climate. What seems to be particularly vital are the extreme shifts, for these moments create more opportunities for marginal adaptations to come into play. Their recurring nature ratchet up our adaptive trajectories to produce, in the words of William Calvin, "major changes in body and behaviour in only a million years or so" – a process he defines as "catastrophic gradualism."[62] Reflecting a similar dialectic in different symbols, Woodworth writes that the Haudenosaunee keep "turning from days of darkness to the knowledge and teachings of light – a kind of cultural breath work which integrates the two ways of the twins in Creation."[63] Flint is much more than an ice bridge that recedes externally; he also has an internal dimension that people have to struggle with continually. In a spirit similar to Calvin's "catastrophic gradualism," Flint's presence returns with disruptive challenges that bring people to a dynamically changed sense of the world and the Good Mind needed to live well within it.

One social adaptation that gradually emerged across the human species in response to climatic changes was a relational etiquette not unlike that represented by Sapling and the Peacemaker. Considering the evolution of altruistic behaviour and the common sharing protocols across hunter-gatherer cultures, Calvin writes that the "repeated fragmentation of large prehuman populations into many smaller subpopulations" during periods of extreme change occasionally created groups focused on sharing as a response to uncertainty.[64] While the more individualistic and self-interested groups wasted "time and effort at fighting over the remaining food, the groups that shared (and otherwise minimized conflict) might have survived better."[65] Also drawing upon the relation of major climate changes to human evolution, anthropologist Brian Hayden and others have noted a pattern which suggests that an emotion-based religious etiquette evolved from these periods of extreme turbulence. Ritual activities in early hunter-gatherer cultures have an emotive connecting capacity that can nurture lasting group bonds and alliances which are helpful for surviving disastrous periods.[66] Signs of these developments are seen in archaeological records and contemporary ethnographies.[67] It also seems that these adaptations were coded in some myths that arose in relation to climate transitions between stability and extreme uncertainty.

This interdisciplinary constellation of ideas resonates with the Haudenosaunee stories that tell of their various descents away from the ways of Sapling and toward Flint, from which came messengers calling for cultural renewal like Skanyiatar:io, the Peacemaker, and before them the Creator. In the earliest days, it is said that the newly created people had forgotten the ways taught by Sapling concerning the dark potential of Flint and how to avert it. He again said to the people, "You have failed in doing what I had asked," and he warned that if they did not continually perform their ceremonial duties then Flint's power would increase as he used them to destroy the creation.[68] As Rice relates, from the moment humans were created, the twins agreed that "their minds would be divided into the right and left sides" so that we are faced with a choice.[69] People can use the right side to follow Sapling's skyward reach from a position firmly rooted on Turtle Island, or engage Flint's hardening way that disrupts the creation through attempting to control it. Would people "only think of themselves in matters pertaining to the world" and thus release Flint's monsters, or would they turn toward a social etiquette that thinks of others and protects the creation?[70] At particular catastrophic moments this choice comes to the fore as a challenge to renew relations with the way of Sapling before Flint goes too far, and each requickening adds strands to the return of a Good Mind.

With night fast approaching, these darkening ravines bring us into relation with a Flint who not only glacially shaped Toron:to but currently seems to be warming the planet, increasing extreme events, and expanding a glacial melt that began with the Bluebird call. Returning to Rice, he writes that when Sapling returned a long time ago to help the people renew a Good Mind, he gave them ceremonies and other gifts to help, including more knowledge about the ways in which his brother would try to undermine them. One thing he explained was that Flint had created beings stored deep "under the Earth" who "are so potent that just the sight of them could overpower someone."[71] They should remain buried, but human beings would arrive one day who will free them and in the process create a poisonous "fire that will be so hot that it could melt anything in my creation." This could give Flint a truly destructive power over creation. For the darkening power of Flint has since the glacial retreat offered a flood of mythic challenges to far-removed people – some of whom are storied ancestors of the modern ship with whom we now need to renew our acquaintance.

———

Late-winter oscillations between warming melt and cold freeze-up make these ravine paths that I have traversed for over a decade seem wilder. With slippery wet ice underfoot, the mind searches for footholds that can give traction to move forward on the winding turns of ravine ridges glazed over by Flint's freezing power. Shale ice and slush can help, though sometimes a concealed slick patch carries you down the slope. Thick slushy snow off the hard-packed icy path provides some better footing, and a strategic slide toward a well-placed oak trunk propels one up the ravine to the less challenging flats above. Though by no means dangerous, there is something about these jogs that leads me to contemplate the attentiveness required when uncertainty prevails. I am reminded of the passage de taronto to the west where Brûlé, coureurs de bois like Leduc, and Indigenous people required knowledge, skill, familiarity, and rituals of thanks to engage uncertain phenomena like a longue sault. Turbulent energies have an engulfing potential which makes it clear that attention is needed. The larger the rapids the greater opportunity there is for a momentary falter of focus. Suddenly a rapidly approaching rock goes undetected or the foot inattentively slips on ice, pulling the body down as the mind hurriedly considers options. Within such quickening "thin" spaces of change, larger powers come to pervade mythic stories, from Flint's massive glacial ice to the floods that follow its melting to our changing position in a climatic longue sault.

As I look down on the labyrinth, the stained glass of Noah's flood brings forth post-Ice Age stories of sudden water transformations not unlike that which threatens island nations and many coastal areas in our current era of fast-melting glaciers. This myth offers a different view on what underlies current and past floods of change than the overly rational vistas of the modern ship. These storied images tell of large cosmological presences that respond to human behaviour, though the reason for the response has different interpretations. In the biblical story, which is iconic even for secular moderns who think that they have long left Christianity and religion behind, it is told that God brings about the flood to purify the world of a pervasive disorder that encompasses all life, not just humans.[72] With a Flint-like darkness increasing, the Creator fears that humanity has become too sinful and thus, "emptied of its seeds of life and creative powers," will gradually waste away in a weakened state.[73] Rather than permit this regression, God brings an all-consuming flood that aims not only at "punishing the guilty but at purifying a polluted earth" so as to renew the ground for human potentiality.[74] The one informed of this forthcoming deluge is Noah, and he is also given the duty of building an ark and ferrying various life forms to the new world.

Just as the Deer's rainbow path connects the heavens with Turtle Island, the biblical flood ends with a rainbow that is a continual reminder of the covenant God made with Noah, humanity, and all living creatures – that there will "be no further cataclysm" in this new world.[75] Despite the similarity, the story is in stark contrast with the Indigenous tale of Sapling's returns to a people who periodically forget to live the good way and thus release Flint's mischievous monsters.

Here Christian purification is replaced with what King refers to as a Creator presence who asks people to do better, and gives duties and teachings that help them follow the rainbow path. In the Middle Eastern region, other interpretations of the human role in disastrous events like the flood also exist. The biblical story is connected to Sumerian records of "a flood so overwhelming that nothing was ever the same again."[76] The oldest such story was written in stone and is found in *The Epic of Gilgamesh*. It tells of a flood brought on not by moral disorder or sin but by the gods' belief that the "uproar of mankind is intolerable and sleep is no longer possible by reason of the babel."[77] The individual who related these events to Gilgamesh was Utnapishtim, who, like Noah, survived. He was saved because one of the gods, Ea, did not agree with this extermination and "warned me in a dream" to build a boat.[78] In these stories, humanity participates in an imbalance that divine beings view in different ways, thus evoking divergent responses.

The source of flood myths appears to be a series of real events that hit different parts of the world as Flint's glaciers melted and brought global shifts in climate, ocean water, and geology. At its largest the Laurentide glacier of Kanatha covered 13.4 million square kilometres. It is an area larger than all present glaciers in Antarctica and contained "enough water to lower the level of the sea worldwide by up to seventy-four meters."[79] With Turtle Island cycling into its regular closer orbit with the Sun, these melting waters came to affect geography, ecology, climate, and myth. The first catastrophic change occurred about 13,000 years ago as the Laurentide receded and, in a relatively short period, a tiny rivulet expanded into a deluge of freshwater entering the Atlantic Ocean, shifting its water circulation and bringing about a lengthy thousand-year cooling in the midst of a warming trend.[80] This and an ensuing series of significant interactions had numinous impacts. As Wilson explains, human populations have for a long time settled on flat land along seashores and when "breakthroughs of the sea occurred, quite possibly suddenly and unexpectedly, they were accompanied by serious localised catastrophes."[81] In the Black Sea region around 5600 BCE, a major flood occurred that was so "devastating that it arguably spawned not only the 'Noah' Flood story" but also a host of flood stories.[82]

On the icy and wet ravine paths of this Kanathian labyrinth, one can almost imagine the texture that was underfoot as the first Indigenous ancestors of Toron:to came north of today's Lake Ontar:io and orally passed down a mythic flood of memories. The Wendat saw the Anishinaabe as the land's elders and this is confirmed in localized stories of floods around Lake Huron and general myths of a deluge. Exemplary of the latter is Basil Johnston's telling of a time when the sky formed great clouds that "spilled water upon the earth, until the mountain tops were covered"[83] and everything was a great sea not unlike the experiences of Noah and Utnapishtim. It is said that all people and land animals perished in an "unbroken stretch of water whipped into foam and wave by the ferocious winds," a set of conditions that persisted for generations.[84] Considering the prevalence of Indigenous flood stories, Lakota scholar Vine Deloria surmises that a global event in the past "gave rise to the legends that recorded memories" of these catastrophic changes.[85] Recent climate and geologic research describes it more as a series of flooding events commonly related to post-glacial sea-level rises in interaction with "localized circumstances such as seismic activities."[86] These mythic glacial melts also transformed Kanatha into a land of more inland fresh water than anywhere else on Turtle Island,[87] and Toron:to into a place of ravines on one of the Great Lakes that together contain almost 5 per cent of the planet's fresh water and regionally supports over forty million people on its shores.[88]

There is sensible in this labyrinth a diversity of flood stories that seem to be climatically related, over several thousand years, to the transformation of Flints' glaciers as signalled by Bluebird. Beyond documenting real events, such myths also, Deloria explains, encode "moral and ethical codes particular to the religion reporting the flood experience."[89] In contrast to the Christian tradition that tends toward a universal apocalypse, the Indigenous stories tell of a moral cosmos that responds to human behaviours and whose cycles mirror "the seasons with its endless circle of degeneration, death, and rebirth."[90] The Anishinaabe stories of Johnston reflect this cycling, since it is following the flood's aftermath that Sky Woman is said to have fallen onto Great Turtle. His version has nuances that differentiate it from the Iroquoian versions but there are also common features, such as

Muskrat, "least of the water creatures," being the one to retrieve the sacred mud.[91] While the Haudenosaunee Good Mind follows the skyward reach of Sapling, Anishinaabe myths tell of Kitchi-Manitou, the great Creator who gave his descendants the land to be listened to as a living book that continually renews knowledge, beliefs, laws, and customs.[92] Anishinaabe "were as keen observers as the scientists of today," though Johnston clarifies that this ecological knowledge is embedded in myths that tell of human "conduct to be imitated and spurned."[93] It is a way to knowledge consistent with the imagination of the Good Mind, for attending to what we do not know requires a knowledge of Turtle Island's changes and ceremonial duties.

Perhaps the most significant difference between the Black Robes and Wendat was the respect the latter showed for their Anishinaabe allies who were seen as their elders because they had lived in the area longer and had followed a hunter-gatherer existence that gave them a closer relationship to the land.[94] While this view clearly is in contrast to the Black Robe concern with wild descents, there is beyond the Christian Bible and the story of Noah a symbolic connection with the Sumerian myth of that region. What led the king, Gilgamesh, to seek out Utnapishtim was a burning question about the transitory nature of life and the way to immortality: "Oh, father Utnapishtim, you who have entered the assembly of the gods, I wish to question you concerning the living and the dead, how shall I find the life for which I am searching?"[95] The search arose from the death of his friend Enkidu, "the strongest of wild creatures" who was reared in the wild hills beyond civilization.[96] These unlikely companions came together when a duel made it clear that the powers of the civilizing king and sauvage were equal, though different, and as such "Enkidu and Gilgamesh embraced and their friendship was sealed."[97] They undertook many adventures and became bolder in eventually disrespecting the power of the gods, with the pivotal moment that led to Enkidu's death being their decapitation of the monster who protected the great Cedar forests from abuse. Haunted by the deathly limits of life, Gilgamesh came to Utnapishtim with his question, to which he received this response: "There is no permanence. Do we build a house to stand for ever, do we seal a contract to hold for all time?"[98]

Wise though this answer was, it was clear to Gilgamesh that Utnapishtim had found some kind of passage beyond death, and thus he asked a more direct question: "How was it that you came to enter the company of gods and to possess everlasting life?"[99] Utnapishtim explained that the silencing of the people's loudness by the gods was brought about by a deluge, though he was warned by one of the gods – who did not agree with this escalation of events – to, like Noah, build a boat. When the waters burst forth they were so expansive that it even terrified the gods, and so "they fled to the highest heaven."[100] After seven days on the boat "I loosed a dove and let her go," but she found "no resting place and so returned." In contrast to the Bible, which tells of the dove finding land, it was when Utnapishtim "loosed a raven" that a land beyond death was found. Both he and Noah have an Indigenous ecological knowledge for navigating these coasts, as reflected in the use of birds to find land, but some wisdom was also needed to find the immortality desired by the king. For Gilgamesh to find passage to a shore beyond death, Utnapishtim advises that he must stay awake for six days and seven nights. He fails this and a second task before returning to his kingdom still mortal and destined to the same death as his wild friend Enkidu.

There is little sense of biblical sin in this story, with the only moral seeming to be a reminder "of our modest dimensions,"[101] like that of Muskrat and Toad. But the humbling task Utnapishtim gave Gilgamesh suggests to William Irwin Thompson that there are deeper teachings here concerned with the humility needed to wisely mind floods of change. The first clue is that Utnapishtim's epithet is Atrahasis, "the exceedingly wise," which indicates to Thompson that he is "more of a yogi than was Noah."[102] The task of not falling asleep is a kind of mystic method for allowing the mind to enter other states of awareness beyond what is rationally known. Thompson describes this as paraduction, letting "the mind wander to gather what associations it may, through dream, reverie, or vision,"[103] which has a certain consistency with the imagination of a Good Mind. This is not the only sign of these spirited dimensions, for the boat Utnapishtim is directed by the gods to make "is no ordinary boat but a perfect cube with seven levels."[104] Thompson sees this as corresponding "to the various vibratory planes of the subtle bodies" that one can engage through meditation.[105] It is a practice for quietly attending our relation with Turtle Island's surrounding messages, for imagining something in the climatic signs beyond what is directly knowable.

People who are too loud tend to not be listening, and thus they cannot hear, are not attentive to the building voice of disturbed divine presences until it is too late. Only Utnipishtam is aware that the commotion has disturbed the gods, and this requires ecological knowledge of the creation coupled with practices for heightened awareness. With the climatic longue sault bringing unprecedented extreme flooding events from Hurricane Katrina in New Orleans to Hurricane Sandy in New York to submerging Bangladeshi coasts to seemingly yearly Red River overflows in Kanatha's Manitoba, one again hears intimations of numinous stories in diverse forms.[106] As with Utnipishtam, it has been suggested by some environmental ethicists that the root of these numinous changes are again people who have lost the capacity to listen. "Swaggering, talking too loud, not knowing how to listen, this very clumsiness" may be exactly what is in need of change.[107] The central lesson of Utnapishtim's flood is, for Thompson, that we cannot save ourselves "from the catastrophes that always come at the end of a world cycle, one cannot simply escape," but rather we must be attuned to the emerging messages of surrounding presences.[108] Seen through such a mythological framework, our ever-growing commotion is initiating all kinds of changes on ecological, climatic, and cosmological scales, changes that call for us to listen more closely as we attempt to respond in turn.

Relating these mythic floods to a kind of paraductive attuning brings my mind back to Notre Dame's numinous Black Madonna. The influence of Saint Dionysius went far beyond the light emanations of cathedral building, for it was a dark mystic practice that informed his approach to the One. To actively recognize the shadows of thought and knowledge that is always limited, he offered what became known in the Catholic mystic tradition as a *via negativa*. Since the Creator is, in Dionysius's words, "the Cause of all beings, we should posit and ascribe to it all the affirmations we make in regard to beings, and, more appropriately, we should negate all these affirmations, since it surpasses all being."[109] In this spirit, the twelfth-century symbolism of Notre Dame elevates images "borrowed from sense-perceptible reality" to mysterious scales through an unknowing of their material references.[110] The potential viriditas of Sapling can be found in Bluebird, Raven, Dove, Heron, Deer, and other presences we become familiar with, but at the same time we need to negate all this knowledge because the orenta behind the creation is clearly more than what is knowable. So the Black Madonna is not simply an external prima materia from which all creation emerges, but also reflects the unknowing taught by Dionysius. In the face of Flint's self-interested swagger, we need to renew a via negativa – from paraduction to imagination – for attending regular cycles and extreme changes that always go beyond the knowable. This does not mean that we disregard what is scientifically known or discontinue searching for the knowable. What the via negativa exactly entails as part of a good climate of mind is more fully contemplated in the next chapter, though a hint of its direction is intimated in the Black Madonna's right hand, which holds a heart as the finger of the Christ child points upward to an arch of golden hearts.

On this spring day, the paths vary from muddy to refrozen ice, and despite the coolness a change of season is in the air with smells of humus and the sounds of bird activity, mostly chickadees, nuthatches, downy woodpeckers, and the odd *purdy, purdy, purdy* of a cardinal. In the ravine below is the fallen onerahtase'ko:wa where I often contemplate Woodworth's Toron:to as a fallen dolon-do, and it now takes on new cosmological dimensions. Walking down the slope I move into Flint's glacial imprint and the subsequent flood of water that released a global chorus of myths from the lands of Kanatha to Sumeria and beyond. In this low spot the twins come into relation with Utnipishtam to teach us the many roles people in the past and present have played in this tree's fallen nature. Sitting down on the lying trunk, I cannot help but mythically connect its current state to the ship's epic conversions and an ever-rising noise that is moving us toward increasing disorder. In the words of Berry, the planet is for the first time "being disturbed by humans in its geological structure and its biological functioning in a manner like the great cosmic forces that alter the geological and biological structures of the planet or like the glaciations."[111] It is not so much a moral sin as a lack of awareness or wilful forgetting that is today fostering the swaggering behaviour so desired by Flint. If I listen closely within the depths of this ravine, I can hear the ancestral presence of Utnipishtam ask us whether we can quiet ourselves long enough to hear what the surrounding cosmological presences say is needed to raise the dolon-do. It is to this mythic challenge of our moment that these weaving passages now guide us.

In the labyrinth's southeastern quadrant, the season shifts toward summer and in this urban place a smoggy heat island gradually slows my pace as I walk across the sixteen lanes of driving severance that opened this book. On the other side, light plays on the ripples of a calm Lake Ontar:io as my eyes are drawn eastward to the CN Tower Prayer Stick and Skyscraper Medicine Wheel of Woodworth's Ashbridges Bay vision. We saw in chapter 2 that the tower stands in his Good Mind as a beacon for "the impending shadow of danger in the darkness" and as a "herald of an emergence from dark into light."[112] The tower's symbolic reference to the CN railroad is a reminder of Kanatha's resourcist history, from fur trade to bitumous sands, while the skyscrapers that house the nation's banks have made much of their ever-growing profits through the progressive conversions of the colonial and now modern ship. On the one hand, the tower and skyscrapers are symbolic beacons of our epic noise that is deepening our turbulent relations with a climatic longue sault. At the same time, they are powerful cultural symbols that have the potential to be renewed. Within Notre Dame de Turtle Island, they can be imagined as vital beacons for, following the guidance of Thomas, our task of lifting the onerahtase'ko:wa once again and putting Flint's poisonous monsters back in their place.

The dual nature of the tower and skyscrapers, seen hazily through the summer smog, calls forth another numinous myth of the Bible that Hulme adds climatic dimensions to. Generations following the flood, an increasingly confident people said, "Let us build ourselves a city, with a tower that reaches to the heavens."[113] The divine response to the Tower of Babel was a destructive humbling of the people. This myth not only highlights for Hulme a persistent human desire for "god-like status, acclaim and personal glory," but more fundamentally offers lessons on the hubris of over-confidence and the urge to dominate that is being replayed today in our climate responses. There is in the modern ship a self-interested swagger that seems to guide people to the extremes of either a denial of the challenge or towering scientific urge to "climate mastery and control."[114] The result is that our Tower of Babel is crumbling from within as conventional political instruments of diplomatic negotiation, economic trade, and international regulation continue to fail, despite the most globally extensive scientific effort represented by the IPCC. Our trajectory into stormy waters seems to progress unimpeded. The observation of Hulme brings my mind back to Stephen Harper's epic characterization of Alberta's bitumous sands and the hubris that has been at the core of Kanatha's conversionary history.

This swaggering confidence is the outward reflection not simply of an addictive bind to resources, fuel, and technology but, more importantly, of a particular mindset whose reach extends far higher than any tower. What built the CN Tower and bank buildings was a Kanathian resource economy based on the idea of unending economic growth. As the ecological economist Peter Victor explains in *Managing without Growth*, this "pre-eminent economic policy objective of government" is rooted in the Enlightenment idea of progress based on individual self-interest.[115] It is a powerful factish that continues to motivate even though it has been tempered by international calls for sustainable development and green alternatives, as well as the emergence of "government departments and agencies charged with safeguarding the environment."[116] These responses have to date, Victor explains, been mixed because the context is unending global economic growth. In the case of Kanatha, the resulting policies from both the left and right of the political spectrum have not ended poverty or unemployment, have widened the inequalities related to income distribution, have "exacerbated, not been a panacea, for environmental problems," and continue to fuel the nation's still growing greenhouse-gas emissions and resulting climatic turbulence.[117] It is these numinous challenges to our tower that are behind the subtitle of Victor's book, *Slower by Design, Not Disaster*, for it seems to him that a choice is before us concerning how, not if, we navigate these changes.

In this labyrinth, a grounding of our modern bind in economic self-interest and growth brings us back into the presence of Flint, who, we recall, was given his name because, in the words of Sky Woman, "you only think of yourself." As we saw earlier, Flint influences the left side of the human mind and, if not held in check, will lead people to be overly self-interested and controlling.[118] When this tendency becomes strong in people, then Sapling or a messenger like the Peacemaker comes to remind them of their role in the darkness. In Sapling's final return, he said that in "controlling your minds" Flint has the objective of destroying all that is good and makes life worth living on Turtle Island. There is a resonance between this mythic lesson and Victor's research which suggests that, beyond a basic level of need, an individualized self-interest in the acquisition of more wealth does not bring more happiness, fulfillment, or stability. As he muses, "if people understood that demonstrating status through consumption is self-defeating it might help relieve the pressure on the environment caused by making, distributing, using and disposing of these goods."[119] But people do not recognize this self-defeating behaviour for the simple reason that a rational explanation for it does not suffice. Our minds are caught in Flint's presence. This is more consistent with an eco-psychological diagnosis of economic self-interest as a kind of narcissistic wound that leaves many on the modern ship with "almost no sense of their own insides, their own bodily felt living," thus binding us to towering self-images and consuming appetites as compensation for inadequacies.[120] A dark web of disorder reaches deep within, something the Haudenosaunee mythically relearned many times, with Skanyiatar:io most recently counselling them "to be vigilant in the ceremonies."[121]

By this point Flint has shape-shifted into many forms, from glacial ice to colonial conversions to a self-interested part of our mind to a climatic longue sault fuelled by "a great smoke" of rising emissions seen by Skanyiatar:io. One can be left to ask what he is. On an energetic level, Flint and Sapling are cosmological manifestations of the otkon and orenta from the Sky World. As Rice related in chapter 5, Sky Woman's fall was initiated by a heavenly imbalance wherein the vibrating light energy, orenta, was giving way to the "freezing dark energy known as otkon."[122] It is an understanding that has a certain resonance with the scientific definition of energy as capacity to work and its two laws of thermodynamics.[123] As Suzuki explains, the first law states that the "amount of energy in the universe remains constant"; no more can be created and it is not possible to destroy existing energy. What the energy can do is convert across different forms such that the Sun's energy can be taken up into the growth of a sapling that can, with other trees, sequester carbon and further transform the world. These trees can also be chopped down and used for heat in our homes, which leads to the second law of thermodynamics. Though energy is never lost, converting wood or bitumous sands into fuel dissipates energy into a low-quality form that cannot be harnessed for work. It is a random and disorderly state referred to as entropy, and can be minded as a dimension of the otkon that Flint further animates in the dark self-interested changes of people on Turtle Island.

Despite the second law of thermodynamics, the orenta in Sapling's creation does not tend toward disorder. This is because the Sun's energy "is constantly flooding our planet, providing high-quality energy to compensate for the steady decay of energy."[124] Ever renewing sunlight is the basis of life, and thus we have seen that, for the Haudenosaunee, Wendat, Anishinaabe, Chartres, and many others, the Sun has its own seasonal ceremony and recognition in daily Thanksgiving. In Suzuki's ecology of mind, the Sun's renewing energy also has an equivalent in human relations as a love that research indicates is essential to a child's healthy upbringing.[125] Parental love must continually give of itself for the child to feel safe and become a well-functioning part of his/her family, community, and world. This kind of attraction may, Suzuki goes on, "be built into the very structure of all matter in the universe," from protons and electrons that are held together by their mass and electric charges to planetary gravity to our biological beginnings in the oxygen revolution some two billion years ago.[126] The evolution of cyanobacteria emerged from two bacteria that came together and eventually "transformed the chemistry of

Earth and primed the environment for new forms of life."[127] This may be the biological and heliocentric basis of a sharing etiquette that is also told in the changing stories of Sky Woman, Budding Flowers, Sapling, and Flint – a widely shared sensibility within which, "however imperceptibly, all matter feels drawn together."[128] From this cosmological perspective, orenta is like a spirit of mutual attraction, with Turtle Island orbiting Sun, child orbiting parents, Haudenosaunee orbiting onerahtase'ko:wa, and Flint trying to freeze a loving mind with a globalizing self-interest.

A spirited pull toward each other is also core to the thought of Teilhard de Chardin's perspective on our planet's evolutionary Omega point, a view that helped us in the previous chapter enter a co-created Notre Dame de Turtle Island. His blending of science and religion leads him to see a movement of life toward a more nuanced, complex, and mysterious consciousness that interacts with biological evolution. This movement is pulled by a spirited energy or orenta that is before and after the creation. He writes that all life, including humanity, is "knit together and convulsed by a vast movement of convergence ... the term of which we can distinguish a supreme focus of personalizing personality."[129] The nightmarish Darwinian struggle for species to blindly and yet naturally select ways of surviving, producing offspring, and passing on a viable inheritance becomes, with Teilhard, "the way leading to personalization, a movement which has direction and aim."[130] Species interactions with local, regional, and planetary ecologies and climates have an effect on the evolution of both physical realities and what he termed a psychic noosphere. This is a much more spiritual sense of the ecology of mind that emerged with Leopold, Carson, Livingston, and others at the dawn of the environmental movement. For with Teilhard and the Haudenosaunee orenta, a broader energetic presence is drawing forth the evolution of life, humanity, and our minding of relations. A climate of mind is thus not simply a non-descript phenomenological way of minding life relations, but rather is moved through participation and familiarization toward a particular quality of relation. We have a choice between a personalizing loving orenta that affirms creation, or a tower-building otkon that potentially undermines it.[131]

In contrast to the Sun's ever-renewing energy, the oil that excessively fuels the modern ship and climatic longue sault is, Suzuki writes, "a once-only gift of ancient life-forms to an energy-hungry industrial civilization."[132] Going one step further, Mircea Eliade connects the industrial myth of unending progress and economic growth to a kind of sorcery that accelerates "the natural tempo of things by an ever more rapid and efficient exploitation of mines, coal fields and petrol deposits."[133] These conversions are of ancient fern-like trees from over three hundred million years ago that sat in warm swamps and over eons became today's coal fields, and hydrocarbons of animal fat that were pushed down "between ten thousand and thirteen thousand feet below sea level where temperatures are high enough (100 to 135°C) to boil organic matter into petroleum."[134] The succession of geological and glacial changes with which Flint and Sapling are associated buried these presences that humans have since the Industrial Revolution released with increasing speed. It is a process that Livingston saw as moving us toward a time when nearly all the "long-dead animalcules" are unearthed "to rejoin their kindred in the great heat sink in the sky."[135] While the tower-building hubris of the modern ship leads in this direction, the eco-mysticism of Berry points to the lack of deep reflection that such acts are based on. In his words, we "can only marvel that scientists generally seem never to have reflected on or explained to the community why the petroleum is buried in the Earth in the first place."[136] A mindful pause clarifies that Turtle Island has carefully buried great amounts of carbon in the form of coal and petroleum, and this has manifested the balanced creation humanity finds itself within.

The good ways of Sapling and Utnipishtam suggest that we cannot deny an evolving knowledge from either the modern ship or Indigenous canoe but rather must engage the spaces being opened by turbulent changes in ways that attempt to lovingly rebalance relations. Though there is obvious resonance between the earlier discussed paleoclimate research of Calvin and the persistence of numinous myths, it can be hard for modern researchers, myself included, to embrace the latter's wise sense of a moral cosmos whose numinous potential requires us, before all else, to act respectfully. It is easy in today's highly anthropocentric culture to assume that the evolution of communal sharing is simply a rational response to climatic uncertainties, but this misses the fundamental point – we are grounded in a responsive and cycling cosmology. The earliest religious impulse was toward surrounds filled with ancestral and divine beings; physicality and spirit were interpenetrated in ways that "may be central to the way of life of tribal and archaic peoples."[137] Even the scientific story of evolution tells of deep ancestral connections to life on Turtle Island.

Sharing is an innovative human ritual that is today attempting to renew itself in interdisciplinary, intercultural, and global calls for collaboration. But there is also a numinous presence confronting the renewal of a sharing etiquette that is difficult for the modern ship to appreciate and yet is central to the old stories emerging from this labyrinth. As Hulme and Thompson relate, myths like those of the Flood, the Tower of Babel, the Peacemaker, and Flint "reflect back to us truths about the human condition – instincts – that are both comforting and disturbing" because they are in such stark contrast to the progressive linear power of scientific reductionism, unending economic growth, and fossil-fuel consumption.[138] The selfish gene popularized by Richard Dawkins is as real as Flint, but so are the broader emanations of Sapling's life-giving orenta experienced in the Sun's rays, Notre Dame's divine reminders of what life is based upon, and duties of a Good Mind.

Fossilized sources of energy are time-limited gifts for which Thanksgiving must be offered, especially in a turbulent time that is highlighting the unanticipated impacts of their release and the need to reduce how and how much we use them. The entropy of these emissions is what creates space for Flint's disorder, and their deep source is our economic self-interest in unending growth. In a context where epic management regimes are what appeals to the ship, deep-set factishes continue to bind many of the proposed alternative routes to a sustainable future. As the "efforts to reign in the damage using conventional human control systems are deemed to be failing," Hulme explains that there is increasing pressure to come up with viable proposals for "a new, but now deliberate, great geophysical experiment with the planet."[139] Some of the grand proportions of such a design challenge are highlighted by Calvin when he considers what our ancient climate-mind relations have to tell us about the present situation. Beyond simply reducing emissions, he proposes that our goal "must be to stabilize the climate in its favourable mode and ensure that enough equatorial heat continues to flow into the waters around Greenland and Norway."[140] This can be designed for "in computer models of climate, just as architects design earthquake-resistant sky scrapers." But there are issues with relying too heavily on modelling more information and data for an even more towering project. In the words of Ophuls, the costs of increasing complexity "grow disproportionately until they eventually reach a point of diminishing or even declining returns," a dynamic that can escalate quickly for a civilization already stressed and becoming less flexible in its responses.[141] Even as the tower reaches higher there is the risk of a cascade of failures. Coming to a renewed sense of balance is central to why the modern mind now needs to re-engage Indigenous wisdom.

We are entering a period of transition from the somewhat climatically stable Holocene to a new epoch on the other side of this climatic longue sault, and the swaggering ship now needs to humbly renew a quieter, more attentive, slower, and more loving etiquette. Though not stated in those exact terms, this is the central message of Victor's slower-by-economic-design alternative to impending disaster. In confronting unending growth, he does not suggest that we "adopt zero growth as an alternative, over arching policy objective," but rather forget about growth or make it secondary to a specific objective like well-being.[142] A one-size solution cannot flexibly mind the diverse relations that manifest uniquely in different regions of Kanatha and the world. There are places where growth is needed, as reflected in the injustices of colonial and modern conversions that continue to play out and need to be dealt with while recognizing the world's ecological limits.[143] Rich nations and pockets of wealth within nations need to make space. At the same time, Victor simulates various no-growth scenarios to clarify a disastrous potential reminiscent of the Great Depression when carelessly implemented, as well as opportunities if these scenarios are used as a tool with clear long-term objectives.[144] A deliberate slowing of growth that leads to stability in a couple of decades can "be consistent with

attractive economic, social and environmental outcomes: full employ-
ment, virtual elimination of poverty, more leisure, considerable
reduction in GHG emissions and fiscal balance." Some of what has to
occur in this design is decreasing consumption and sharing work so
that people move more slowly and have more time to enjoy their
relations. We are being asked to moderate Flint's self-interested tower
building through emphasizing more loving attentiveness to where we
live, who we live with, and how we live.

Many of these changes are not to be found without but are within
the modern ship, for making these shifts will transform our relations
with Notre Dame de Turtle Island. The difficulty with this kind of
change is not lost on Victor, who is aware of the institutional binds
that continue to block significant action on regional, national, and
global scales. Though government policies are needed and innovative
ideas like localization will help, he concludes that the entrenchment
of power structures means that "any movement towards managing
without growth will have to be driven from the grass roots."[145] He
specifically highlights the old historic tradition across cultures of
"simple living" that moves in a different direction than self-interested
consumerism and unending growth. From Saint Francis to Thoreau,
many people have sought "the simple life as a solution to the stresses
and strains of the mainstream." This is a call for the kind of ecological asceticism and frugality that Saint Tekakwitha helped us contemplate in chapter 4. But here Victor makes the important point that,
while a grass-roots push is needed, it will not produce the desired
result without connections to institutional and policy change on
regional, national, and global levels. Focusing on climate change,
Hulme likewise argues that we cannot yield to one-dimensional goals
like no growth, simple living, or even the idea of restabilizing the
climate.[146] In fact, he writes that "setting the overarching goal of
humanity as the restabilization of climate will, I believe, lead to disil-
lusion." The question for us now is how we approach the climatic
longue sault's emerging nature in relation to the changes we have to
make, not out in the world but in our etiquette.

There is an accumulation of evidence that Flint's self-interest as an
orienting way of living was a failure during past periods of climate
instability, and it is highly likely that the current global approach to

trickle-down self-interest will fail our common Kaniatarowanenneh. Reflecting on the continuing prevalence of an "Everyone for themselves" attitude in global climate politics, Naomi Klein warns that if we do not "radically change course, these are the values that will rule our stormy future, even more than they already rule our present."[147] In light of the ship's addictive binds, a deluge of pain may well mark the passage to reconsidering our choice of direction beyond Flint's self-interest. But that is only the pessimistic side of the story, for we have seen that there is also a more ancient connecting etiquette that we can draw upon in our search for an alternative path. That this alternative is not out there in some other culture but in our common human and Kanathian ancestry is an important point. For evolutionary transformations do not arise as miraculous mutations but rather "work most effectively where there is already some incipient expression of a trait that can become adaptive under changed conditions."[148] Perhaps we can see the current rise of interdisciplinary, intercultural, and global dialogues as just such an incipient trait that is in the spirit of Teilhard's loving Omega Point – a cosmological etiquette that we can broaden and deepen as self-interested maladaptations crack in the face of quickening shapes of violence.

These labyrinthine passages are initiating us into a socially dynamic reality that always transcends our towering knowledge because it is rooted in felt relations. This does not mean that our searching for knowledge and sustainable ways of living ends, just that this search needs to be conducted with an active recognition of the numinous presences that permeate our myriad relations. In contrast to those who deny climate science for various religious, economic, political, and personal beliefs, the cycling mystery of Notre Dame de Turtle Island does not refute the empirical knowledge of interdisciplinary research or Indigenous experience, though such knowledge is wisely transmuted in imaginative ways that have the cosmological reach of Sapling rather than Flint's freeze-framing tower. We should employ a principle of precaution not simply because we recognize the limits to our knowledge but because the world is social, animated, and responsive. Our precaution primarily comes from respectfully and lovingly engaging the world as the place in which we want to be, and is only tangentially a conservative stance in the face of a numinous reality.

The difference is in the spirit of a mythic precautionary etiquette that tries to emulate the grounding and skyward reach of the onerahtase'ko:wa, which can mind the changing movements of Flint and Sapling without and within.

———

The smoggy summer view on Woodworth's CN Tower Prayer Stick across this Lake Ontar:io bay intimates a mythic challenge to renew a wild wisdom. What is at its core? That acting from a passionate, loving, sense of what connects us to each other is vital, and that the universalizing ideas of economic growth and self-interest need to be imbued with an etiquette of dutiful service and loving devotion to Notre Dame de Turtle Island. So as not to recapitulate the hubris of another towering epic, the modern ship needs to actively renew the orenta of past cultural adaptations in relation to present realities and possibilities. This approach is in accord with Bateson's call for a move toward service that can heal our binds by reaffirming our interdependent relations with powers "greater than any of us."[149] This is a return "of" and not "to" sauvage wisdom, for what is unique now is our capacity, through technology, interdisciplinary knowledge, and global networks, to evoke cosmological responses, and the potentiality of humans to respond communally in turn. It is clear that all the modern options, including policies based in economic self-interest, will be needed to navigate a sustainable and just way forward in times of increasing turbulence, but they can no longer be the driving measure of all things. This factish needs to be jettisoned from the ship before it pulls us down into Flint's engulfing darkness.

Beginning to walk westward along the beach toward the mouth of the passage de taronto, I am stopped in my tracks by the stillness of a great blue heron reposing with one leg on the lake's break-wall some twenty-five metres into the water. This descendant of the winged being who first recognized Sky Woman's fall from above and was companion of Saint Tekakwitha on her isle has also been a vital guide through this book. The heron's countenance speaks to the kind of calm attentiveness that is needed in the midst of our quickening changes. As counterintuitive as it seems within turbulent times like ours, the most important thing we need to do is slow down our driven lives, our insatiable consumption, and our obsession with technological development and limitless economic growth. The numinous myths of this chapter ask us to attend more closely to "what we really want to achieve for ourselves and for humanity," though Hulme adds that their irrational dimensions are disturbing because they suggest that "we are limited in our abilities to acquire or deliver it."[150] The image of a Chinese finger lock is an apt analogy for the way out of this dark bind we seem caught in, for the shift away from Flint and toward Sapling is as counterintuitive in a time of high stress as releasing fingers by bringing them together rather than instinctively pulling them apart.

We are being drawn into a paradoxical challenge that the Heron calmly bears witness to on the lake's shimmering waters, as its close relation Bittern did almost a century ago across the city with a young Livingston. A short eastward flight of the wings to Ashbridges Bay takes us beyond the wobbling CN Tower to an outcrop of rocks looking out at a seeming sea of blue. With the city out of sight and the traffic hum barely audible because of the shore's slight edge of trees behind me, the water lulls my mind toward a solitary peace that momentarily opens me out to this chapter's flood of glacial stories. In these myths the sacred wild comes forth, unbeckoned, with Muskrat lessons focused on humility, modesty, and the etiquette of being quiet so as to listen. Similar capacities are needed today, something that becomes apparent in these places where the urban edges out to a numinous mystery that surrounds our lives. It is not the question of "distance from society" that is problematic with a romantic "thin" space, for that is valuable at times. Rather, it is the sense of it being a permanent, totalizing, and pristine sojourn.

Wild openings occur in Toron:to, such as in the ravines of the passage de taronto where a winter snowstorm restricts one's vision to a few metres and mutes the surrounding noise. This weather keeps most urbanites away from the woods, and so there are some paths where I will be the lone walker for two to four months, a perverse joy in this city. Meanwhile, at the height of summer the forest viriditas comes together with birdsong to put the urban rush in the background, if not totally hush it. On these lake shores of Livingston's youth the wild opens momentarily even on a busy summer day, though winter quiets the human presence even more as Flint's ice extends, the cold winds increase, and the lake's waves roughen. These temporally shifting "thin" spaces remind me that the human world, even urban Toron:to, is surrounded by profound mysteries, and that the climatic longue sault has numinous teachings for how we need to renew our lives.

The scale of uncertainty that is confronting our wobbling self-interest with unexpected extreme events like the flooding of torrential downpours draws me again to the paraductive imaginings of Thompson. He writes that "great ages of transition, with their attendant changes of climate – as we ourselves seem about to discover – are difficult times of decreasing culture."[151] Though these labyrinthine passages suggest that even in times of great uncertainty core parts of culture remain with us to be wisely renewed, Thompson is referring to a decreasing amount of useful energy, a kind of entropy, behind prevalent ideas and practices that have become frozen in their relation to ecology, climate, and cosmology. They cannot sense the numinous presences as the dominant stories become brittle. But yet it is in this brittleness that Berry saw a hopeful potential for change. We need to remember, he writes, that "the dark periods of history are the creative periods; for these are the times when new ideas, arts, and institutions can be brought into being at the most basic level."[152] Our darkening milieu of change is bringing a choice inexorably closer: Do we wisely set aside some beliefs, factishes, and practices? Or do we wait to be forced into a requickening not of our choosing? What can be taken across these rapids, and what needs to be left behind? Not to respond with orenta is to evoke Flint's hard otkon and, eventually, experience an utter powerlessness and vulnerability not unlike that felt by many during the deluge of Noah's and Utnipishtam's time or, more recently, Indigenous people like Skanyiatar:io in the face of colonial conversions.

With this thought, the Great Blue Heron takes me on one more imaginative flight beyond Ashbridges Bay to the Kaniatarowanenneh of my childhood and Saint Tekakwitha's sacred isle. Her presence is a reminder that we cannot be romantically flawed in imagining what the pressure of quickening changes looks and feels like. Marked on her body and canonized spirit are the mixed histories of Kanatha's colonial conversions. She lost her Algonquin mother and Mohawk father to a dark pox that she also barely survived. She was left with a scarred face and, for the rest of her life, was physically weak and suffered poor eyesight.[153] These afflictions brought her to Kahnawake, the Catholic faith, the Île-aux-Hérons, and a Lenten death in her early twenties. But in the midst of Flint's colonizing darkness, a "thin" space also opened for weaving an ocean passage that was not thinkable decades or even years earlier – the adoption by a Mohawk-Algonquin woman of Catholic mysticism in the midst of what was seen by Black Robes as a wild Kanatha. Without denying the violence of towering conversions, can we find the orenta to follow the Protectress of Kanatha's healing weave before we are forced that way by a catastrophic flood of change? Can a humbled ship be renewed so as to help in lifting the fallen dolon-do? It is to a concluding contemplation of what this dark healing passage may entail that we now turn.

7

A Life That Is Real

These ceremonies were handed down through the visions of a people who lived the Creator's way, and the spirit of those ceremonies can only be regained through the visions of a people who again live a spiritual life, one that goes beyond the power of words, a life that is real.

John Mohawk (2010), 6

From the sandy beach the voice of a small girl exclaims with joy to her friend that the "water is so shiny and sparkly." This little excited voice pulls me back from the repose of the Heron and my flights of imagination to the reality that I am not in the stillness of a labyrinth, but on a beach during an early summer weekend that brings many people for family barbecues, sunbathing, and the abundant recreation of volleyball, boating, swimming, and biking. This is the watery place to be on a sunny Toron:to day that, when in Notre Dame, also symbolizes life's joyful viriditas. The waters of Lake Ontar:io have a sparkling beauty that can draw forth such childish enthusiasm, a spirit that is perhaps the natural root of Thanksgiving in the summer sunlight and even those more numinous Flint-inspired seasons of cold stormy winds. Yet, as with the towering beacon, the joy of this beach often seems loud and largely focused only on other humans, tending to forget even the possibility of a duty to give thanks to the diverse presences that make this possible – from that which is cosmologically given to that which has been taken. Overt signs of swagger are seen in the garbage left behind in expectation that someone else will do the clean-up. In more secluded places the trash builds over the warm season and highlights a forgotten etiquette for becoming respectful of life on this sacred lake. Despite the joy on these waters,

17 Great blue heron on a rock. The heron that has flown through this book
stands perched on an isle-like rock here.

there is a painful strangeness to our relations that makes a state of
reverence seem quite distant.

Leaving the beach and walking into the labyrinth's southwest
quadrant, I head back toward the Howard tomb where we began,
across Lakeshore Boulevard, under the Gardiner Expressway and
railroads, and up to the Queensway and streetcar tracks. Stopped by
a light, I can hear from the open window of an idling car the latest
news report on what seems like a rising apocalyptic sense of our position. Earlier I had woken up to this story on the CBC radio morning
show that was describing the latest IPCC report and its central warning that "Global warming dials up our risks" and "the wild climate
ride has only just begun."[1] In the words of the IPCC's chairman,
Rajendra Pachauri: "Nobody on this planet is going to be untouched
by the impacts of climate change." Not long after the news, the CBC
Radio program *The Current* offered a more detailed examination of
the subject in an episode titled "IPCC Climate Change Report:
Official Prophecy of Doom."[2] The emerging scientific and environmental consensus is that our situation is much worse than expected,
there is no turning back from the direction in which we are headed,
the spectre of climate-related food shortages will expand far beyond

18 Rose cross on Howard tomb (winter 2015).
The rose cross found at the Howard tomb
offers a way of minding the heart that is central
to ceremony and healing, the central concern of
this final chapter.

the injustices of the modern period, damages related to extreme
events will become more normal, and everyone will be affected, from
seemingly remote areas to urban centres like Toron:to.

With the light changing, the vehicle's grim reminder of our numinous moment recedes as I make my way up the ravine toward the
Howard tomb and a dark presence that these labyrinthine passages
seem to continually cycle me back to. Places feel; all you need is to
slow down and become familiar with their patterned textures, beings,
and moods. There is something about coming around the ravine bend
toward this tomb that often evokes in me sadness and a sense of grief.
These feelings are consistent with the death marked here, though they
are clearly not connected to the Howards, who are on personal,
familial, and cultural terms largely strangers to me. What I feel here
instead is in fact connected to my current participation in Livingston's

19 Turtle Rattle (summer 2015). A turtle rattle was traditionally used to help people rhythmically enter into a ceremonial space of healing. As this chapter contemplates, it is simultaneously a symbol of Turtle Island and a tool for shaking those afflicted with illness to an experiential sense of the orenta underlying creation. Turtle Island is being climatically shaken today in a way that calls for ceremonial healing.

paradox. The modern Kanathian culture in which I have grown up and which I love in many ways appears to be coming into collision with limits it once thought it could progress beyond. Despite the pain of all its colonial, ecological, and climatic severances, there is a part of me that feels a need to grieve that death as a vital step toward transforming the ship's approach to Toron:to, Kanatha, and the planet. The drive below is the culture I grew up in – it is of my family, recent ancestors, friends, and country – and thus has an emotional bond that must be recognized before any renewal can occur.

Over the years, this tomb has become for me not simply a memorial to the Howards and the land they bequeathed, but a symbol of ancestors who can offer vital wisdom during times of numinous and painful change. We are haunted today by death and the suffering it brings as not only an integral part of life but a dark reality that is being heightened and displaced onto others by the modern ship's release of Flint's poisonous monsters. This is part of what I see here, though there is something about this darkness that simultaneously highlights the capacity for ancestors to remind us of other ways and duties. As Robert Pogue Harrison stated in the Introduction, dialogue with the dead needs to be ongoing so as to "keep open the possibility of a 'reciprocative rejoinder' that never simply denies but freely avows or disavows the will of the ancestors."[3] We need to be in relation with the past so as to have a wiser sense of who we are and where we are going. The common reverence for ancestors across cultures and time is in contrast to the modern ship's continuing disregard of elders, honoured dead, and a wild wisdom that goes in the opposite direction of the progressive belief in self-interested economic growth and unending energetic consumption. Revering ancestors is an act that reminds me of the prophetic insights from the previous chapter about the need for an etiquette when approaching a social cosmos of numinous potentiality. Today, the darkening paths of this labyrinth seem to draw our modern minds relentlessly toward a renewal of this ancient sensibility.

Dark climate prophecies reverberate through my mind as dusk falls upon this high spot, restating a message that in many ways began with the first ancestor who initiated me down the passages of this book, Livingston. As prominent environmentalist Farley Mowat wrote, "the breadth of his grasp of what ails this planet, and the ruthlessness of his conclusions combine to give him the stature of an Old Testament prophet."[4] In an appreciation following Livingston's death, Graeme Gibson similarly described his critique as requiring us to "rethink everything" in recognition that it is we who "have to change."[5] Such strong messaging is what prophets do, at least in the Old Testament tradition of the Bible where anger "denotes what we call righteous indignation, aroused by that which is considered mean, shameful or sinful; it is impatience with evil."[6] Though Livingston did not use the religious terms of sin and repentance, Mowat, Gibson, and others saw in him a resonant message concerning an approaching cataclysm if the ship's behaviour did not change. His capacity to recognize and confront these tendencies required an emotional commitment, one that was instilled by his early Ashbridges Bay initiation into a bird's-eye view and subsequent severance that felt like something being cut "out of my stomach." It was a painful experience that marks many lives on the modern ship, though it is often forgotten and denied in our driven lives.

In the Haudenosaunee Good Mind, people cannot enact the reverence of Thanksgiving if they are outside of Sapling's creation because of a disorder, illness, or grief. At such times, Woodworth explains that another sacred duty taught by Thomas needs to be engaged, the Ceremony of Condolence, *Kanianerenkó:wa*, that is traditionally offered to "those found wandering far from home and seeking a place with us on our land."[7] Though we saw with Saint Tekakwitha's time of mass requickenings that this ceremony has its own mixed history based on the pain of colonial conversions, it was originally another gift handed down from the Peacemaker through the line of royaner as a way of renewing the Good Mind.[8] As Longboat writes, the whole story is "a succession of episodes in which the Peacemaker encounters people who are not behaving rationally," whereupon he teaches a Condolence Ceremony to heal those who are dispirited and far from home.[9] It is a devotional duty that Woodworth, following Thomas, says must be resumed if "the healing of the profound rift between the Indigenous and the Western mind" is to occur. To heal this severance will require significant change in how the modern ship relates with the canoe and our common waters. In a sense, that is what the pain of Livingston is prophetic of.

What has become clear is that a good climate of mind is grounded in ceremonies that can raise enough orenta to heal deeply painful severances. As Mohawk says, such duties can be renewed only by "people who again live a spirited life, one that goes beyond the power of words." It is a wisdom consistent with Shawn Wilson's description of Indigenous research as "a life changing ceremony" that aims to raise our insights about the world and thus foster stronger relations between people and cosmos.[10] This is what Thanksgiving is for Woodworth when he engages Ashbridges Bay or offers his dissertation as "A Prayer for Recovery and a Spiritual Explanation of the Process for Receiving Duties."[11] The duties are "not just the period at the end of the sentence" or an occasional methodology engaged for a specific end, but rather a continual process of listening to ancestral and cosmological presences.[12] From Haudenosaunee Thanksgiving to Saint Tekakwitha's eco-asceticism to the braided myth-modelling of the last two chapters to Livingston's bird-inspired scales of consciousness, each of these can be conceived as ceremonial acts for engaging different dimensions of the climatic longue sault. This last chapter weaves healing ceremonies like Condolence with the aim of further contemplating what is required to reactivate a felt bond with Notre Dame de Turtle Island. At the Howard tomb, the surrounding presences have much to teach about healing a disorder that cannot continue if we are to move toward "a life that is real"[13] rather than be flooded by another deluge.

—————

The labyrinth's southwest quadrant unfurls into a central rose that I not only stand upon but look upon as the Rose Window above the Chartres' Royal Portal fuses with the rose cross on the tomb. These flowers are the symbolic source of the joy and pain that is part of all life. The tomb's rose cross derives from the Rosicrucian-inspired Masonic tradition of Howard, and represents "the chalice into which Christ's blood flowed, or the transfiguration of those drops of blood or, again, the symbols of Christ's wounds."[14] It is the Sacred Heart on the cross that symbolizes the mystery of tears, of an inspiration that comes from experiencing "the joy of sorrow and the sorrow of joy."[15] Meanwhile, at the centre of the cathedral's Rose Window, "sits the Virgin, crowned, the sceptre of world rule in her right hand and her left supporting the infant Christ."[16] As with the Blue Virgin, she is the dynamic ever-cycling support for our ecological, climatic, and cosmological relations, from birth to death and back again. Just as the thorns of a rose prick, the severing wounds of another can, if attended to, pierce and inspire a Good Mind to move toward compassion. Empathy for the pain of another can transform the estranged ways of life on the modern ship into the first ceremonial steps of healing; that is what the rose centre asks us to contemplate.

"Look with reverence" – the line has become a call to recognize these Budding Flowers as not simply roses but our never innocent participation in the interplay between Sapling's vivifying orenta and Flint's disorderly otkon. This is the deeper cosmological source of Livingston's painful paradox that cycled through his life, leaving an open wound that led him to conclude that "preservation is a cata-strophic, heart-breaking disaster."[17] While lamenting that the "over-whelming losses can instantly, easily, be attributed to the urban-industrial process of scarification, to world human populations, to institutional greeds," he added that it was more advisable "to stay closer to home" and recognize that the "fault is not and never was 'out there.'"[18] Echoing his words is Leopold's lament that a penalty "of an ecologi-cal education is that one lives alone in a world of wounds."[19] The death of Livingston in 2006 called forth remembrances of a life haunted by such an education. Though commonly seen by students and friends as a man who advocated a deep emotional relation with the world as the basis for change, some spoke of a ghostly pain that over time intensified his prophetic tone. One wrote that a "lifetime of frustration led to a kind of internal exile" and that a "constant sim-mer of rage arose from psychic pain – something I realized only after several months in his class."[20] It clearly was not easy to be an early prophet of Kanathian environmentalism while in a modern ship focused on unending progress.

Because there is in Livingston's environmental tradition a tendency to bypass what it means to engage such pain consciously and deeply, I am led back to the Haudenosaunee Good Mind as it manifests itself in the Condolence Ceremony or Kanianerenkó:wa. The understand-ing Woodworth has of this duty "is based on witnessing the last com-plete recitation of the *Kanianerenkó:wa* in September 1996 by Cayuga Chief Jacob Ezra Thomas."[21] It occurred a two-hour drive east of these Budding Flowers on Lake Ontar:io's Bay of Quinté where the Mohawk community of Tyendanegea and Eagle Hill is to be found. This land is, we saw earlier, where Toron:to's land transfer with Anishinaabe was first negotiated and, more importantly for the Haudenosaunee, where the Peacemaker was born of a Wendat woman. Upon arriving, Thomas and Woodworth placed "tobacco in a fire offered in Thanksgiving" and then invoked the "Memory of the Ancestors" as Thomas recited over nine days the Peacemaker's story, teachings, and ceremonies. Condolence is a storied ceremony that takes people back to a time of disorder and suffering, thus bringing those present to a renewed sense of how returning to our duties requires going through a period of dark grief and healing with the loving support of others.

The Peacemaker's move south of Lake Ontar:io is said to have been out of compassion for all those affected by the brutal wars and violence that was afflicting the Haudenosaunee in this time before colonialism and the confederacy. A pivotal moment in the creation of the Condolence Ceremony occurred when the Peacemaker comes across his eventual helper Ayenwatha. The latter was alone at the forest's edge, beyond the village, grieving the loss of his three daughters who died in quick succession of each other. Carrying a heavy sorrow and sitting alone amidst elderberry bushes, Rice writes that Ayenwatha spoke: "If anyone was burdened with grief like me, I would make sure that someone was there to console him."[22] With that Ayenwatha heard a rustling from the bushes as the Peacemaker, who had spoken in his village earlier, came forward and said: "If anyone was burdened with grief like you, Ayenwatha, I would wipe away the tears from their eyes … so that they could see clearly again." In the presence of the clear-minded Peacemaker, Ayenwatha then spoke of his loss and grief. Doing this, he began to feel better and then they went over the steps of what is needed to heal one who suffers. Once Ayenwatha felt better, the Peacemaker said that "now we must use this to comfort nations."[23]

The Condolence Ceremony is given to the people as a way of healing minds that are in pain and have become more susceptible to Flint's dark power. As Longboat explains, the ceremony consists of fifteen "matters." It begins by the speaker for the clear-minded ones addressing the physical manifestations of a grief-stricken or troubled people, and then moves to putting things back into their proper places, and from there to restoring the people's minds and spirits, and then to reminding them to resume their responsibilities and obligations, and to carry on with life."[24] Those who have grieved and come back to a clear mind can approach the wood's edge of communities still in mourning to begin the process of clearing their eyes, ears, and throats.[25] It is a healing protocol that Woodworth says expresses "compassion for the trauma of dissociation from Creation"; condoles for the grief of wandering rolling heads; requickens people into meaningful roles that were once filled by an ancestor; and finally "dignifies the shifting phases of our lives" that inevitably end with a return to the Sky World.[26] After so much disorder and killing, it is through Condolence that people can come into the peace of the skyward-reaching onerahtase'ko:wa. It is a ceremony that needs renewing whenever Flint's dark cycles become too engulfing, such as at the dawn of the Haudenosaunee Great Peace, during the colonial period of Saint Tekakwitha and Skanyiatar:io, and in our present time.

While Livingston carried the wound of an ecological prophet to his grave, the pain Woodworth carried to the Peacemaker's home on the Bay of Quinté was only partially informed by the loss arising from colonial conversions. He draws on a more intimate suffering to appreciate what the healing of Condolence entails, one that seemingly recapitulates the colonial pain of disease and extensive loss in a modern world of denial. As a gay man, the years from 1989 to 1995 were a time when he "became bloated by the weight of so much experience" as he watched his "long time companion and lover slip into the ravages of time and loss which is AIDS."[27] With so much death around his community of friends and violent denial in surrounding society, something died in Woodworth that eventually brought him to the Haudenosaunee teachings of Thomas's Good Mind and Condolence. Looking back on this grief, he writes: "My entire life is being lived in the liminal state where I am busy gathering the experience and learning the lessons with which I will return to the home from which I came."[28] His life gave him these experiences, and to come home in a deep ancestral sense required looking at his pain. It was with the Peacemaker's condolence for his tears that Woodworth's wound began to heal and his mind became clearer. From this humble start, the Toron:to skyline Woodworth called home for five decades began to change into a place suffused and shaped by the continuing presence of the ancestors in their myriad ecological, climate, and cosmological relations.

There is a potential bridge from Woodworth's AIDS loss to Livingston's stomach-wrenching pain that is offered by the Toron:to-based queer-ecologies thinker Catriona Sandilands. She asks: "How can the overtly politicized understanding of melancholia located in the midst of AIDS illuminate unrecognized losses in the midst of environmental destruction?"[29] Whether it be the queer experience of the AIDS epidemic or the ecological suffering of, drawing upon Leopold, living "alone in a world of wounds,"[30] pain that cannot be openly grieved can foster a melancholic condition. This is, Sandilands writes, "a state of suspended mourning in which the object of loss is very real but psychically 'ungrievable' within the confines of a society that cannot acknowledge" the death in their midst, from queer relations to non-human beings and ecologies.[31] We live in a milieu of melancholy that, through our denial of it, binds moderns as we struggle "to hold onto something that we don't even know how to talk about grieving."[32] A semi-conscious grief afflicts the modern ship and maintains it in a numbed position of denial that reflects a generalized refusal to move toward anything akin to condolence and healing. Following Woodworth, we are kept in a rolling-head position that is painfully far from being at home on Turtle Island.

The AIDS-inspired lesson of queer melancholia offers Sandilands insights on the value of not getting over or denying our traumatic losses, of nurturing a firm awareness on the death that surrounds us.[33] It is an approach that attends the mixed places of our immediate relations as the intimate, often forgotten, context of the loss of pristine or wild realities. From "the glut of aspen-loving birds in the clear-cut" to "bald eagles near the landfill" to an urban drive past the Howard tomb, there needs to be a commitment to maintaining awareness on the interpenetration of life and death where we are. In some ways this approach recalls the life of Livingston as the initial pain came to inform a closer look at the death around him. It was from this position that his writings moved over time toward bolder critiques and simmering anger at what is being done, a kind of dark prophetic view about the depth of the turbulent change and the long-delayed passage to an adequate response. He would boldly write that the "last time I looked, there was an endless tumbling cataract of identical managers, administrators, and technicians" intent on rationally planning the management of all the planet's resources.[34] Within the deep corridors of the ship, he lamented that there was not "a freedman in sight." Despite the progressing world of wounds, a queer melancholia seems to be easily bypassed and denied by rationalizing tendencies.

It must be noted that there is little engagement of feminist, let alone queer, perspectives in the thought of Livingston. Elsewhere I also critiqued his engagement of Indigenous knowledge as not recognizing the evolving nature of these cultures and their relation to the land, thus displaying a degree of participation in the modern ship's freeze-framing tendencies.[35] As with any of us, our minding of relations can get snagged by the powerful assumptions of our times. Yet, limited though Livingston was in engaging Indigenous, feminist, and queer voices, his fundamental insights about our technological and economic colonization are being borne out as each year passes without significant action. In a sense, he presaged contemporary critiques of ecological modernization and before that sustainable development as compromised approaches to a good future. Bringing this critical spirit into the present, Sandilands describes the popular nature-nostalgia of "ecotourist pilgrimages to endangered wildernesses, documentaries of dying peoples and places, even environmentalist campaigns to 'save' particular habitats or species" as the expansion of a modern melancholy that simultaneously denies the pain at its core.[36] It does this by incorporating our ecological crises into the practice of capitalism. The environmental movement often seems to drive past grief with the heroic urge of its modern milieu. Consequently, the natural losses all around us largely remain invisible and thus ungrievable.[37]

Situated deep within the ship's rationalizations, many environmental responses speed past the rose-centred pain and thus miss the first attentive step toward condolence, healing, and homecoming. That said, there is a danger in being suspended in loss and melancholia, from the extremes of pervasive ambivalence to Livingston's frustrated anger. The boldness of words like "tumbling cataract of identical managers" and not a "freedman in sight" are emotive signs of what makes him more prophetic naturalist than traditional academic thinker. While some could embrace his approach, others could not and indeed were sharply critical. In the words of Mowat, "he is revered by true believers but feared and rejected by those who are preoccupied with worldly self-interest."[38] His assessments could even bring forth the claim that he was a misanthrope. As Livingston's former students and friends Leesa Fawcett and Connie Russell wrote in their remembrance: "He was appalled when someone called him a misanthrope," for how could this be true of "a man who went home every day and loved his wife and children, who was a great friend, and a dedicated teacher."[39] Yet it is a characterization I have also heard when drawing upon his ideas.[40] Interestingly, these critiques come from those who are environmentally minded. Though he was widely respected, one obituary noted that he had disagreements with various environmental colleagues because of his "reputation as a nay-sayer, and a misanthrope who was against almost any form of commercial or resource development."[41] His opinions were seen as becoming more despairing and "entrenched as he retreated from the opportunistic and pragmatic world of commerce and public policy into a rarefied and idealized philosophical atmosphere."[42] The space of Hekademos fostered a tone in Livingston's writings that could turn some people off, thus marginalizing his pain-induced thought in the shadows of a perceived misanthropy.

Others had more nuanced recollections about specific points of conflict that raised concerns, though not total rejection. While co-working on the documentary series *A Planet for the Taking* in 1985, Suzuki remembered one such argument with Livingston: "I was involved in the anti-nuclear movement and his attitude was if humans were stupid enough to develop nuclear weapons and to drop them, well, so be it, the rest of nature would be better off for it. I had a hard time with that."[43] Despite this difference, Suzuki is clearly supportive of the ecology of mind advocated by Livingston, Leopold, and others discussed here. To him, a rejection of courageous positions is the modern ship's internal way of denying the extent of human change that is required. In response to the critics who "often accuse deep ecologists of being misanthropes, caring more for other species than for our own fellow beings," he offers the thoughts of the American poet Gary Snyder: "A properly radical environmentalist position is in no way anti-human. We grasp the pain of the human condition in its full complexity, and add the awareness of how desperately endangered certain key species and habitats have become."[44] The needed changes are significant and simple tinkering will not be enough. If we are in denial or simply wish to distance ourselves strategically from that sense of our uncertain moment, then it is natural to paint Livingston as a misanthrope rather than a thinker with vital insights about our web-like binds. Such an approach symbolically highlights a denial that keeps us far from the healing that is needed to find our way out of the modern ship.

When I read Livingston's body of work and consider the influence he had on friends, students, colleagues, and distant admirers like me, I have to conclude that at his best moments he did not believe some of his more extreme statements. We have all had loss in our lives, and most of us may be able to remember rash words said in the searing pain of an open wound that we would later wish to take back. At the core of his prophetic assessments was the multi-scalar trauma of the Ashbridges Bay severance that was continually reopened over a lifetime in what was the most ecologically destructive century since the dawn of humanity. To be consciously tethered to other cosmological scales of the creation was then and now to be in the perpetual pain symbolized by the roses of this labyrinth. Such melancholic awareness rather than denial reflects, Sandilands writes, "a psychic state of being that holds the possibility for memory's transformation into ethical and political environmental reflection."[45] The prophetic rhetoric of Livingston and other environmentalists is about how to "change the future rather than to predict it."[46] Maintaining mindfulness today is an act of staying with our losses in defiance of so many imperatives to forget and progress to the next greener market possibility.

From within these wounded roses there arise a starker sense of our situation and the kind of response that is needed. Rather than misanthropic prophet, Livingston and other ancestors of this eco-lineage can be seen as people whose pain led them to have a stark view on human behaviour and a desire for us to renew something like a bird-inspired flight of mind. But being in perpetual melancholy can take its toll over time, for loss and pain that is not condoled can eventually leave anyone, even prophets, estranged from a peaceful state. This was true of Skanyiatar:io before his alcohol addiction brought him to the brink of death and a Good Mind vision of renewal for the Haudenosaunee. We are now likewise facing a moment of choice. Do we follow the queer path of Woodworth toward acknowledging the pain-evoking denials of the modern ship, or will we wait until we are forced into this painful awareness? The difficulty with the latter option is that there will be less space for Sapling's healing orenta to manifest itself and help support less turbulent and thus painful changes. Recognizing this moment of decision amidst these Budding Flowers, I am drawn back to another significant moment in the Peacemaker's story of Condolence.

Before meeting the grief-stricken Ayenwatha, the Peacemaker came across a cannibal who was "carrying a human body over his shoulder" on the way back to his lodge.[47] Climbing onto the lodge's roof and looking down through a smoke hole, the Peacemaker observed the cannibal dissect, prepare, and cook the body in a large cauldron. While testing his meal, the cannibal "saw a face looking back at him" from the boiling water. Three times he looked, and the last time he noticed a radiance reflecting from the face. The sight left him so awe-struck that, Thomas tells us, he "sat down and thought how beautiful the face was" and began wondering "why would he eat such flesh" that could birth a being like that.[48] Recognizing he was moving toward true remorse, the Peacemaker climbed down and explained that Flint had deceived his mind, and that in understanding this he could now be more vigilant in his actions.[49] For Longboat, there is an important lesson in the way he jolted "the man into a moment of clarity, and then reasoned with him."[50] When we meet someone whose behaviour has become irrational, we must determine whether the behaviour is permanent or only due to particular circumstances. Is the estranging sickness of the modern ship treatable or not? Is there a capacity for us to be jarred, as Livingston was, into a recognition of the wound and healing that is needed? Can we return home to Notre Dame de Turtle Island? These are open, yet vital, questions for us today.

Contemplating the heart-felt roses before me, I feel pulled downward into the dark ravine, far below the Howard tomb and the driven ways to the south. This descent leads to a marshy pond of ducks, swans, herons, and red-winged blackbirds. Kneeling down to cup some water over my head, I can almost imagine the young Livingston rapt in wonder at such a place, as well as his stomach-wrenching pain of having it taken away. In a sense, his life's work attempted to heal a progressive operation that had cut into the core of his being and left a technological prosthesis, one that he saw proliferating everywhere. Illnesses and disorders that inspire calls for renewal are an old story not unlike the role alcoholism played in Skanyiatar:io's awakening to a Good Mind. There is in Indigenous traditions, Catholic mysticism, and many cultures an appreciation of this dynamic.[51] A wounded call to mind relations differently is another way for us to contemplate Livingston's childhood pain, Woodworth's queer losses, and the colonial severances that have initiated the climatic longue sault. It is to this strand of a healing braid that we now turn along these waters near sacred Lake Ontar:io.

In front of this pond within Notre Dame de Turtle Island, I am simultaneously drawn into an ancestral French Catholic approach to the passage that led me here. Near the Chartres' south portal one can go below the labyrinth to enter the crypt where a subterranean well of healing known as the Well of the Saints-Forts is found. The way to these waters takes one by the carved communion of saints who offer various healings, including the mercy of Saint Lawrence and the praise of Saint Dionysius. In this Kanathian labyrinth, we can also imagine Saint Tekakwitha as someone who, less than two years after her death, had a regional healing cult dedicated to "the Good Catherine," as she became known in the rural French communities of La Prairie and Lachine near Kahnawake,[52] places from which the coureurs de bois began their journeys into the pays d'en haut. Healings came to be attributed to Tekakwitha's intercession, giving rise to a local healing cult that three centuries later led the Church to canonize her as the Protectress of Kanatha – an ancestor who carried the pain and healing of colonial conversions within and without. From her Île-aux-Hérons to a sacred well to a marshy pond, these merciful waters can be engaged as "thin" places for attending our wounds and melancholy. Here we can ceremonially contemplate what is required to renew an ancestral healing energy that is beyond the capacity of simply people to enact.

At the time of Saint Tekakwitha's death, people on both sides of the Two Row took diverse approaches to healing that often crossed the

Kaniatarowanenneh. There was a tendency to, Greer writes, "invest the Other with valuable healing qualities" such that Indigenous peoples often "sought out the medicines and arcane religious rituals of the French, while many French turned to Indians."[53] It was in this spirit that Father Cholenec wrote of how Tekakwitha had attracted the prayers of poor Canadien habitants. An example was a crippled boy who was brought to her grave and dedicated to her care while a local Haudenosaunee woman was hired to recite a novena that consists of a nine-day cycle of prayers. Not long after this ritual, the boy was reportedly walking. It is important to note that Tekakwitha's healing cult did not arise among Haudenosaunee converts. As Greer explains, they tended to show "much less interest in saintly curing than did their French-Canadien neighbours" because there was no tradition of seeking cures from ancestors in their culture.[54] Ancestors are, we have seen, important and death itself was in Tekakwitha's time loudly mourned in ceremonies. But in contrast to French Catholic saints, the souls of the dead were seen as remaining close to the place of death for only a few days before moving to "the Country of the Ancestors."[55] It was, Greer adds, unthinkable to ask for support from a soul who had moved on and now had their name and communal role requickened to another living person as part of Condolence. And yet, in between embrace of and resistance to Saint Tekakwitha, some Haudenosaunee have found middle ground by recognizing the resonance between her healing practices and "our ceremonies, our herbs, our rituals, our traditions."[56] These two approaches to our numinous surround do not need to discount, prioritize, or convert differences; in fact, they cannot do so if they are to nurture a healing braid.

Though the understandings from the Kaniatarowanenneh's two sides may differ, there are points of resonance that I hear when someone like Woodworth talks of the continuing influence of Thomas and other ancestors like the Peacemaker and Skanyiatar:io in his thought. Ancestors may not be directly called upon by Indigenous peoples, but we have seen that the Haudenosaunee and Wendat do recognize the cosmological being of Sapling and Flint within our minds. To engage the Good Mind is to requicken oneself, it is to participate with vital ancestors like the Peacemaker in raising the onerahtase'ko:wa here in Toron:to, Kanatha, and elsewhere. It is not a far leap from this wisdom to the French Catholic veneration of saints who participated in other kinds of skyward reaches, ones that through prayer, novenas, and other ceremonies could heal and support life here. A "reciprocative rejoinder" with significant ancestors puts us into the healing gravity of their Good Mind. Such a participatory awareness is exactly what the modern ship wants to forget, as epitomized in those social forces that surrounded the life and thought of Livingston. And yet his rose-centred pain that brought us down to these healing waters can be seen all around us today not simply as an external climatic longue sault but as a kind of dark modern sickness that has globalizing dimensions. Attempting to heal this disorder seems preferable to the prophetic alternative that Livingston's melancholic pain sometimes led him to envision – that is, of an eternally damned people and human species in the midst of a totalizing flood.

The healing braid offered by Saint Tekakwitha needs to be seen as grounded in ancestral Haudenosaunee and Algonquin traditions that linked people marked in unique ways with a special spiritual role. In her Algonquin tradition, a shaman or shamaness was often called by spirit through an illness that limited their social interaction and took them on an inward trajectory that also, paradoxically, was the point of entry into the cosmological relations that surround human life. There is an urgency to this kind of sickness, as reflected in this description: "I became a shaman only to escape illness."[57] She could have likewise been spiritually responding to the colonial illness that afflicted her short life, or perhaps she inherited from birth a recognized predisposition toward this path. As Greer writes, there is "no way of knowing whether Iroquois people might have seen her as a *dehninotaton*, a dangerous/vulnerable child prodigy, signalled at birth by the ominous presence of fetal membrance over the head and valued for her purity after reaching puberty."[58] Though she may have inherited a spiritual inclination, it also seems that something significant occurred in Tekakwitha's last two years such that she came to stand out among many converts in the mind of her Jesuit guide.[59] Even her death seemed to affirm this change, with the hagiography telling of how within an hour of her last breath her face became luminous and "appeared more beautiful than it had been while living."[60] These words return me to the radiance of the Peacemaker's face that looked back at the cannibal from his deathly cauldron, a story that Tekakwitha surely intertwined with her Catholic meditations on the Île-aux-Hérons. The context of her emergence as a healing presence is a cannibalizing colonial illness that was in her being, a painful Budding Flower that afflicted Woodworth, Livingston, and many others to varying degrees and in varying ways.

Not far from the sacred well is a replica of the Black Madonna – known as Notre-Dame-Sous-Terre or Lady of the Underground – that brings me back to a dark contemplation of a Catholic strand in the Protectress of Kanatha's weave. She holds in her hand a heart that in the statue on the cathedral's main floor is emphasized by a surrounding arch of sacred hearts. These are symbols of the loving energetic potential from which saintly ancestors derive their healing power, the Blue Virgin cycles forth viriditas, the Peacemaker manifests peace, and the Bluebird calls forth spring as Flint recedes. These ravine passages are, perhaps surprisingly, the kind of place we must go to experience Saint Dionysius's via negativa as not simply a ritual practice but a transforming heart-felt presence who can be ceremonially engaged by entering the Black Madonna's darkness. As we saw in chapters 5 and 6, this saint's approach to the Blue Virgin's light recognizes that "we cannot know God in his nature, since this is unknowable and is beyond the reach of mind or of reason," though semblances of the divine are somewhat knowable from how everything is arranged.[61] To mind the divine mystery requires staying with the heart's light-giving orenta while recognizing the surrounding darkness. "With a clear eye," Dionysius writes, we look "upon the basic unity of those realities underlying the sacred rites ... which he had undertaken out of love for humanity."[62] Bringing us face to face with the Black Madonna, this ancient tradition teaches that, rather than wallow in the ship's rational denial of the pain, death, and melancholy surrounding us, the bird-like angels and ancestral saints can help lift our minds to a heartfelt sense of the world we live within.

The loving via negativa of Dionysius is a set of practices for mindfully initiating one into a kind of "charismatic healing" with cosmological presences that always transcend our thought, beliefs, and towering idols. As former Jesuit priest and Kanathian eco-poet Tim Lilburn explains, his words are meant "to draw the imagination of initiates into the beneficence and compassion of the artefact of divine eros" so we can be more receptive to "cleansing enlightenments."[63] The practice entails using affirmations of everything in our worldly relations followed by negations that are meant to quiet the rational mind and thus open space for good imaginations of the energetic source that underlies all forms. In other words, all our interdisciplinary knowledge must be affirmed as a reflection of what can be known of Notre Dame de Turtle Island, but then it must be negated because the surrounding mystery is before human thought and constructions. The words of Dionysius are a ceremonial theurgy that guides one into practices for transforming relations through knowledge that is lovingly aware of its limits. This practice creates, in Lilburn's words, an "apt strangeness" to our immediate knowledge of reality by imbuing it with mystery.[64] Though I began using Indigenous terms like Toron:to, Ontar:io, Kanatha, and Kaniatarowanenneh out of respect for this land's ancestors, I now understand that doing so is a via negativa practice for fostering an "apt strangeness." It is a ceremony that helps to make spaces thin by bringing our familiar knowledge into felt relation with ancestral and cosmological presences that are so beyond it.

Nightfall around this marsh and ravines often evokes in me numinous feelings. It does so in a kind of dialogue with Livingston's prophetic anger and the climate messaging I come into contact with, such as when the morning radio announces the latest cycle of dark news on another record-breaking hurricane, flood, and so on. The headlines then reverberate through my head with the dread of a modern mind screaming: "Ohhhh! Do I have to listen to this first thing?" Later in the evening, another environmentalist on the television describes our situation as akin to a culture that is speeding toward a brick wall that is simultaneously speeding toward us. Being concerned about the rapid climatic changes we are experiencing leads me to listen, but I cannot shake the feeling that there is something in such prophetic words that make me want to follow much of the population in tuning it out. It is partly because I am a father of two young children, and such numinous visions are hard to accept. But perhaps it is also the Christian heritage of this prophetic tone that has grown tired. It was just such a preachy predisposition that guided the ship

toward its colonial conversions and continued to limit its engagement of Kanatha's Indigenous Good Mind, seeing all the suffering as
evidence of its own universalizing story. This is a significant part of
what needs unknowing, of what needs to be made strange to us
before we can attend the pain in our midst and potentially move
toward healing.

We need to become that stranger the Howard tomb welcomes so as
to live better within the mystery of Notre Dame de Turtle Island, to
put Flint's freeze-framed idols back into the flow of orenta. One way
to do this is through ceremony, and what Lilburn found was that the
via negativa was one such ritual practice from the modern ship's
ancestry that can return us to our home. When he moved to
Moosewood Sandhills, Saskatchewan, some worries arose. The first
six months and winter were marked by the modest presence around
him of "a little grass, blow-outs where there was bare sand below and
overhanging thatch of jumper roots – there wasn't enough for the
eye: I thought I'd starve."[65] His words remind me of my unwilling-
ness for a long time to consider moving to Toron:to, a place seen
from a distance as asphalt, concrete, and not much else. For Lilburn,
there was a need to find a way of relating to his new home, and
Dionysius was a helpful ancestor. As he writes, "the contemplative
focus described with such phenomenological rigour in the texts from
antiquity and the Middle Ages could be turned to a place … without
any change in its form or diminishment of its vitality."[66] A mystic via
negativa can help make us strange enough to the places in which we
live so as to be able to relate with them again. This approach has
potential connections with Livingston's bird-inspired flight of mind
that led him to state that the nature of mystic experiences that "cannot be force-fitted into convenient categories" is something "the naturalist knows – profoundly."[67] Others of this environmental tradition
have explicitly recognized the need for bringing interdisciplinary ecological knowledge into a kind of unthinking, which suggests "that
'thought' and 'beyond thought' are not as opposed as we might
think."[68] Thought is brought into the rose-centre so that it is made
strange enough to recognize the wounds without and within, thus
ceremonially connecting us to an orenta for reimagining our relations, starting in those most intimate places where we live.

Though it may be that such mystic traditions can be brought to the lands of Kanatha as they are, these pages suggest that the process of opening up to a place's Indigenous ancestry will come to colour an emerging braid in untold ways. For the practice of seeing a Moosewood Sandhills, Toron:to, Long Sault, or Kanatha as places far from the sacred has deep colonial roots, as we have uncovered here. In Lilburn's words, the descendants of the colonial and modern ship have barely begun the task of "learning to be spoken by the grass and cupped hills."[69] What we need to learn is not so much ecology, geography, an environmental ethic, history, or a new sustainable economics, "but a style that is so much ear, so attentive, it cannot step away from its listening and give a report of itself." It is in such attentive spaces that the land can bring forth to us stories like those of a fallen dolon-do in Toron:to, the Peacemaker's birth along this lake, the Protectress of Kanatha on her Île-aux-Hérons, Livingston and Woodworth at Ashbridges Bay, and Notre Dame de Turtle Island as the sacred context of it all. Without this capacity "to take in the genius of the place, let it say its piece through you," Lilburn writes that "the place will throw you out."[70] This is what the increasingly turbulent climatic longue sault is doing, throwing us out of our ship to see if we can come back to a good minding of what it is to be here with each other. Ceremony, it seems, is central to this renewal.

Though Dionysius offers us a via negativa for wisely engaging our dark pain and melancholy before we have to, the Black Madonna also symbolizes a numinous potentiality that can engulf us in an experience of what needs to be unknown, cleansed, and healed. Even for Catholic mystics who often followed Dionysius's guidance, there was commonly an experience of coming to a point where they felt helpless, arid of any feeling, alone, and utterly out of control. It has been called the "dark night of the soul" and described as a spiritual "state of disharmony; of imperfect adaptation to the environment."[71] The unique unfolding of this painful experience is well documented in the life of Saint Tekakwitha's Catholic namesake. Before the mystic marriage that was the pivotal act in Catherine of Siena's transformation from "an illiterate daughter of the people" to a leader who tended those afflicted by the plague and reformed the Church, she underwent a series of trials that tempered her faith.[72] As we saw in chapter 4, her thin space was the seclusion of a little room in her home. Within that island, Catherine was "afflicted by dark visions of sin and a felt absence of anything divine, all of which haunted her until that moment when she made this affirmation: 'I have chosen suffering for my consolation.'"[73] Catherine then heard a divine voice explain that "to raise the soul from imperfection ... I withdraw Myself from her sentiment," and it is this that cuts to the root of spiritual self-love and self-interest, thus ensuring that one's preceding limited sense of being does not "turn again in its death struggle." The sense of abandonment, helplessness, and loss of control is due to a limited awareness that falters in the face of the ceremonial participation with such prodigious orenta.[74]

Traversing the dark night of the rose cross requires a degree of self-surrender that is the essential mystic act in the Catholic tradition, which is not, as is often assumed today, simply a permanent wild seclusion from Turtle Island.[75] The healing pain brought Saint Catherine into a greater engagement with the surrounding social, political, and religious disorders of her time, all before her death in her early thirties. Through the via negativa practices that brought her into relation with the Black Madonna's dark night, she saw these corruptions and offered alternatives that were closer to her sense of how spirit worked. Exemplary is a passage in *The Dialogue* that recites what she is told about the difficulty of taking on a religious mission while not being "humbly clothed" in the will of divine relations. People in this position, she is taught, "may often sin against their very perfection by setting themselves up as judges over those whose way is not the same as theirs. Do you know why this happens? Because they have invested more effort and desire in mortifying their bodies than in slaying their selfish wills."[76] Is this not a way of contemplating the mistakes of the succeeding colonial missions to Kanatha, starting with the Black Robes and progressing to residential schools? What seems to have been at work was a desire to engage these wild lands as something in need of conversion to a particular view of the human place in creation, rather a spirit of humbly engaging the experience of new relations as a via negativa ceremony for learning what is needed to live well with others on our common Kaniatarowanenneh.

The dark night makes that which we think we know about the world aptly strange, opening up a thin space to move beyond our preconceived missions. A few centuries after Saint Catherine's life, and across the ocean, the Black Madonna in many ways encompassed Tekakwitha's short life through disease, family death, dislocation, and then a spiritual calling to her isle. The stories of Tekakwitha and her Catholic namesake Catherine commonly tell of people who actively engaged spirit in the world, which is also the central purpose of Dionysius's via negativa and Thomas's Good Mind. This is what significantly differentiates our situation of being passively pulled toward change from the dark nights of the Catholic mystics. These ancestors attempted to attend to an encompassing spirit that renewed their sense of who they were. But ultimately what the dark night of the soul highlights is that effort alone can take one only so far, that a spirited orenta also enacts a kind of painful purge from without which can feel dark and devoid of love. For the Catholic mystic, a bright light comes to shine on every dark corner of their being as they recognize their helplessness and dependence on something more than themselves. They are brought to a state of surrender where before "a new form is to come into existence, the old must of necessity be destroyed."[77] It is the rose-centred cross that Jesus Christ dies on before being put in the dark tomb and then resurrecting. Another resonance arises here with the healing of Condolence that wipes away tears before requickening or renewing the person to a new communal role. Surrender to something larger than ourselves is common to these braided Kanathian passages.

From the firewater illness of the royaner Skanyiatar:io to the colonial disease that afflicted Tekakwitha, these dark nights are initiations into a renewal of ancient ways. It is the rose-centred pain that makes the world aptly strange enough to open up a thin space for imagining something else. But more is reflected in these stories. What makes these dark nights transformative is the wisdom, humility, and love that arises from a ceremonial practice for finding a way through, not around, this pain. Our present situation is quite different, for the numinous challenge is confronting us even as we continue to deny the wound and melancholy. We think that more rational solutions will allow us to manage our way out of the labyrinth. Not surrendering is what makes those on the modern ship feel comforted and in control, and thus the surrounding darkness grows in proportional response to that wound denied deep within.

The passage toward healing for people in denial of their mysterious context is difficult to imagine or predict, though the previous chapter's flood of stories displays how past social binds, captured in mythology, hold lessons for the present. Whether it is Condolence or a contemplative labyrinth, stories told in the context of ceremony are never simply about ancestors from the past. These presences are to be engaged in a spirit of dialogue with honoured elders, and this is quite difficult for a modern ship that still fundamentally believes that spiritual ways of minding relations are of the past – that we have progressed beyond ancestors, angelic theophany, mythic cosmologies, and ceremony. Yet, from within our state of denial, the climatic longue sault seems to be shaking these beliefs to their core, and thus opening space for attending multi-scalar presences who can help us imagine alternative passages. Our dark night is a numinous force breaking in upon us exactly because we have forgotten old teachings of how to ceremonially engage the Black Madonna, and these now need some kind of renewal.

As I look one last time into the marshy water, the reflections of Livingston and Saint Tekakwitha recede as my imagination draws forth the shock of the cannibal seeing a radiating face in the broth of his cauldron. The Peacemaker is in many ways looking at us right now, seeing if we can be jolted out of our consumptive addictions and denial of the severing pain that is being forced on so many. A cosmological determination is being made, for, as Longboat relates, if the cannibal is "permanently irrational, then one must prepare to effectively deal with him, because eventually that person will disrupt the peace."[78] Alternatively, if the mysterious jolt leads to actions which suggest that real change is possible, then one is obliged to find ways of bringing "that person back to a Good Mind." Mythic events like a fall, deluge, epic explosion, or Peacemaker reflection ultimately have much to teach us about how we should live and about the changes we need to make. The dark night is within and without, thus the reason for aptly strange ceremonies that can evoke heartfelt healing and renewal. More vital than prophesizing, a good climate of mind must proceed as if the estranged ship can be brought back to a reverence for our home on Notre Dame de Turtle Island. This leads us to contemplate one last strand of this braid that is concerned with reconciling us to the orenta underlying all life.

———

Leaving these waters that are the source of this chapter's rose centre, we reverse the labyrinth passages and bypass many now familiar presences: the tomb's call to "Oh, Stranger," Lake Ontar:io, a fallen onerahtase'ko:wa, and ravine paths that lead up to the staghorn-sumac grove where we began weaving this braid in Toron:to. Our late-summer return to this spot has a different feel than what we encounter here during winter's barrenness, as deep red pods reach antler-like toward the sky amidst a green canopy. Their calming presence draws me back to the Tahontaenrhat, people of the Deer, and legends of them as vital healers. They survived the seventeenth-century conversions and requickening somewhat intact, and their village in Seneca country "achieved renown throughout Iroquoia for their effective curing rituals."[79] While the Wendat, like the Haudenosaunee and Canadiens, had many natural remedies known by all, Sioui explains that when it came to ailments of a more supernatural origin, then specialists were called in. These healers were chosen and apprenticed as children to become ones with *arendiwane*, which derives from *arenda* (orenta) and *wane* or "great."[80] They guided people through ceremonies "disguised as hunchbacks and who carried sticks and wore wooden masks," thus suggesting a link with the False Face (Hadoui) Society of the Haudenosaunee.[81] It was these masks that Thomas carved and referred to as Hadouis, with one hanging on his bedroom wall across from "a small portrait of Jesus, holding a lantern knocking at a door."[82] The healing ceremonies associated with these "aptly strange" masks offer us insights on responding to turbulent change.

While there are physical ailments and disease, in the Wendat Tahontaenrhat tradition suffering and pain often have a deeper source that calls forth the ceremony of these masked healers. The disorder was thought to arise from a "secret or unconscious desire of the soul that the Wendats called 'ondinonc,' 'a desire inspired by the spirit.'"[83] To cure someone of an illness that has more dimensionality than what can be physically observed, the masked society fasts, sings, and engages in purifying rituals aimed at understanding the spiritual source of a disorder that is in need of healing.[84] It is about physical health, though it extends to personal happiness, fulfillment, and fostering communal relations.[85] The Hadoui Society can conduct individual healing at any time, though Woodworth was taught by Thomas that there are also communal healing ceremonies in spring, when the plant medicines are coming back to life, and fall, when these plants are gathered.[86] These are times on the edge of Flint's freezing power, one leaving and the other entering. With "terrifying cries, the sounds of rattles, and singing," the seasonal ceremonies of the Hadoui Society attempt to frighten disease from the community by acting as conduits for Sapling's orenta.[87] Such masked strangers aim to get at the dark spiritual source of disorder that manifests itself in a variety of ways.

The diagnosis of a soul illness pulls my mind back to the Howard tomb's Christian intimations of severance as grounded in a sinful "Fall" from grace, one that seemingly fuels an unquenchable desire to convert the Kanathian wild. As was briefly touched upon in chapter 5, the popular story is that Eve's desire for knowledge occurred in the natural world. This act led to "the Fall" that introduced death not as that which regenerates life, as in the cycles of a Good Mind, but as a total end.[88] With this earthly sin, humanity and Turtle Island fell into a degraded state that could be saved only through conversions. In the words of Livingston, there is "no conceivable reason for the existence of the blue planet apart from the needs of God and man," needs that are oriented to somewhere else than here.[89] This is in contrast to the Sky Woman's fall, which originates in a light-dimming disorder, not guilty sin, in the Sky World.[90] Such mythic "falls" are, Thompson explains, common across the planet, and tend to be about "the conditioning of time-space out of which all events arise," the fall of an ancestral soul into time and the creation of the many.[91] There are Christian traditions related to Chartres' Neoplatonic sensibility that similarly describe the Fall in terms of a soul rather than a bodily event. From this view, locating the fall on Turtle Island is a materialistic interpretation of the biblical story that inevitably leads one to see nature and our flesh as enemies of spirit that need to be managed and converted.[92] Rather than graciously recognizing the gift of orenta in creation, the modern ship has inherited a deep cosmological severance that, following the Wendat sense of ondinonc, is calling forth the climatic longue sault's shapes of violence.

The story of the Hadoui mask's origin as a healing power directed at just such a soul sickness takes us back a long time ago when a stranger, Woodworth writes, "challenged the Creator to move a mountain."[93] In a deep ravine-like valley cradled by a large mountain, Sapling met this stranger, who was unaware of the being in his presence and boasted that he could move the mountain. Other versions talk not of the Creator, Sapling, but of a great medicine man who was "granted supreme powers by the Great Good Spirit because of his true love of people, birds and animals"[94] – he embodied the spirit of Sapling. With their backs to the towering rocks, "the stranger tried twice to magically move the mountain but succeeded only in causing it to tremble."[95] Recognizing the effort, Sapling said: "Your medicine is powerful" but what you lack is "complete faith, not skill." Then he brought the mountain forward with a power that shook Turtle Island, leading the stranger to spin round quickly and smash his face against the mountainside. With the now crooked-nose stranger looking with reverence, Woodworth writes that Sapling healed him and compassionately taught him the way of complete faith.[96] The stranger's face came to adorn those masks that also have various names: False Face, Old Broken Nose, Great Hunchback, and Thomas's Hadoui or the Hunchback's "real face."[97] In ceremonially becoming the stranger, shaking his Turtle rattle, singing his songs, and performing his dances, Sapling's healing orenta can come into the creation to heal.

Great power is needed to heal a soul illness like the climatic longue sault, and the Hadoui Society symbolically represents and enacts this in various ways. Its masks are carved "from a living tree, so that the power inherent in the tree becomes part of the mask."[98] While the mask is vital to the ceremony, the individuals wearing them also need to act, dance, and sing forth Sapling's healing orenta into the presence of others.[99] Everyone who participates in the ritual must also dance, for it is through this that the masks are further "infused with the supernatural power to heal."[100] Scholars in the evolution of religion now propose that ceremonies like these, Condolence, and Catholic traditions initially emerged as an embodied "musilanguage" that allowed the bonding of "larger groups than was possible using the conventional primate mechanism of physical grooming."[101] Such practices became more complex as people further ritualized their relations with a changing sense of communal, ecological, and cosmological presences, a dynamic resonant with the Haudenosaunee stories of successive renewals of Good Mind rituals. Powerful emotions can arise during ritualistic communication with deities and ancestors, thus fostering communal bonds, enabling the survival of communities in the face of destructive changes like colonial conversions, and fostering the global appeal of religious experiences to this day.[102] Of this nature, masked rituals from around the world are "dedicated to the attainment of ecstasy at the moment when it contains within it the god or spirit."[103] Paradoxically, they enact a "motionless ecstasy." This seems to be symbolized in the Hadoui mask's carving from a living tree, communal dancing, and a turtle rattle that not only moves the mountain of a shaking Turtle Island but ritually calls us to maintain awareness of the ecstatic shaking brought by a spirited approach to healing.

The Hadoui Society enacts healings that go far beyond the often frozen categories of a modern mind. This freeze-framing tendency seems to play out in the academic approach to the masks themselves. The foundational study of these masks by William Fenton classified the forms of the carved faces[104] and was an important and respected analysis that even Thomas referenced as a carver.[105] This is part of Thomas's Good Mind approach that was always inquiring about research on "the past that could corroborate his knowledge of the oral tradition."[106] That said, some have critiqued Fenton's approach in relation to the modern "need to classify – which was considered unimportant or even irrelevant to the function of the Faces in Iroquoian society."[107] The storied cosmology of Sapling's meeting with a stranger becomes secondary to the needs of an objective study that empties the forms of meaning. It thus becomes possible to fill these masks with different meanings, a colonizing tendency referred to as decontextualization.[108] So, while there may be a common evolutionary basis for musilanguage rituals and masked ceremonies across the human species, such categorical understandings are secondary to the more vital lived and storied context. This takes on other dimensions in the collection of Hadoui masks for museums, which many Haudenosaunee traditionalists believe become trapped, "restless and agitated, and need to be 'fed' or renewed with tobacco and sunflower oil."[109] If we return to Latour, it is as if a static face without gets freeze-framed to its core, emptying it of the orenta which, through ceremonial performance, offers people the basis for healing and renewal. The modern stranger's soul illness is maintained as the humbling sense of spirit is severed from our experience.

The academic urge to classify and purify the orenta of Hadoui masks has another colonial connection to the earlier-discussed Black Robe history and what some see as the emergence of this Hadoui Society in the midst of the seventeenth-century disease carried by Saint Tekakwitha. It has been suggested that the Hadoui Society arose at this time as communal responses to the extensive illness brought over on the colonial ships.[110] This can be seen in the origin story where the stranger's collision with a mountain is used as a symbol "powerful enough to represent the profound devastation of European disease."[111] It may not have been a totally new development as much as a renewing of established practices in light of extensive changes. The legendary Tahontaenrhat healers were part of a Wendat culture where shamans and curing societies played a big role.[112] In fact, these shamans were in conflict with the Black Robes, who were concerned about their power, especially when the rituals worked, since success suggested "the assistance of supernatural forces that, because they did not derive from God, must be diabolical."[113] This attitude is one more sign of that soul sickness at the core of the ship. Perhaps what made the Tahontaenrhat curing society so renowned was its longer experience with the colonial illness; the Hadoui Society became more established at this time just as Skanyiatar:io and Saint Tekakwitha brought distinct spirits of renewal through their colonial relations. Anytime the dark pain of Flint increases, a renewal is called for. This seems to be a budding-rose feature of the Good Mind healings that stretch back to the Peacemaker and Sapling.

From masked healers to Condolence ceremonies to a mystic via negativa to mythic presences, these are not freeze-framed images or beliefs but dynamic ways of engaging the numinous forces underlying our lives. Such practices recognize the presence of something much stronger than people. They allow us to move toward a "thin" space where there is greater potential for surrendering to an orenta that, as with Thomas's picture of Jesus holding a lantern, can light our way out of Flint's darkness. The paradox of valuing ancestors and elders is that outwardly it seems like a conservative act, but it is fundamentally grounded in attending to a dynamically cycling and shaking Turtle Island. Donning the Hadoui mask or a via negativa requickens us to the ancestral pain of having our nose misaligned. It is an act that makes us strange to our deep-set beliefs and assumptions, like those concerning a globalizing individual self-interest. Despite its historic limitations during the early colonial era, Longboat explains that such requickening has the intent of placing a newly energized person in the place of one who has recently passed. Ancestral relations are affirmed so that family and community can, in the words of Condolence, *enhowa'nikonhketsko* or "raise up the minds" and enable "the spirits of the people to soar once again, to rise up from sorrow and grief to the high purpose of the Great Law, the Great Peace."[114] Because Notre Dame de Turtle Island is ever-renewing, we engage ancestors whose lives can tell us something about what it means to foster our relation with numinous presences that bring our pain into greater awareness and provide signs of how to heal. This is the ancestral healing Thomas has in mind when he asks Woodworth to help lift the fallen dolon-do by renewing Thanksgiving on these Toron:to shores.

From my French side of the Kaniatarowanenneh, taking on a saint's name and giving forth devotion can be engaged as another way of bringing the ancestral wisdom of Saints Tekakwitha, Catherine, Lawrence, Francis, or Dionysius into renewed dialogue with the present. It is interesting to note that Dionysius was rejected during the Enlightenment mainly because new scientific procedures revealed that he was not a New Testament figure who had dialogues with Saint Paul, but rather a fifth-century Eastern Christian monk who came to be charged with misrepresentation for taking on this pseudonym. What was being forgotten at this time of the dawning modern mind was the ancient practice of cloaking one's identity within a "trusted ancient name" based upon the conscious recognition that one's culture was borrowed from the past.[115] This mask was becoming a rigid category to be studied, not a dynamic ritual for opening us out to an ever-changing cosmology. If I return to such a view, then perhaps I can link John Livingston with the prophetic revelations of Saint John, the celestial theophany of Jon Scotus Eriugena, and the dark web of John Mohawk, ancestors connected in first name and varied mindings. Such relations are vital in times of crisis not only because, in Harrison's words, "the dead possess a nocturnal vision ... to certain insights" that life in our materialistic age blinds us to, but also because in moments "of extreme need one must turn to those who can see through the gloom."[116] Adding a relevant insight is Sandilands, who describes queer melancholy as a kind of internal mourning that "may be a *prerequisite* for letting the object go."[117] Those "objects" that need to be surrendered by the modern ship are the beliefs, factishes, and practices that are bound to the intensifying climatic longue sault. Thanks to the ancestors of our Two Rows, we have many aptly strange teachings, stories, and ceremonies that can be donned to help us let go and renew our way home to reverence.

Ancestors show our connection to a stranger reality than we could ever rationally model. This is true for both those who inspire and those of a darker, perhaps mixed, character whose masks can fearfully jolt our hearts. As I mentioned in chapter 2 when outlining the personal ancestries the Good Mind was leading me to contemplate, there is a need to remember the positive potential of a coureur de bois braid as well as the twisted rotten branches. The stories of Flint, Hadoui, Gilgamesh, and the Black Robes remind us of the way to disorder. A feature of "less dangerous stories" is that they make "heroes of characters who cooperate" while "giving antagonists names, faces, and purposes that cannot be immediately dismissed."[118] On today's national and global stage, former prime minister Harper was one such antagonistic character because of the way he effectively moved the debate from the climatic longue sault to the economic necessity of building epic pipelines from Alberta's bitumous sands to wherever viable. Closer to home, the former Toron:to mayor Rob Ford began his 2010 term by declaring that "the war on the car is over" and in the process brought uncertainty to the city's mass-transit plans and climate response. Though very different leaders, both Harper and Ford approached politics with a "my way or the high-way" attitude that aggravates divisions and conflict. In our time of change, a large proportion of the population, though not a majority, elected such leaders who tend to act unilaterally and are primarily concerned with driving the Kanathian economy via an outdated rolling-head disorder. They highlight the addictive bind of the modern ship's painful denial even as the turbulence rises. On a mythic scale, our current political climate may in the future be turned into stories that can teach about signs of the seemingly strong, yet brittle, grasp of a soul illness that wants to continue but can sense the approaching dark night of the modern ship and on some level knows its days are numbered.

The dominant environmental stories of our climatic longue sault are too securely situated in a rationalizing mind focused on ecological, climate, or energetic modernization; there is little that is "aptly strange" in them. As we manoeuvre through this labyrinth, questions arise as to whether environmental and academic critiques can ultimately inspire the loving effort needed to unknow and transform our web-like binds. Once again, we are in the company of Livingston, who asked, "Why are we, the unique possessors of THE WORD, so pathetically inept in adding to our number?"[119] The "we" are those who are ecologically minded and, despite our time of crisis, seem unable to move people toward significant change. It is something that Jane Bennett also considers when asking whether environmentalism "is the most persuasive rubric for challenging the American equation of prosperity with wanton consumption, or for inducing, more generally, the political will to create more sustainable political economies."[120] The tendency is toward scientific pronouncements and policy options that speak rationally when people need inspiration and imagination for attending the death, pain, and melancholy that confront us, the necessary first step in any move toward healing. A paralysis of guilt about not doing more in the midst of dark prophecies seems to pull us into Flint's dark mind. To get out of this bind, we need the numinous jolt of a Hadoui mask, the Black Madonna, or the Peacemaker to help distance ourselves from many pervasive modern idols and factishes.

In recounting the origin stories at the heart of the Hadoui and Condolence ceremonies, I do not mean to suggest that the modern ship needs healing at the hands of Indigenous people like the Haudenosaunee, Wendat, or Anishinaabe. As the Truth and Reconciliation Commission on residential-school abuse and the colonial conversions of these pages attest, there is enough pain and grief in Indigenous communities that needs healing – not to mention that this pain has arisen in relation to colonial and modern Kanatha.[121] In the words of Justice Murray Sinclair, the chair of the commission, the "loss of language and culture, the impacts on family function, the devastation to self-identity, the loss of respect for education, and the loss of faith and trust in Canada's government will take many years to overcome, and will only be achieved with a focus on a vision for a new relationship and a commitment to behavioural change and positive action."[122] We need to become clear-minded enough to recognize our continued participation in these severances and the need to participate in the healing of relations. This is the other side of the reconciliation process that is clearly understood in the Good Mind tradition but hardly recognized in popular-media portrayals of this process as an Indigenous issue. As Sinclair states, "commitment to change will also call upon Canadians to realize that reconciliation is not a new opportunity to convince aboriginal people to 'get over it' and become like 'everyone else.' That is, after all, what residential schools were all about and look how that went."[123] Perhaps reconciliation cannot be fulfilled until it renews the spirit of a braided Condolence Ceremony, one that the modern ship actively participates in, and is concerned with what the ship needs to attend to and let go of on the way to good relations and reverence.

Everything we bring into ceremonial dialogue with the land's Indigenous elders and ancestors has personal, familial, and cultural dimensions. As with Livingston's conclusion that "the fault was never out there," I have something in common even with leaders like Harper. I have the large crooked nose of the stranger who is trying to learn how to live here in good ways, and their politics is a caricature of what I, we, need to bring into reconciliation. But what gives these labyrinthine passages relevance beyond me or any other individual is their participation in the common waters of the Two Row. A good Kanathian climate of mind must be situated in each of our experiences and yet recognize the ways in which the personal and ancestral opens out to broader shared experiences of community, culture, nation, ecology, climate, and cosmology. Healing, condoling, and reconciling can never be done alone. Our leaders today dance in front of us on our ever-proliferating flat screens, displaying an epic mountain-moving desire that should guide us back to Sapling and some alternative questions: Do we have "absolute faith" in something better than what we currently are? Can our innovative tower-building skills be transformed toward a greater service? Does the climatic longue sault need to break our nose and more before we compassionately understand what conversionary severances bring to Turtle Island, including those on the modern ship? Can we lift the healing orenta of Toron:to and Kanatha higher than the darkness of our past and current shapes of violence?

As I walk the sandy savannah paths beyond the sumacs, the haunting songs and turtle rattles of the masked Tahontaenrhat slowly recede into an autumn season that sees the tall grass just beginning to yellow amidst colourful wildflowers and black oaks whose brilliant red, orange, and yellow displays are immanent. Moving out of the labyrinth's weave, I am struck with the feeling that these Kanathian passages are indeed *toronton*, abundant, another Wendat term sometimes associated with the origin name of Toron:to.[124] This abundance underlies not only the beautiful ecologies of savannah, marsh, and ravine that were once more expansive here, but also, as we have seen, the Kanathian resource web that is connected to the mixed Toron:to name, the CN Tower Prayer Stick, Medicine Wheel Skyscrapers, and the Kaniatarowanenneh Two Rows. The beacon observed by Woodworth asks whether we are satisfied with following Flint in simply taking and consuming Sapling's abundant viriditas, or whether we can find the heart to give thanks and thus participate in raising the orenta that continually gives birth to all this. In Thomas's teachings, this means coming back to the energetic source of our lives, a place that "is known as the Sky World to my indigenous ancestors, heaven to others, space to the scientific mind."[125] But strange as it sounds, this mysterious place is not to be found somewhere else beyond

Toron:to or wherever we live, for we are, Woodworth adds, "always on our way back 'home' to this place, and so it is that I now find myself conscious, awake to, the grand journey of returning."

When I started out on the passages of this book I had no idea that the Haudenosaunee and Wendat Good Mind would become so central, but each step opened in unexpected ways to a Budding Flower at the centre of my being as a Kanathian. The severance of the stranger has come to have a meaning very different from the Christian view of some totalizing Fall from a once sacred world. I now primarily mind it as a temporary or temporal ceremony for becoming strange enough to the world so as to again experience its reverential beauty. This severance calls us to recognize our pain out of compassion for the impacts we are having on others and ourselves; to feel the grief of loss, and desire for something better. If we cannot recover this, then perhaps we are permanently irrational and Livingston's prophetic anger is well founded. That said, he also wrote that, while he could not "know whether your or my qualitative experience of nature will ever benefit wildlife," his life suggested that "the joy of self-recovery and liberation from cultural tyranny is contagious."[126] A love for being here on Turtle Island can potentially draw the modern mind out of the ship and into an aptly strange, yet familiar, reality that is more vital and liveable. It is consistent with Woodworth who, after recognizing his pain, says, "I am left reverent with inspirations to perform spontaneous ceremonial speaking and doing to acknowledge the spirits and ancestors who accompany me constantly... I am filled with love at last."[127] This is what arises in me as well when I sit in the "thin" space of a Kaniatarowanenneh isle with the Protectress of Kanatha, hear the Peacemaker amidst onerahtase'ko:wa, or walk with Brûlé and Tahontaenrhat on the passage de taronto.

Ceremonies of Condolence healing would, Thomas told Woodworth, someday have to include all Kanathians, for only then will we be able to join together to take up our human duties of lifting the fallen onerahtase'ko:wa. A ritualistic return can lead to small changes with deep roots that have the potential to collapse the "distinction between the differing roles I fulfill, either academically or personally."[128] Beyond the need to consume, drive, and fly less, such devotional acts will ask us to remember the many presences that surround us when we turn the key of an automobile or flick the switch in our home, thus limiting how, how much, when, and what is consumed. What inherited ceremonies and stories can inspire us to slow down, to contract our economy while dealing with injustices, and shift our focus from ever-growing bank accounts, GDPs, and energy webs to questions about the quality of relations we want? What will it take to re-envision renewable energy and no-growth economics as not simply policy plans or broad-scale detoxification programs that inevitably manifest various resistances, but as duties arising from healing ceremonies? Before such possibilities can even be entertained by a modern ship still rationally focused on mountain-moving epics, we need to learn to grieve that which is being lost with the coming changes.

We have become too strange to the world; that is the central lesson of the climatic longue sault. To heal this spreading pain we need to become aptly strange to the ways of the modern ship so as to again return home. For Thomas, the issue is not whether we are Indigenous to this place or not, but rather whether are we "real people" who are willing to take up our duties. He does not call himself Native or Indigenous, but rather "Ongwehónwe, meaning 'a real person.'" To be a "real person" is, he counsels, to recognize our position in the creation of and dependence on all that surrounds us. It means to take up those duties that have been given us based on where we have come into the creation. To return to Mohawk, "a life that is real" is based in practising the Good Mind ceremonies that the ancestors have handed down to the present, and then acting on the arising insights. As a royaner, Thomas had the duty of maintaining open spaces where the community could remember ceremonially who they are and what they need to do in creation. In a spirit similar to Woodworth and Sioui in chapter 5, he teaches that we are all living on "the land of the Creator and that we are all natives of this Earth," that we all have an inheritance for being "real people in our own ways."[129] In the pages of this book, braiding stories of the Peacemaker, Skanyiatar:io, Saint Tekakwitha, Utnipishtam, Brûlé, Livingston, and others has revealed the joys and struggles of finding passage to being a "real person" rather than Flint's "rolling head." The alternative, it seems, is to become so estranged that the Kaniatarowanenneh is no longer hospitable for the ship, canoe, and many presences that have lived here longer than all of us – to be engulfed in a dark night not of our choosing and for which we are ill-prepared to navigate.

The gift of an emblazoned sumac grove, a fallen dolon-do, a numinous labyrinth, the sacred sparkling waters of Lake Ontar:io, and the Kaniatarowanenneh of my original home to the east – there is so much here to be grateful for, to feel Thanksgiving well up from within. It is not simply that there is no other place to go to reconcile ourselves with each other and the sacred; there are also no new stories or ceremonies needed to help renew our orientation on such numinous passages. What we have in our diverse and mixed ancestries simply needs renewing in relation to the common waters of our historic moment and place. Even as the turbulence of the climatic longue sault intensifies, we need to unknow, heal, and requicken our relation with the cosmologies of Kanatha wherever we happen to live. Modern science, economic policy, and technologies still need to be part of our response, though all this must be engaged with humbler methods which recognize that knowledge, beliefs, and factishes are always limited. Ceremony and mythic storytelling have a vital role to play in weaving us toward Saint Tekakwitha's Île-aux-Hérons where we can experience the reverence of a good Kanathian climate of mind.

Some will discount the labyrinthine passages that have led to these words as a mixture of interdisciplinarity run amok, romantic excess, or simple delusion, all of which misses the central lesson I came to learn in braiding the passages of this book. Because we co-participate with so many other ecological, climatic, cultural, ancestral, and cosmological presences in the creation of Notre Dame de Turtle Island, numinous times of change like ours require diverse ceremonies and stories for lifting our energy, the orenta, higher than the flood of darkness that can engulf us. While we each need to find our unique ancestral ways to a diversified Two Row, to share the common Kaniatarowanenneh in a good way means recognizing the central role of the land's Indigenous elders wherever we live. Then we may begin to recognize the climatic longue sault as a call from Sapling, the One, or the great mystery, a call to attend to our pain and renew our duty to offer Thanksgiving. With that we may still be able to lift the onerahtase'ko:wa above Flint's poisonous monsters, within and without. We are being asked to return home, to renew a life that, echoing Mohawk, "goes beyond the power of words, a life that is real."[130]

Look with Reverence

Oh! Stranger look with reverence.
Man, Man! Unstable Man,
It was thou who caused the severance.

Coming to the edge of the northeast savannah, a red-tailed hawk swoops down in pursuit of a black squirrel that tightly circles the trunk of an oak. Unable to imitate the manoeuvre, the hawk flies widely around the tree and soars up to a branch on a neighbouring oak. Meanwhile, the squirrel stops on the trunk and gives an excited call before proceeding to run back and forth between the two trees, directly under the hawk's branch. As it climbs the latter oak with an intent that brings it to the same branch as the hawk, the two keep a steady eye on each other and me. The squirrel then suddenly runs directly at its dark predator to within a metre before turning back quickly. This occurs twice before the hawk seemingly grows tired of this game, jumps toward the squirrel with wings spread and a guttural sound, but leaves it untouched. It then takes flight and glides off in a northeastern direction toward the bush and urban traffic farther in the background. For the ten minutes that this strange sequence occurred, my mind stopped in a kind of spontaneous reverence and joy for what was before me.

On many occasions I have seen hawks chase, capture, and, sometimes, lose their prey, often squirrels within the tree cover of Toron:to's ravines, but never anything like this. With the hawk's departure and the squirrel's seeming victorious call, an energetic flurry of impressions flood my mind. I am struck by a seeming agency in these animals to go beyond not simply the expected but the evolutionary sense

20 CN Tower and White Pine. Wherever we are is sacred land; that is what this Kanathian beacon teaches.

of self-preservation and predatory survival instinct to eat what is given. At first, the movement of the squirrel toward its shadowy predator seemed like a kind of sacrifice, and yet the hawk did nothing but watch. There was a playfulness in how the squirrel got closer, shifting my impressions of the interaction from a sacrifice to a child-like game of chicken with this potentially death-bringing presence. And, while

the hawk closed this set of relations by spreading its wings and showing what it could do, its departure felt like a nod of respect for the
courage of this small squirrel – modest in stature not unlike Muskrat
and Toad in the chapter 5 telling of Sky Woman's fall. The thought
that something bigger, with the power to bring death, can act beyond
evolutionary instinct or an ecological food chain has the feel of something miraculous.

Situated as I am in today's Hekademos, it is clear that many of the
thoughts and questions buzzing around my head suffered from a
romantic anthropomorphizing. But in the moment of Squirrel-Hawk
interaction there was an "apt strangeness" that jolted my modern
mind's critique of such romanticism into silence and allowed me to
feel the possibility of something else. Walking these winding paths for
over a decade has gradually instilled in me a sense that this place and
world is much smarter than I or any individual can ever be. This is a
central idea behind a good climate of mind: that the consciousness
we embody as human beings arises from a complex and numinous
reality that goes so far beyond it. If the world is smarter than us and
the climatic longue sault is one of a host of interconnected environmental mysteries that are requickening the modern ship to an old
sensibility, is it not possible that the best features of our humanity can
also mark the participatory consciousness of animal life and the
world itself? What if the world has the capacity to respond to other
beings, and to us, in compassion and not simply as a survival instinct?
That orenta infuses all life. Such questions and thoughts seem absurd
in today's popular flat-screen milieu, and yet the preceding pages suggest that the Good Mind that underlies them will be vital if we are to
navigate our way to a sustainable shore.

A Canadian Climate of Mind is animated with a seeming anthro-
pomorphism not to put humanity up on an evolutionary pedestal or
within a towering religion, but to help us recognize that human ways
of minding relations are gifts from Notre Dame de Turtle Island.
Compassion, forgiveness, love, wisdom, awe, and humility, as with
Flint's darker selfish binds, are ways of being that humans have particular talents for manifesting. But the mythic stories, interdisciplinary
research, and ceremonial duties that were renewed on these weaving
passages suggest that such capacities participate in something much

bigger. As the last three chapters displayed, our minding of relations can cosmologically lean toward either Sapling's orenta or Flint's otkon. To unknow a human-centred tendency is to recognize that we are not the reason for creation, but a vital species that has arisen from an energetically spiritual surround. We are healthy when we are devoted to peacefully raising the fallen onerahtase'ko:wa, the CN Tower Prayer Stick, rose-centred cross, or whatever other culturally appropriate symbol and ceremony that can be woven here to renew the energy in us for change. A good climate of mind recognizes that our beautiful relations, even in a deeply severed Toron:to and Kanatha, can encompass us in a wisdom that is smarter than any of us.

Being humble about human and cultural limits is not to deny Thomas's teaching that people have a duty to side with Sapling in raising the dolon-do. As I was writing these words, a national initiative emerged that seems to epitomize the braided passage we need to navigate humbly. On 4 September 2014 Canadians for a New Partnership inaugurated a "broad-based, inclusive, leadership initiative to engage Canadians in dialogue and relationship building aimed at building a new partnership between First Peoples and other Canadians."[1] This new partnership is being built upon the national apology for the residential schools and the heart-breaking stories recounted to the Truth and Reconciliation Commission. In its words: "Canadians for a New Partnership is not here to bury the past, no matter how harmful it was, but to use it as the foundation upon which the new partnership is built." There is something in these steps that feels like a renewal of the Haudenosaunee Condolence Ceremony as we communally come to terms with the pain imposed upon Indigenous peoples but is also the inheritance of all Kanathians. As we saw in chapter 7, this ceremony moves from the recounting of sorrow following a long period of violence, communally wiping our tears away, and then helping each other renew the duties of a Good Mind. The move from the tears of Truth and Reconciliation to requickening a New Partnership follows old ancestral grooves that can perhaps be a helpful guide.

The proponents of this new initiative include two former prime ministers, Indigenous leaders and elders, and other influential figures in media and business, though its original inspiration arose from a former premier of the Northwest Territories and president of the Northwest Territories Dene Nation, Stephen Kakfwi. What moved Kakfwi in this direction was a challenge that arose from dialogues he had with his children as the Idle No More movement was receiving national attention at the end of 2012 and into 2013. In the wake of this Indigenous movement that was calling for significant change, "I expressed to my adult children the urgent need for a renewed Nation-to-Nation relationship, built upon mutual respect and understanding between Indigenous and non-Indigenous governments and people across Canada." This family discussion was also influenced by testimony before the Truth and Reconciliation Commission, for his partner, Marie Wilson, was one of its three commissioners. Bringing light to this abuse is a vital first step that needs to be followed by a healing of the historic severance between Indigenous and non-Indigenous peoples.[2] In talking of this vision, Kakfwi was asked by his children "to take the initiative and I began making calls to former leaders of all political stripes and backgrounds and was surprised and gratified by the very immediate and positive responses."[3] This diverse leadership symbolizes the broad action that is needed across political parties and nations, for renewing the roots of peace requires Kanathians to declare their intention to participate across the country in a diversity of places and relations.

Something in the behaviour of the strangely courageous Squirrel and magnanimous Hawk also posed to me questions about the humility, bravery, and respect needed to make such a turbulent passage toward healing after so much pain. Our present moment is putting Kanathians in a position to recognize themselves as that cannibal who, in chapter 7, was jolted by the Peacemaker's beautiful reflection into asking if there was a different way. The ship is called to a humbler sense of how to be in relation with the Indigenous canoe as part of the creation, and thus participate in the healing of a pain that has been forced on many. As Canadians for a New Partnership declares: "For too long our relations have been marked by misunderstanding, betrayal and neglect. While the sins of the past can never be erased, by acknowledging Aboriginal and Treaty rights and forging a new partnership we can stop the cycle of negativity that past wrongs have wrought on the generations that followed." Retracing the steps of Condolence, we need to wipe away the tears of our Indigenous brothers and sisters, and those of us on the ship need to ask them to wipe away our tears for what we have done. A renewed partnership must be "based on the principles of mutual respect, peaceful co-existence and equality." The expectation is that such a changed way of relating will give us the good orenta to deal with historic injustices in real ways, to tangibly hold open "the promise of better living conditions, education, and economic opportunities for First Peoples." The spirit of Kakfwi's initiative resonates with the Peacemaker's message to respect our differences while affirming our common source. It also sits close to the heart of the Protectress of Kanatha on her Île-aux-Hérons as she holds open a "thin" space of potentiality between the Kaniatarowanenneh's two shores for braiding a new future.

As I follow the red-tail's flight to where the savannah gives way to some limited forest cover beyond which a driving hum is increasingly audible, my peripheral vision is drawn by orange streaks back across the tall grasses. A setting Sun fills the sky to the west with warm hues of light radiating through the ravine trees from the busy street to the north to the distant Lake Ontar:io in the south. Farther up toward the Sky World the colours change from orange to purples and pinks as the Sun's rays meet some clouds. Such encompassing spectacles have been here long before humans came to this place with the ancestors of the Anishinaabe, Wendat, and Haudenosaunee. They will also continue to reveal their beauty long after the ways of the modern Kanathian ship have been eclipsed by those good partnerships which are emerging from the climatic longue sault and other challenges brought by Flint's ever-changing cycles of darkness and Sapling's renewal. This timeless cosmological context that was the focus of chapters 5 and 6 is the deep source from which a peaceful initiative like Canadians for a New Partnership can potentially requicken our minding of relations. In the words of its declaration: "For over four centuries we have shared the same land, water and air that form one of the most bountiful and prosperous countries in the world. In this way Indigenous and non-Indigenous people are bound together in an inseparable bond." A common reverence-inspiring beauty still flows in Kanatha despite all the violent severances that the past seven chapters have described as deeply wounding Notre Dame de Turtle Island and its inhabitants.

The quieting of the mind by numinous forces and behaviour that inspire respect, awe, and gratitude can be understood as an inherent feature of creation. It is from these "thin" spaces that seeming miracles can burst forth like the viriditas of spring. From the modest Squirrel to a glorious-setting sun, there are emergent potentialities beyond control, management, and prophetic prediction that the modern ship needs to relearn how to appreciate. This will partly entail bringing our extremely detailed scientific knowledge, political skills, and technological capacities into felt relations that have the potential to miraculously transform how we imagine relating with each other and the Kanatha we want to live within. The miracle of a magnanimous Hawk is from this perspective not something that happens to us from on high without change, but rather phenomenally reflects a sacred response from Notre Dame de Turtle Island and the Sky World to good ways of relating. We are in need of a miraculous transformation that is akin to the emergent properties of deeply interconnected relations like the beautiful colours before me. Such a participatory miracle is the response of Sapling's creation to people who have dutifully come together in partnership to heal and renew their devotion to being here in, following Mohawk and Thomas again, "real" ways.

While Canadians for a New Partnership offers a beacon of hope that the Sun is setting on colonial Kanatha, the vision of Woodworth that began unfolding in chapter 2 highlights the need to find places in our local relations for raising such a peaceful dolon-do. For much of the past decade, Woodworth has attempted to initiate the building of a beacon that is concerned with partnering the place's Indigenous ancestors and the diverse settlers of the Toron:to he has called home for decades. Just west of Ashbridges Bay is a sandbar that once held its marshy waters in place and "is still intact at the foot of Cherry Street, along the only stretch of original shoreline in the downtown core."[4] It is here that he proposed to the city of Toron:to the need to build a sacred architectural site in the form of a "round lodge in the Anishinaabe tradition woven among the branches of a tree" that is at its centre. Within this building inspired by the land's elder Indigenous nation would be found models of Toron:to that move through twenty thousand years, from ten thousand years into the past as we follow Flint's receding glaciers to ten thousand years into possible futures that "remind us of the cyclical nature of our occupation here." Extending from the lodge's central cosmic tree that can be found in so many traditions of this planet would be two longhouses and beyond each "staghorn sumac bushes, the traditional place of teaching" in Haudenosaunee and Wendat traditions. As with Canadians for a New Partnership, this sacred place would remember the past while reaching out to something better.

The central purpose of this round lodge is for Woodworth to create a communal place where we can heal "the profound rift between the Indigenous and the Western mind." Being Haudenosaunee, he talks of it as a place for fulfilling the Peacemaker's prophecies that we would "welcome people from the four directions."[5] Under the lodge's central tree Woodworth envisions burying the weapons of the colonial era as we attempt to reconcile and move toward a new "spirit of peace." These weapons, we have seen, range from remains of residential schools to policies like the Indian Act to those tools that support an unquenchable addiction to resources and energy. A communal healing of this past will entail resuming traditional duties like Thanksgiving and a Condolence Ceremony whose purpose is, in his words, to "clean you up before inviting you back to our longhouse community."[6] Bringing this sacred architecture into manifestation was the goal of Woodworth's foundation, A Beacon for the Ancestors. His dialogues with municipal officials, corporate bodies, and interested academics were filled with starts and stops that gradually made it clear to him that the city was not ready to acknowledge formally the continued presence of its ancestors. In the absence of a sacred lodge, there has emerged a virtual Indigenous beacon called "First Story Toronto" that has been developed by the Native Canadian Centre of Toronto. Its mobile app named "First Story" maps research and stories on the Indigenous heritage of Toron:to "with the goal of building awareness of and pride in the long Aboriginal presence and contributions to the city."[7] While Indigenous-inspired initiatives like these, Idle No More, and Canadians for a New Partnership come forward, official governmental institutions seem to remain resistant to fully recognizing both the past it is situated within and the strides toward a good future that may come about if we formally renew relations with Kanatha's ancestral elders.

The moment we live in is one where beacons for change clearly shine forth and yet old institutional structures, beliefs, and factishes progressively hold on to inhibit changes that are inevitable. Wherever we live in Kanatha, this interplay of Flint and Sapling can be observed in a myriad actions and denials. In this time of cosmological change, we need to do something that is totally impractical, useless, and aptly strange to the modern ship; that is the only way to lift up enough orenta to imagine something different and yet familiar enough to inspire. I am reminded of Joseph Campbell's description of Chartres and all the twelfth-century French cathedrals dedicated to Notre Dame as works of passionate faith that in Europe are unparalleled "by any single economic effort, except war."[8] Continuing, he adds that, from the perspective of the modern rational mind, the early stages of many great civilizations tend to be marked with signs of seeming insanity; contrary to modern beliefs, "economics, politics, and even war are, in such periods, but functions of a motivating dream."[9] Though we have seen in relation to Chartres and Kanatha that civilizing epics can also perpetuate unjust ecological and social conversions of grand scales, I am left wondering whether new dreams in the spirit of Kakfwi and Woodworth can energize a different braided passage down the Kaniatarowanenneh that goes beyond our current impasse.

It is clear that the time and space for epic models, solutions, and conversions is from the past, and that a sustainable and good way forward will need to symbolically renew that which has been built in our midst. That said, there seems to be a need in the heart of all human cultures for sacred spaces where we can communally hear stories and participate in ceremonies that remind us of where we have come from, who we are, and where we are going. This is what Woodworth envisioned for a sacred Toron:to lodge, a potential "thin" space that could have benefits for many communities across this nation. In a sense, bringing Canadians for a New Partnership and Truth and Reconciliation into fruition in such a vast country will require a diversity of regionally based partnerships in spaces where people can come to be familiar once again with each other and our common waters – from the Kaniatarowanenneh and Lake Ontar:io to the Northwest Passage to the Fraser River Delta. Such sacred and yet tangible spaces of partnership can act as reminders for recognizing our mixed histories, giving thanks for the sacred offerings of the past to the present, and modifying our being in relation to emerging braided stories of what the climatic longue sault is teaching us. Linking together such regionally diverse beacons is one way of conceiving the peaceful spirit of the Kanathian initiative imagined by Kakfwi and, perhaps even more aptly in its built form, the National Centre for Truth and Reconciliation.[10]

Down I go one last time into the ravine's wet ground, passing another oak and small sumac grove before exiting onto a busy Toron:to intersection, the cars getting louder with each step. The drive of this Kanathian culture impresses on me the miraculous change that is now needed; it is of a scale not unlike the severance that occurred over the past three centuries here, though of a different reverential quality. A planet for human taking and consuming has not required us to pause much for giving thanks. We live in a time of "doing." Pipelines must be built, the economy must be turned around, people must shop, jobs must be created, we must work more, and so on. Even among the ecologically and climatically minded who critique unending economic growth, humans as unquenchable resource consumers, and the social inequities of recent political calls for austerity, the overall tone often rings familiar. Ecological and climate systems must be accounted for, sustainability policies must be economically viable, and we must buy green. It is in this context of limited action that another kind of partnership has emerged between environmentalists and Indigenous nations who are utilizing treaties and land claims to slow down the dark spider web of pipeline developments that Mohawk helped us observe in chapter 3, while also building a national movement aimed at limiting bitumen and fracking extractions.[11]

Not only can we no longer be idle about renewing our partnerships with the Indigenous elders of Turtle Island, but we also can no longer be idle about our energetic relations with the cosmological presences made increasingly apparent by today's climatic longue sault. Our historic moment is a beacon itself, one that is revealing the limits of many modern factishes about how knowable and manageable our world is. In the midst of all our doing, rapid changes are calling us to slow down and do less, relying on whatever partnerships and ceremonies can help us move beyond Kanatha's history of colonial severances – from Indigenous to land to climate relations. With the Hawk watching closely, waiting to see how we will now respond, perhaps it is possible to see today's crisis as a reflection of what happens when a great number of human communities forget how to pause and give thanks for each other and our common Kaniatarowanenneh. From such a perspective, ending our energetic idling has to be central to any peaceful partnership in the twenty-first century.

While we saw in chapter 4 that the Earth Hour can be criticized as caught in the modern ship's state of thankless amnesia, it can also be optimistically conceived as the beginning of humanity's global return to Thanksgiving after a modern hiatus. Ceremonies of thanks are found in various states of use or disuse across cultures, and what they commonly ask is for people to pause and deeply consider that which our life depends on and thus is worthy of our thanks. From Christian Lent to Islamic Ramadan to many other spiritual ceremonies across the planet, there are a host of acts that humans can access to deepen, broaden, and renew our expression of gratitude in climatically real terms. The inspiration of these diverse traditions will be needed for us to reduce our carbon footprint, challenge our participation in unsustainable energy politics, respond to climate injustices, and, most important, remain firm with a good Peacemaker intent during the difficult but necessary transformation that is upon us. I am reminded once more of Chief Thomas, who taught that in his Good Mind tradition everyone arrives into this world with a purpose. This is why our common Creator "gave us a culture," and it is our duty to keep it going with its particular gifts and talents.[12]

We have all inherited stories, ceremonies, and teachings about what it is to be here on Notre Dame de Turtle Island, just as each individual has his and her own unique positions and gifts. The difficulty, we have seen, is when one culture deems its gifts as the way for all, thus breeding painful severances everywhere it goes. Returning to

Thomas, he said that "I have a talent to sing and you have a talent to dance or you have a gift of knowledge about ceremonies," and these can be brought together for a common purpose both within and across cultures. Such gifts are given by the Creator so we can help one another, and when we do this, he believed, we become "real people." The beacon of his student Woodworth, the initiative of Kakfwi, and the National Centre for Truth and Reconciliation have the potential to renew such a spirit by creating spaces where we can reconcile our diverse cultural gifts and histories so as to return to good relations based on the hope for a better common future. To become "real people" requires sacred spaces and time for lovingly renewing ceremonies that have their own seasonality to them. Together with our inheritance of unique ancestral rituals, Earth Hour can be engaged as a common yearly beacon for minding our national, continental, and global participation in the unfolding revelation of an intelligence that informs every Earth Hour, day, season, year, and life.

We need ceremonies for renewing Sapling's orenta in our everyday ways of living, since this good energy is required to confront our addictive binds and thus further those emerging "real people" partnerships aimed at putting Flint's dark presence back in its place. Ritually engaging our fuel usage is one way of slowing down that can potentially create space for small Squirrel miracles to emerge and grow. The Thomas Berry–inspired Toron:to eco-theologian Stephen Scharper writes of just such "silver linings" in the ecological and carbon asceticism that I began connecting with Saint Tekakwitha in chapter 4. He tells of how he and family were originally stuck in a kind of calculus logic concerning the idea of giving up their car in an urban reality where that is viable because of public transit. Beyond saving money and taking one car off the urban road, with its average annual emission of 4,480 kilograms of carbon monoxide and 20 kilograms of carbon dioxide, something unexpected was gained.[13] Silver linings included the more relaxed pace of public transit that offered space to talk as a family, and "a marked decrease in impulse buying" related to the sacrificed convenience and functionality of the car.[14] Of course, this is not possible across many regions of Kanatha, though it is so for those who live in urban centres like Toron:to. It is also not to suggest that the burden of change is on individuals when, as was discussed at various points, broad political, economic, and cultural changes are needed as well.[15] The point is that silver linings and miraculous possibilities abound if we make a move toward becoming aptly strange to modern ways, from our individual and familial lives to supporting local beacons and national partnerships aimed at reconciliation and peace.

A Catholic spirit of renewed responsibility toward each other and the common waters of Turtle Island took on a global dimension in June 2015 with the release of the first papal encyclical focused on changing human ways of relating with environment and climate. Pope Francis expresses concern with the modern belief in unending economic growth and consumption as the basis of a good life that is simultaneously having devastating impacts on many people and increasingly the planet itself. As his encyclical states in blunt terms, "the rejection of every form of self-centredness and self-absorption are essential if we truly wish to care for our brothers and sisters and for the natural environment."[16] Commenting on this radical document, Scharper writes that "Francis takes aim at a 'throwaway culture' of unbridled consumerism."[17] The human role in today's climate changes is revealing the need for "a seismic shift away from 'rapidification' and 'irrational confidence in progress and human abilities' toward a culture of social and ecological inclusion, to protect, in the words of St. Francis, 'Mother Earth.'" The fact that Francis is the first Jesuit pope and took the name of the patron saint of ecology connects him with a host of ancestors from the ship who have been central to these Kanathian passages – from the Jesuit Black Robes to the Franciscan Sagard to Saint Tekakwitha. With his encyclical confronting the progressive delusions of the modern ship, calling for social justice, and affirming our dependence on creation, Pope Francis offers us another opening to slow down and consider the ancestral wisdom that can help us renew relations on Notre Dame de Turtle Island.

Moments of pause in our driven lives can give us space to reflect on our mixed relations with others and mind the impacts of our actions. Sharing work, making and consuming less, double-clicking less, going offline more, reducing our epic predispositions, coming together in "real people" partnerships – generally doing less and being with those we love more – all of these acts and others discussed here can allow possibilities to emerge that are impossible to model objectively from within the walls of Hekademos. For what the mythic stories, interdisciplinary thought, and ceremonies of these pages teach is that creation has the capacity to respond to our relations with Flint's numinous otkon or Sapling's miraculous orenta. But, more than anything else, what I take from Scharper's silver linings, Woodworth's beacon, and Livingston's flight is that ceremonial Thanksgiving cannot simply be about critical reflection and austere personal sacrifice, for that will not inspire the deep cultural change needed today. We need to project and share an onerahtase'ko:wa's calming peace, covenant rainbows, graceful Heron flights, humbling Squirrel courage, Hawk-like compassion in the midst of death, and a toronton of myths and ceremonies meant to unfreeze our violent past and inspire partnerships for a renewed future.

As I take the last steps up the path to the busy Toron:to intersection, the sacred contours of this weaving labyrinth can still, with firm attention, be sensed cycling through the everyday concrete passages of this twenty-first-century city. The dipping roads leading from the park to my home are paved-over ravines, the glacial marks of Flint that chapter 6 displayed as being deeply related to the floods of numinous stories. On my front yard, a staghorn sumac in the sandy soil reminds me of the savannah ecology this urban reality is partially paved over, the Tahontaenrhat rainbow path to the Skyworld, and Woodworth's vision of a sacred round lodge flanked by these trees. From home to Hekademos, every northward-moving subway trip to the York University campus where Livingston wrote of his bird-inspired flights of mind and I began contemplating the subject of this book submerges me in Turtle Island's dark recesses. It is a deep space within Turtle Island that opens up all the debates on the future of sustainable transportation. The last leg of this commute transfers from subway to a bus that traverses a transit-only road through an energetic corridor of hydro towers and the controversial Line 9 pipeline that is projected to bring Alberta's bitumous oil to the eastern seaboard. With this passage, the binding lessons of the climatic longue sault come to the fore as the oft-cited greener way of public transit rides the dark spider web seen by Mohawk, Banyayca, and Livingston. Then there is the railway along Lake Ontar:io that often transports me eastward to my familial home along the Kaniatarowanenneh that was the focus of chapters 3, 4, and 5, opening up other stories about Kanatha's colonial roots, Thomas's Two Row teaching, and Saint Tekakwitha's healing braid.

On the other side of this book's passages, everything has become sacred, even a Toron:to that for a good portion of my life was seen as an urban wasteland. The wild fissures of this place have taught me reverence for ecological, climatic, ancestral, and cosmological presences that happen to be wherever we are in Kanatha, even this city. I am filled with a renewed spirit that brings to mind the late-nineteenth-century Indigenous vision of Black Elk as his Lakota people were rushed into America's colonial conversions. While terribly ill as a young boy, he had a vision of himself standing on the sacred Harney Peak in the Black Hills of South Dakota, "the highest mountain of them all," from which he could see beneath him "the whole hoop of the world."[18] In his words: "I saw that the sacred hoop of my people was one of many hoops that made one circle, wide as daylight and as starlight, and in the center grew one mighty flowering tree to shelter all the children of one mother and one father. And I saw that it was holy."[19] From Harney Peak to Chartres to Lhasa to Mecca to the Île-aux-Hérons to the National Centre for Truth and Reconciliation to the Fraser River to Toron:to to whatever community we live within, sacred presences are waiting to see if we choose to align with Sapling or Flint in our ways of minding relations. The choice to lift the orenta of a fallen dolon-do is open to all of us, though the potential passages to a peaceful partnership are as diverse as the cultural, ecological, climate, and cosmological relations that suffuse Notre Dame de Turtle Island's many places.

We need more Île-aux-Hérons moments of repose where we can take time with Saint Tekakwitha, the Peacemaker, Thomas, Livingston, and other vital ancestors to remember the sacred dimensions of where we live, and the hard choices required to lovingly engage a numinous climatic longue sault rather than simply survive it. The feel of such quiet and transitory openings takes me back to the calm joy that arose from simply lying on the labyrinth's fallen onerahtase'ko:wa in early spring; taking in the Sun's rays as the last remnants of snow melted away; sitting in a darkened ravine with wind blowing overhead, leaves rustling, birds singing, and a downy woodpecker drumming away; and late-summer savannah walks with the Sun shining down, the smell of grass in the air, and goldfinches flitting about. Such experiences inspire a sense of wonder, spark a recognition that the intense pace of modern life makes them a rare privilege, and reminds me to do something that is rare in our consumer culture – give thanks. It was through relating with Woodworth's Good Mind vision of Toron:to that I gradually came to appreciate the profound implications of this largely forgotten, yet common and simple, act.

Can modern Kanatha find a way of renewing this peaceful spirit of partnership as we undertake the next turn in the cycling passages of life within Notre Dame de Turtle Island? This is an open question, though the surrounding presences are becoming increasingly numinous in their pressing for a good response. We are being asked to co-create the kind of world that moves us to want to live within it; that is what it is to participate in raising orenta. It is an act of humbly recognizing our limited nature, as embodied by Squirrel, Muskrat, and Toad, and yet follows their courageous spirit in digging deep below the dark waters for the sacred mud that is needed for a good way forward. From Canadians for a New Partnership to a sacred lodge in downtown Toron:to to our common global passage over the climatic longue sault, there is much work and healing to be done that will require modesty, courage, and inspired vision. We have the stories and ceremonies in our collective ancestries to help us along the way, but they will need to renew the knowledge, technologies, and political possibilities of our moment if the fallen dolon-do is to be lifted again. Requickening the ancestral past to present and future ways of living on this sacred Turtle Island is the central good duty of *A Kanathian Climate of Mind*, and our turbulent time is calling us all to that task.

Glossary of Indigenous and Braided Terms

Aataentsic: The Wendat name for Sky Woman, introduced in chapter 5. Also see *Sky Woman*.

Ayenwatha Wampum: This beaded wampum belt represents the peace brought by the Peacemaker, as discussed in chapters 2 and 7. It has a purple background that symbolizes the sky or cosmos, with white lines connecting two squares, a tree (or heart), and two more squares. These five figures symbolize the original Five Nations that were brought together by the Peacemaker and Ayenwatha. From east to west, they are Mohawk (Keeper of the Eastern Door), Oneida (People of the Standing Stone), Onondaga (Keepers of the Fire), Cayuga (People of the Swamp), and Seneca (Keepers of the Western Door).

Black Madonna: A darkened representation of Notre Dame de Turtle Island whose statue is found at the Chartres cathedral and is discussed in chapters 6 and 7. Also see *Notre Dame de Turtle Island*.

Black Robe: This is the English translation of the term that the Wendat and Haudenosaunne used in referring to Jesuit missionaries, largely because of the black robes they wore. These missionaries are discussed primarily in chapters 4 and 5, and then the term is used as a historic reference to colonial conversions in Kanatha.

Blue Virgin: An east-facing stained-glass window at Chartres cathedral that represents the promise of birth and life's viriditas. She is the light-bearing complement to the Black Madonna. Also see *Notre Dame de Turtle Island*.

Budding Flowers: She is the daughter of Sky Woman who had relations with Turtle and then gave birth to twins, Sapling and Flint. The latter killed Budding Flowers when he attempted to enter this world through her armpit. Her story is told mostly in chapter 6, though it is also touched upon in chapter 5.

Canoe: See *Two Row Wampum*.

Climatic longue sault: Climatic longue sault is a recasting of today's climate changes as akin to a long set of rapids (in French, longue sault) that are asking us to increase our attentiveness to the numinous and powerful world we live within. The image arises in chapter 3 at the place on the Saint Lawrence River where the longue sault was once situated and is now harnessed for its energy. The term climatic longue sault is engaged to help us mind the turbulent and violent dimensions of today's climate changes, as well as the kind of human knowledge and behaviour that is needed to navigate.

Dolon-do: The term dolon-do arises in chapter 2 as a Mohawk oral understanding of the origin of Toron:to's name. It means a damp wet log, and for Woodworth it is imagined in his Good Mind tradition as a great white pine or onerahtase'ko:wa because of its symbolic references.

Factish: This is a term coined by Bruno Latour to refer to beliefs of the modern mind that are deemed objective facts but actually are grounded in particular cultural assumptions. I use the term across many chapters to help destabilize some of the ideas central to Canada's historic relations with Indigenous people, the land, and climate.

Flint (Sawiskera): Flint is the second twin whose birth killed his mother, Budding Flower. He subsequently becomes associated with various dark changes that try to undo the creative work of his brother, Sapling. The okton energy he brings into Turtle Island is a destructive force that destroys the creation and also disorders the Good Mind of the people. Flint arises in chapter 6 and then continues to be discussed in chapter 7.

Hadoui Masks: These masks, which are discussed in chapter 7, are also known as False Faces, and they are associated with the False Face healing societies in Haudenosaunee and Wendat culture. The term Hadoui is employed because it is the one used by Jacob Thomas.

Île-aux-Hérons: This is the island on the Saint Lawrence River between Kahnawake and Lachine where it is said that Saint Tekakwitha and her friends engaged in a kind of spiritual hermitage. It is discussed in chapter 4 in relation to various other ideas around sacred spaces, and is connected to the idea of "thin" spaces where in some traditions it is said that spirit is close to the material world. The island is referred to in all subsequent chapters as a way to remember the interaction of sacred places found in Nature with particular spiritual practices. It represents an approach to minding relations from particular "thin" places where Saint Tekakwitha helps us come to a braided understanding of our place on Turtle Island.

Kahnawake: This is a Mohawk village on the Saint Lawrence River or *Kaniatarowanenneh* that is associated with Saint Tekakwitha's story in chapter 4. It is from this village that the inhabitants of Saint Regis and Akwesasne, communities farther west along the river across from Cornwall, Ontario (discussed briefly in chapters 3 and 4), originated.

Kanatha/Kanathian: Iroquoian term for village that comes up in chapter 5 in relation to Jacques Cartier's approach to the kanatha of Stadacona on the Saint Lawrence River. The term replaces the contemporary "Canada" and "Canadian" as a way of remembering the roots of the Canadian nation in colonial misunderstandings that later progressed into conversions. In the last few chapters it also becomes a way of conceiving the nation of Canada on a human relational scale that simultaneously recognizes its grounding in a cosmology of powerful presences.

Kaniatarowanenneh: Mohawk term for Saint Lawrence River that means great waters and in this book is used as a symbol of the common waters of the Two Row Wampum. It is in between the two rows/shores of this great river that Saint Tekakwitha is imagined as sitting on her isle, thus offering a symbolic space for braiding stories and practices for a good future.

Lorette: The village or kanatha of Wendat Christian converts found on the Saint Lawrence River or *Kaniatarowanenneh* following the mid-1600s violent displacement of Wendat from southern Ontario. This history is discussed in chapters 4 and 5.

Longue Sault: See *climatic longue sault*.

Notre Dame de Turtle Island: The Haudenosaunee/Wendat creation story of Sky Woman tells of her falling onto Turtle Island after an illness in the Sky World that was her home. In chapter 5, this story is brought into dialogue with the cosmology of the French cathedral Notre-Dame de Chartres, a dialogue mediated by a late-1600s letter sent from the

Wendat of Lorette to Chartres. This letter is engaged as a way of bringing these two different cultural creation stories into a braided dialogue,
symbolized from that point on as Notre Dame de Turtle Island.

Onerahtase'ko:wa: This is a great white pine that is the central tree of the
Haudenosaunee Good Mind tradition as discussed in chapter 2 and further elucidated in subsequent chapters. Its white roots connect the original five nations that are also symbolized in the five needles of this tall
and straight tree. With the help of Thomas, it is connected by
Woodworth to the fallen dolon-do along the passage de taronto, and
thus becomes a way of minding the origins of Toron:to.

Ontar:io: Mohawk term for shining sparkling waters that are today
known as Lake Ontar:io.

Orenta: See *Sapling*.

Otkon: See *Flint* and *Sapling*.

Passage de taronto: Oldest written reference to the Humber River and
associated portage trail that went up the Oak Ridges Moraine, and
another Mohawk or Wendat reference to the Toron:to's original name.
The varying perspectives on this term are discussed in chapter 2. Also
see *Toron:to*, *dolon-do*, and *onerahtase'ko:wa*.

Peacemaker: The central Haudenosaunee ancestor associated with the
peace of the Confederacy and the Good Mind tradition discussed in
chapter 2 and further examined in subsequent chapters. Also see
Ayenwatha Wampum and *onerahtase'ko:wa*.

Protectress of Kanatha: This is the designation given to Kateri Tekakwitha
when she was canonized as a saint, and she is the central ancestor
engaged here for considering the violent power of conversions in French
Catholic and subsequent Canadian tradition, the resulting divide
between Indigenous and non-Indigenous peoples that exists because of
this history, and the value of significant ancestors/saints for offering
stories pointing to an alternative future.

Rolling head: This term comes up in chapter 2 as a Haudenosaunee Good
Mind view on colonizing ways of minding relations. It refers to the tendency to always be moving from where we are living, or converting that
which is in our midst based on preconceived ideas that are not in relationship with creation. It is braided with Northrup Frye's conception of
a garrison mentality in Canada (chapter 2), John Livingston's technological imperative (chapter 1), and John Mohawk's energetic web

(chapter 3), and then in subsequent chapters it takes on other energetic and technological dimensions.

Rotea:he: He is the husband of Sky Woman who was ill in the Sky World and thus precipitated her fall, though there are different interpretations as to why this happened, as discussed in chapter 5.

Sapling (Teharonhia:wako): Sapling was the first twin born of Budding Flowers, and he becomes the Haudenosaunee creator who represents everything good in the creation. He manifests a light energy known as orenta which is the source of creation and in which people need to participate through Good Mind practices to keep Flint's dark destructive otkon at bay. His story arises in chapters 6 and 7, though it is also discussed briefly in chapter 5.

Ship: See *Two Row Wampum*.

Sky Woman (Otsi:tsia or Mature Flowers): See *Aataentsic* and *Notre Dame de Turtle Island*.

Tahontaenrhat: The particular Wendat nation that was in the Toron:to area at the time of Champlain's arrival in the early 1600s.

Thin space/time: This term refers to particular places or times where spirit is felt to be close to the creation and where people are thus able to come to different ways of minding what their life is about. It represents a spiritual way of minding the ecology of mind tradition's valuing of wilderness experiences (first arising in chapter 1), and we are brought to this older perspective through the coureurs de bois in chapter 3 and then Saint Tekakwitha in chapter 4. Subsequent chapters consider the turbulence of our climatic longue sault as a kind of thin time that is initiating us into a more social world and underlining the need for a more social way of minding creation. Also see *Île-aux-Hérons*.

Toron:to: This is the Indigenous spelling that Woodworth uses for the city of Toron:to, and which I follow starting in chapter 2. Also see *passage de taronto*, *dolon-do*, and *onerahtase'ko:wa*.

Two Row Wampum: A treaty that is first discussed in the Introduction and then used symbolically throughout the book, with more dimensions added in chapter 4. The canoe mentioned in the treaty symbolizes the Haudenosaunee culture and the ship the colonial (and then modern) Canadian culture. They have their own ways which need to be respected, and they travel down a common river that must also be respected for a good way forward. These common waters are reconsidered in relation to our common water, land, and climate, all of which are becoming increasingly turbulent because of the ship's colonizing disrespect of the canoe and the water.

Via negativa: This is a mystic method associated with Saint Dionysius for unknowing knowledge in an attempt to participate with a divine reality. It is primarily discussed in chapter 7 in relation to the Hadoui masks, though some of the ideas associated with it begin to be highlighted in chapter 6.

Viriditas: Latin for greening of life and associated at Notre-Dame de Chartres and particularly the Blue Virgin stained-glass window with the baby Jesus Christ's halo. This term arises in chapter 5 and then in later chapters is braided with the peace of Sapling and the Peacemaker.

Notes

FOREWORD

1 Although there will always be controversy over the actual date, conveyed as it is in an oral tradition that did not use the current calendar, the earliest projection comes from Barbara A. Mann and Jerry L. Fields, "A Sign in the Sky: Dating the League of the Haudenosaunne," *American Indian Culture and Research Journal*, 21(2) (1997): 105–63. It dates the founding of the Confederacy by the occurrence of a total eclipse of the sun over the territory of the Seneca, the last nation to accept the Great Law, on 31 ugust 1142.

2 Safa Motesharrei, Jorge Rivas, and Eugenia Kalnay, "Human and Nature Dynamics [HANDY]: Modeling Inequality and Use of Resources in the Collapse or Sustainability of Societies," *Ecological Economics*, 101 (2014): 90–102.

3 Based on personal communication with Jake Thomas and Michael K. Foster, author of "Jacob Ezra Thomas: Educator and Conservator of Iroquois Culture," *History of Anthropology Annual*, 1 (2005): 219–45. Michael was a close friend of Jake and mine.

4 A traditional term that was used, according to Jake Thomas, to describe those who wandered and were unwilling to join a community.

INTRODUCTION

1 Schmidt and Remiz, "High Park Waterways"; May, *Emerald City*.
2 Root, Chant, and Heidenreich, *Special Places*, 260.

3 Much of these contemplations on northern warming were published in Leduc, *Climate, Culture, Change.*

4 Klein, *This Changes Everything,* 25.

5 Ibid., 460.

6 Otto, *The Idea of the Holy,* 7.

7 *Oxford Canadian Dictionary,* 173.

8 See Spitzer, "Milieu and Ambiance: An Essay in Historical Semantics," 206.

9 Intergovernmental Panel on Climate Change, Climate Change 2013, *Fifth Assessment Report.*

10 Hulme, *Why We Disagree about Climate Change,* 361.

11 Latour, *On the Modern Cult of the Factish Gods,* 97.

12 Ibid., 97.

13 Thoreau, *Walden*; Leopold, *Sand County Almanac*; Carson, *Silent Spring*; Louv, *Last Child in the Woods.* Also see Asfeldt et al., "Wolves, Ptarmigan, and Lake Trout"; Curthoys, "Finding a Place of One's Own"; and Orr, *Earth in Mind.*

14 Livingston, *The John A. Livingston Reader,* 129.

15 Livingston, *Rogue Primate,* 196–7.

16 Livingston, *The John A. Livingston Reader,* 12. Also see Carrothers, Kline, and Livingston, *FESKIT: Essays into Environmental Studies.*

17 Bateson, *Steps to an Ecology of Mind,* 454.

18 Livingston, *The John A. Livingston Reader,* 163.

19 Ibid, 164.

20 Carson, *Silent Spring,* 297.

21 Arthur and Otto, *Toronto: No Mean City,* 6.

22 Woodworth, "Urban Place as an Expression of the Ancestors," 226.

23 Ibid., 226. Also see Woodworth, *The Morning Star.*

24 Freeman, "Indigenous Hauntings in Settler-Colonial Spaces," 235.

25 Woodworth and Leduc, "Recovering the Tribal Identity of Toronto."

26 Freeman, "Indigenous Hauntings in Settler-Colonial Spaces," 230.

27 Ibid., 213.

28 Innis, *Staples, Markets, and Cultural Change,* 127.

29 Klein, *This Changes Everything,* 25.

30 Mohawk, *Thinking in Indian,* 277.

31 McGregor, "Coming Full Circle," 388.

32 Woodworth, *The Morning Star,* 159–60.

33 Saul, *A Fair Country*, 3.

34 Saul, *Reflections of a Siamese Twin*, 98.

35 Ibid., 104.

36 There is a need to differentiate the potential of a métis in Canada's history and future from the particular historic and regional realities that led to the Métis Nation. It is for this reason that the terminology I primarily utilize in this book is "braid" or "weave" rather than "métis," though the latter term is used in relation to particular historic relations that highlighted a potentiality that was often violently marginalized in the creation of Canada as a nation-state. This issue is taken up in detail by Chris Anderson and has many resonances with issues that are further discussed in chapter 3 and particularly chapter 4. See Anderson, *"Métis"*; Anderson, "'I'm Métis, What's your Excuse?'"

37 Gruenewald (now Greenwood), "The Best of Both Worlds," 9.

38 Robinson, "Being Undisciplined"; Ford, *Beyond the Modern University*; M'Gonigle and Starke, *Planet U*.

39 Mohawk, *Thinking in Indian*, 275; Woodworth, *The Morning Star*.

40 Harrison, *The Dominion of the Dead*, 102–3.

41 Livingston, *The John A. Livingston Reader*, 110.

CHAPTER ONE

1 Livingston, *Rogue Primate*, 197. Over the following chapters I draw from my journals between the years 2005 to 2014 to situate and ground the analyses. I will not be referring to specific dates, though I usually give seasonal, climate, and ecological context for the reader.

2 Livingston, *The John A. Livingston Reader*, 129.

3 Livingston, *John Livingston: The Natural History of a Point of View*.

4 Gibson, Appreciation, ix.

5 Livingston, *The John A. Livingston Reader*, 164.

6 Ibid., 164.

7 Leopold, *A Sand County Almanac*, 138.

8 Ibid., 140.

9 Ibid., 239.

10 Baskerville, *Ontario*, 195.

11 Root, Chant, and Heidenreich, *Special Places*, 83.

12 Ibid.

13 Grady, *Toronto the Wild*, 6.

14 Livingston, *The John A. Livingston Reader*, 164.

15 Livingston, *Rogue Primate*, 129.

16 Livingston, *The John A. Livingston Reader*, 164.

17 Bonnell and Fortin, *Ashbridges Bay*.

18 Cited in Robertson, *Walking into Wilderness*, 110–11.

19 Cited in Bonnell and Fortin, *Ashbridges Bay*.

20 Livingston, *The John A. Livingston Reader*, 163.

21 Gibson, in Livingston, *The John A. Livingston Reader*, ix.

22 Ibid., ix.

23 Archer, "I Remember," S7.

24 Wadland, *Ernest Thompson Seton*, 52.

25 Ibid., 54.

26 Seton, *Wild Animals I Have Known*, 47.

27 Sousa, "Re-Inhabiting Taddle Creek," 238.

28 Ibid., 239.

29 Benn, "Colonial Transformations."

30 Ibid., 69.

31 McNeill, *Something New under the Sun*, 12.

32 Root, Chant, and Heidenreich, *Special Places*, 83–4.

33 Ibid.

34 Cited in Sousa, "Re-Inhabiting Taddle Creek," 234.

35 Cited in ibid., 239.

36 Gibson in Livingston, *The John A. Livingston Reader*, x.

37 Ibid., x.

38 Louv, *Last Child in the Woods*, 36. Also see Selhub and Logan, *Your Brain on Nature*.

39 Livingston, *Rogue Primate*, 134.

40 Livingston, *The John A. Livingston Reader*, 151.

41 Carrothers, Kline, and Livingston, *FESKIT*, Foray Four 9.

42 Ibid., Preamble 1.

43 E.g., Livingston, *The John A. Livingston Reader*, 12; Livingston, *John Livingston*.

44 Shepard, "Ecology and Man," 2.

45 Shepard, *The Only World We've Got*, 61.

46 Bateson, *Steps to an Ecology of Mind*, 454.

47 Ibid., 484.

48 Gibson in Livingston, *The John A. Livingston Reader*, ix.

49 Code, *Ecological Thinking*, 50.

50 Ibid., 50.

51 Martin, "John Livingston, Naturalist 1923–2006."

52 Ibid.

53 Suzuki, *The David Suzuki Reader*, 279.

54 For an analysis of Suzuki's *The Sacred Balance* in relation to a kind of spiritual ethic, see Taylor, *Dark Green Religion*.

55 Suzuki, *The David Suzuki Reader*, 8.

56 Ibid., 8.

57 For analyses of issue and place-based interdisciplinarity, see Robinson, "Being Undisciplined"; Ford, *Beyond the Modern University*; M'Gonigle and Starke, *Planet U.*

58 Harrison, *Gardens*, 60; Dillon, *The Heirs of Plato*.

59 Harrison, *Gardens* 67.

60 Harrison, *Gardens*, 61; also see M'Gonigle and J. Starke, *Planet U.*

61 Livingston, *The John A. Livingston Reader*, 133.

62 Ibid., 298.

63 Ibid., 298.

64 Bai, "Reanimating the Universe," 140. For comparable critiques, see Berman, *The Reenchantment of the World*, 60; Primavesi, *Exploring Earthiness*; Saul, *On Equilibrium*, 45; Williams, *Problems in Materialism and Culture*, 75.

65 Bai, "Reanimating the Universe," 141.

66 Soulé and Press, "What is Environmental Studies?" This issue is also covered in Leduc and Morley, *Five Decades of FES at York*.

67 Carson, *Silent Spring*, 13.

68 Esbjörn-Hargens, "An Ontology of Climate Change," 150; Esbjörn-Hargens and Zimmerman, *Integral Ecology*; Hulme, *Why We Disagree about Climate Change*.

69 Hulme, *Why We Disagree about Climate Change*, 351; Leduc, *Climate, Culture, Change*.

70 Haraway, "Situated Knowledges," 590; Haraway, *How Like a Leaf*.

71 E.g., Abram, *Becoming Animal*; Abram, *The Spell of the Sensuous*; Code, *Ecological Thinking*; Morton, *The Ecological Thought*; Morito, *Thinking Ecologically*.

72 Code, *Ecological Thinking*, 41.

73 Fawcett and Russell, "Remembered: John Livingston." Also see Gibson, "Appreciation."
74 See Cronon, "The Trouble with Wilderness"; Taylor, *Dark Green Religion*.
75 Livingston, *The John A. Livingston Reader*, 342; Also see Cronon, "The Trouble with Wilderness."
76 This analysis is made in Leduc, *Climate, Culture, Change*.
77 I first talked about this in "A Climate for Wisdom?" 3.
78 Leduc, *Climate, Culture, Change*, 30–2.
79 Shepard, *The Only World We've Got*, 63.
80 Cruikshank, *Do Glaciers Listen?*; Wilbert, *Mindful of Famine*; Calvin, *A Brain for All Seasons*; Burroughs, *Climate Change in Prehistory*.
81 Calvin, *A Brain for All Seasons*, 184.
82 Crate and Nuttall, eds., *Anthropology and Climate Change*; Cruikshank, *Do Glaciers Listen?*; Rappaport, *Ritual and Religion in the Making of Humanity*; Wilbert, *Mindful of Famine*.
83 Atleo, *Tsawalk*, 117.
84 Mokuku, "Lehae La Rona," 167.
85 Wilson, *Research Is Ceremony*, 54. Also see Alfred, *Peace, Power, Righteousness*, 23.
86 See Leduc, *Climate, Culture, Change*.
87 Cited in Leduc, "Inuit Economic Adaptations for a Changing Global Climate," 27.
88 Atleo, *Tsawalk*, 118–19.
89 Esbjörn-Hargens and Zimmerman, *Integral Ecology*, 355.
90 Livingston, *Rogue Primate*, 196.
91 Ibid., 196.
92 Cronon, "The Trouble with Wilderness," 171.
93 Ibid., 175.
94 Mills, "The Living Machine."
95 Root, Chant, and Heidenreich, *Special Places*, 238.
96 Ibid., 220.
97 Grady, *Toronto the Wild*, 228.
98 City of Toronto, *High Park: Restoring a Jewel of Toronto's Park System*; City of Toronto, *High Park Woodland and Savannah Management Plan*.
99 Boudreau, Keil, and Young, *Changing Toronto*, 149.
100 Mills, "The Living Machine," 217.
101 Cited in Boudreau, Keil, and Young, *Changing Toronto*, 149.

102 Livingston, *Rogue Primate*, 119.

103 Suzuki with McConnell and Mason, *The Sacred Balance*, 261.

104 Livingston, *Rogue Primate*, 133–4.

105 Haraway, "Cyborgs and Symbionts," xix.

106 Campbell, "Technology and Temporal Ambiguity," 269.

107 Selhub and Logan, *Your Brain on Nature*, 30.

108 Ibid., 32.

109 Fisher, *Radical Ecopsychology*, 158.

110 Selhub and Logan, *Your Brain on Nature*, 30; Fisher, *Radical Ecopsychology*.

111 Desfor and Kiel, *Nature and the City*, 72.

112 Evernden, *The Natural Alien*, xi–xii.

113 Hulme, *Why We Disagree about Climate Change*, 340. Though climate change is not usually seen as a problem "concerned with questions of ultimate reality," some researchers have suggested this is because modern thought is grounded in an "implicit ontology" that is difficult to perceive, let alone question. See Bonner, "Ontology and Climate Change," 2; Chen, "Why do People Misunderstand Climate Change?"; Esbjörn-Hargens, "An Ontology of Climate Change."

114 Timmerman, "The Human Dimensions of Global Change," 9.

115 Latour, *On the Modern Cult of the Factish Gods*, 97.

116 Interview with Farley Mowat in Livingston, *The John A. Livingston Reader*, xxii.

117 Livingston, *The John A. Livingston Reader*, 130.

118 See Desfor and Kiel, *Nature and the City*; Sandberg, Wekerle, and Gilbert, *The Oak Ridges Moraine Battles*.

CHAPTER TWO

1 Woodworth, "Urban Place as an Expression of the Ancestors," 229.

2 Woodworth and Leduc, "Recovering the Tribal Identity of Toronto."

3 Petrides and Wehr, *Eastern Trees*; Root, Chant, and Heidenreich, *Special Places*.

4 Hayes, *Historical Atlas of Toronto*, 12–13; Hayes, *Historical Atlas of Canada*, 169; Johnson, "The Indigenous Environmental History of Toronto."

5 This and subsequent chapters are informed by the thought of Woodworth as expressed in his writings, a co-authored conference presentation with

me, and various public presentations by him. These can all be found in the Bibliography. No interviews were done as part of this research.

6 Williamson, ed., *Toronto*; Woodworth, "Urban Place as an Expression of the Ancestors"; Root, Chant, and Heidenreich, *Special Places*.

7 Williamson, ed., *Toronto*, 52.

8 Johnson, "The Indigenous Environmental History of Toronto."

9 See Johnson, "The Indigenous Environmental History of Toronto"; McIntosh, *Earliest Toronto*; Robinson, *Toronto during the French Régime*.

10 McIntosh, *Earliest Toronto*.

11 Woodworth, "Urban Place as an Expression of the Ancestors," 223, 229.

12 Williamson, ed., *Toronto*, 52.

13 King, *The Truth about Stories*.

14 Woodworth and Leduc, "Recovering the Tribal Identity of Toronto."

15 E.g., Williams, *Problems in Materialism and Culture*; Cronon, "The Trouble with Wilderness"; Sachs, "Environment"; Latour, *We Have Never Been Modern*; Hulme, *Why We Disagree about Climate Change*; Bennett, *Vibrant Matter*; Merchant, *The Death of Nature*; Shiva, *Monocultures of the Mind*; Plumwood, *Environmental Culture*.

16 Chief Thomas with Boyle, *Teachings from the Longhouse*, 10.

17 Woodworth, "Urban Place as an Expression of the Ancestors," 227.

18 Ibid., 225.

19 Woodworth, *The Morning Star*, 171.

20 Woodworth, "Urban Place as an Expression of the Ancestors," 225.

21 Rice, *The Rotinonshonni*, 255.

22 Woodworth, "Urban Place as an Expression of the Ancestors," 51.

23 E.g., Wilson, *Research Is Ceremony*; Harrison, *The Dominion of the Dead*; Cruikshank, *Do Glaciers Listen?*; Rappaport, *Ritual and Religion*. Other references to ancestors are made in subsequent chapters.

24 Leduc, *Climate, Culture, Change*, 4.

25 Freeman, "Indigenous Hauntings in Settler-Colonial Spaces," 212.

26 Ibid., 210.

27 Fawcett, Russell, and Bell, "Guiding Our Environmental Praxis."

28 Sheridan and Longboat, "The Haudenosaunee Imagination and the Ecology of the Sacred," 375.

29 Cited in McGregor, "Coming Full Circle," 388.

30 See chapter 6 of Leduc, *Climate, Culture, Change* for a dialogue on Inuit hunting practices and animals in a context of changing northern climate patterns.

31 Mohawk, *Thinking in Indian*, 274.

32 Morito, *Thinking Ecologically*, 9.

33 Mohawk, *Thinking in Indian*, 277.

34 Longboat, *Owehna'shon*, 278.

35 Rice, *The Rotinonshonni*, 24.

36 Eliade, *The Sacred and the Profane*.

37 Longboat, *Owehna'shon*, 278; Woodworth, *The Morning Star*; Rice, *Rotinonshonni*.

38 Woodworth, *The Morning Star*; Longboat, *Owehna'shon*; Rice, *The Rotinonshonni*.

39 Thomas with Boyle, *Teachings from the Longhouse*, 15.

40 Woodworth, *The Morning Star*, 156.

41 Longboat, *Owehna'shon*, 113; Woodworth, *The Morning Star*; Rice, *The Rotinonshonni*.

42 Longboat, *Owehna'shon*, 273; Woodworth, *The Morning Star*; Rice, *The Rotinonshonni*.

43 Boudreau, Keil, and Young, *Changing Toronto*, 161–2.

44 Ibid., 166.

45 Kennedy et al., "Methodology for Inventorying Greenhouse Gas Emissions," 4.

46 Williamson and MacDonald, "A Resource like No Other," 44.

47 See Leduc, *Climate, Culture, Change* for analysis and related references.

48 Smith, *Decolonizing Methodologies*, 137; Wilson, *Research Is Ceremony*.

49 Hayes, *Historical Atlas of Toronto*, 11–13.

50 Desfor and Kiel, *Nature and the City*, 78.

51 Ibid., 78; Sandberg, Wekerle, and Gilbert, *The Oak Ridges Moraine Battles*.

52 Trigger, *The Children of Aataentsic*; Sioui, *Huron-Wendat*.

53 Johnson, "The Indigenous Environmental History of Ontario," 61; Williamson, ed., *Toronto*.

54 Desfor and Kiel, *Nature and the City*, 76.

55 Williamson and MacDonald, "A Resource like No Other," 50; Freeman, "Indigenous Hauntings in Settler-Colonial Spaces."

56 Johnson, "The Indigenous Environmental History of Toronto," 61; also see McIntosh, *Earliest Toronto*.

57 See Johnson, "The Indigenous Environmental History of Toronto," 61–2; McIntosh, *Earliest Toronto*; Robinson, *Toronto during the French Régime*.

58 Trigger, *The Children of Aataentsic*; Engelbrecht, *Iroquoia*.

59 Trigger, *The Children of Aataentsic*.

60 Innis, *Staples, Markets, and Cultural Change*, 135. Also see MacLennan, *Seven Rivers of Canada*.

61 McIntosh, *Earliest Toronto*, 7.

62 Schmalz, *The Ojibwa of Southern Ontario*, 32.

63 Trigger, *The Children of Aataentsic*; Engelbrecht, *Iroquoia*.

64 Johnson, "The Indigenous Environmental History of Toronto," 59.

65 Cited in Roberts, "Whose Land?" 18.

66 McIntosh, *Earliest Toronto*, 8–9; Robinson, *Toronto during the French Régime*.

67 Cited in McIntosh, *Earliest Toronto*, 19; Fryer and Dracott, *John Graves Simcoe (1752–1806)*; Simcoe, *The Diary of Mrs. John Graves Simcoe*.

68 Simcoe, *The Diary of Mrs. John Graves Simcoe*, 179.

69 Ibid., 179.

70 Ibid., 179.

71 Bobiwash, "The History of Native People in the Toronto Area," 5; Sault cited in Roberts, "Whose Land?" 18.

72 Cited in Innis, *The Fur Trade in Canada*, 89.

73 Berchem, *The Yonge Street Story*, 10; Also see Arthur and Otto, *Toronto*.

74 Fryer and Dracott, *John Graves Simcoe*, 133.

75 Berchem, *The Yonge Street Story*, 10.

76 Fryer and Dracott, *John Graves Simcoe*, 182.

77 Ibid., 182.

78 Frye, *The Bush Garden*, 225, 217.

79 Ibid., 217.

80 Berchem, *The Yonge Street Story*, 12.

81 Ibid., 14.

82 Ibid.

83 Frye, *The Bush Garden*, 224.

84 McIntosh, *Earliest Toronto*, 17.

85 Berchem, *The Yonge Street Story*.

86 Cited in Roberts, "Whose Land?" 18; Bobiwash, "The History of Native People in the Toronto Area."

87 Longboat, *Owehna'shon*, 128.

88 Bobiwash, "The History of Native People in the Toronto Area," 18.

89 Ibid.

90 Ibid., 266.

91 Razack, "Colonization," 265–6.
92 Cited in Roberts, "Whose Land?" 18.
93 Baskerville, *Ontario*, 195.
94 Ibid., 28.
95 Sperling and Gordon, *Two Billion Cars*, 38.
96 Kennedy et al., "Methodology for Inventorying Greenhouse Gas Emissions from Global Cities," 5.
97 Bruce and Cohen, "Impacts of Climate Change in Canada," 81.
98 City of Toronto, *Change Is in the Air*.
99 Woodworth and Leduc, "Recovering the Tribal Identity of Toronto."
100 Woodworth, *The Morning Star*, 177–8.
101 Johnson, "The Indigenous Environmental History of Toronto," 62–3.
102 Robinson, *Toronto during the French Régime*.
103 Hennepin, *A New Discovery of a Vast Country in America*, 77.
104 See Robinson, *Toronto during the French Régime*; Leduc, *Antoine*.
105 Johnson, "The Indigenous Environmental History of Toronto," 62–3; Schmalz, *The Ojibwa of Southern Ontario*; McIntosh, *Earliest Toronto*.
106 Cited in Robinson, *Toronto during the French Régime*, 31.
107 Leduc, *Antoine*.
108 Parmenter, *The Edge of the Woods*, 196.
109 Woodworth and Leduc, "Recovering the Tribal Identity of Toronto"; Parmenter, *The Edge of the Woods*.
110 Woodworth, *The Morning Star*, 75.
111 Sheridan and Longboat, "The Haudenosaunee Imagination," 378.
112 Cited in King and Stefanovic, "Children and Nature in the City." Also see Louv, *Last Child in the Woods*.
113 Hough, *Cities and Natural Process*, 101.
114 Vanderburg, *Living in the Labyrinth of Technology*, 307–8.
115 Ibid., 377.
116 Ibid., 421. Also see Livingston, *Rogue Primate*.
117 Woodworth, "Urban Place as an Expression of the Ancestors," 224.
118 Freeman, "Indigenous Hauntings in Settler-Colonial Spaces"; Johnson, "The Indigenous Environmental History of Toronto."
119 Wilson, *Research Is Ceremony*, 127.
120 Grady, *Toronto the Wild*, 3.
121 Ibid., 3.
122 Bobiwash, "The History of Native People in the Toronto Area," 22.

123 Saul, *A Fair Country*, 253.

124 Ibid.

125 Glickman, *The Picturesque and the Sublime*, 45.

126 Ibid., 59.

127 Ibid.

128 Woodworth, "Urban Place as an Expression of the Ancestors," 228.

129 Hillman, *The Soul's Code*, 62.

130 Ibid., 62.

131 King, *The Truth about Stories*, quoting Ben Okri, 153.

132 Thomas with Boyle, *Teachings from the Longhouse*, 15.

133 Rice, *The Rotinonshonni*, 196.

CHAPTER THREE

1 Cited in Grady, *The Great Lakes*, 68–9.

2 Simcoe, *The Diary of Mrs. John Graves Simcoe*, 254.

3 Cited in Grady, *The Great Lakes*, 68–9.

4 Sagard, *Sagard's Long Journey to the Country of the Huron*, 171.

5 Ottawa *Citizen*, *The Lost Villages*.

6 Ibid.

7 Harper, *Harper's Index*.

8 Cited in Indigenous Environmental Network, *First Nations Responds to Suncor Tar Sands Tailings Breach*.

9 Mohawk, *Thinking in Indian*, 90.

10 Cited in ibid.

11 Ibid., 103.

12 Parham, *From Great Wilderness to Seaway Towns*, 94.

13 Ibid.

14 McKay, *Electric Empire*, 33.

15 Ibid., 33, 206.

16 Ibid., 206.

17 Abley, *Spoken Here*, 187.

18 Cited in Longboat, *Owehna'shon*, 137.

19 Cited in ibid., 137.

20 Cited in ibid.

21 Senior, *From Royal Township to Industrial City*, 15.

22 Cited in Parham, *From Great Wilderness to Seaway Towns*, 8; Senior, *From Royal Township to Industrial City*, 15.

23 Parham, *From Great Wilderness to Seaway Towns*, 8.

24 Simcoe, *The Diary of Mrs. John Graves Simcoe*, 344; also see Bassett, *Elizabeth Simcoe*.

25 Senior, *From Royal Township to Industrial City*.

26 Ibid., 19.

27 Ibid., 20.

28 Legget, *The Seaway*.

29 Harris, *The Reluctant Land*.

30 Baskerville, *Ontario*, 135.

31 Parham, *From Great Wilderness to Seaway Towns*, 66.

32 Baskerville, *Ontario*.

33 Parham, *From Great Wilderness to Seaway Towns*, 66.

34 Legget, *The Seaway*, 59.

35 Senior, *From Royal Township to Industrial City*, 435.

36 Judson, *St. Lawrence Seaway*, 106, 118.

37 Swift and Stewart, *Hydro*.

38 Ibid., ix.

39 Cited in ibid.

40 Innis, *The Fur Trade in Canada*, xxxi; Harris, *The Reluctant Land*, 31.

41 Innis, *The Fur Trade in Canada*, 401.

42 Winks cited in Innis, *The Fur Trade in Canada*, xxxi; Harris, *The Reluctant Land*, 31.

43 Innis, *The Fur Trade in Canada*, 400–1.

44 Swift and Stewart, *Hydro*, 206–7.

45 McKay, *Electric Empire*, 35.

46 Hessing, Howlett, and Summerville, *Canadian Natural Resource and Environmental Policy*, 16.

47 Livingston, *The John A. Livingston Reader*, 166; also see Leopold, *A Sand County Almanac*, 262.

48 Mohawk, *Utopian Legacies*, 11.

49 Ibid., 44.

50 Ibid.

51 Mohawk, *Thinking in Indian*, 274.

52 Mohawk, *Utopian Legacies*, 9–10.

53 A local example is one 2050 projection of a decline of 0.5 metres in Lake Ontar:io and 1.3 metres for the Kaniatarowanenneh's Montreal Harbour, which will affect shoreline infrastructure and eco-systems while diminishing the longue sault's power. See Bruce and Cohen, "Impacts of Climate Change in Canada," 82.

54 Mohawk, *Utopian Legacies*, 12–13.

55 Parham, *From Great Wilderness to Seaway Towns*, 111.

56 Senior, *From Royal Township to Industrial City*, 345.

57 Hudema, *Tar Sands Spills*.

58 Cenovus, *Rising to the Challenges*.

59 Winfield, "Opinion: Lac Mégantic Disaster Caused by a Series of Failures."

60 Cited in Klein, *This Changes Everything*, 345.

61 See chapter 4 of Leduc, *Climate, Culture, Change*.

62 Ibid., 131.

63 Ibid., 132.

64 Ibid.

65 Bateson, *Steps to an Ecology of Mind*.

66 Nikiforuk, *The Energy of Slaves*, 120.

67 Livingston, *The John A. Livingston Reader*, 139.

68 Bateson, *Steps to an Ecology of Mind*, 320.

69 Ibid., 326.

70 Homer-Dixon, "The Tar Sands Disaster."

71 Nikiforuk, *The Energy of Slaves*, 182.

72 Homer-Dixon, *The Ingenuity Gap*; Homer-Dixon, *Environment, Scarcity, and Violence*.

73 Homer-Dixon, "The Tar Sands Disaster."

74 Ibid.

75 Nikiforuk, *The Energy of Slaves*, 172.

76 Innis, *Staples, Markets, and Cultural Change*, 473, 475. Also see Berland, *North of Empire*, 283.

77 Hulme, *Why We Disagree about Climate Change*, 351.

78 Ibid., 352.

79 Bateson, *Steps to an Ecology of Mind*, 330.

80 Bennett, *Vibrant Matter*, 25.

81 Ibid., 24.

82 Bateson, *Steps to an Ecology of Mind*, 332–3.

83 Bennett, *Vibrant Matter*, 25.

84 Ibid.
85 Ibid.
86 Trigger, *The Children of Aataentsic*, 261.
87 Sagard, *Sagard's Long Journey*, 160–1. Also see Podruchny, *Making the Voyageur World*.
88 Cited in Podruchny, *Making the Voyageur World*, 63.
89 Ibid., 55.
90 Ibid.
91 Cited in ibid., 58.
92 Ibid., 72.
93 Ibid., 54–5.
94 Greer, *The People of New France*, 12.
95 Ibid.
96 Ibid., 13.
97 Ibid.
98 Miller, *Skyscrapers Hide the Heavens*, 16; Morissonneau, "The Ungovernable People," 17.
99 Podruchny, *Making the Voyageur World*, 57.
100 Leduc, *Antoine*, 106.
101 Behringer, *Witches and Witch-Hunts*, 29; Moogk, *La Nouvelle France*.
102 Trigger, *Natives and Newcomers*, 195.
103 Cited in Greer, *The People of New France*, 84.
104 Morissonneau, "The Ungovernable People: French-Canadian Mobility and Identity," 17–18.
105 Dickason, "From 'One Nation' in the Northeast to 'New Nation' in the Northwest," 23.
106 Cited in Moogk, *La Nouvelle France*, 45–6.
107 Sagard, *Sagard's Long Journey*, 160–1.
108 Trigger, *The Children of Aataentsic*, 263.
109 Saul, *A Fair Country*.
110 Podruchny, *Making the Voyageur World*, 308.
111 Ibid, 293, 305.
112 Berland, *North of Empire*, 76, 97; Innis, *Empire and Communications*.
113 Innis, *The Fur Trade in Canada*, 392.
114 Podruchny, *Making the Voyageur World*, 306.
115 Ibid., 14.
116 Morito, *Thinking Ecologically*, 259.

117 Podruchny, *Making the Voyageur World*, 14.

118 Bennett, *Vibrant Matter*, 36.

119 Ibid., 28.

120 Swift and Stewart, *Hydro*, 3.

121 Vanderburg, *Living in the Labyrinth of Technology*, 307–8.

122 Bateson, *Steps to an Ecology of Mind*, 322.

123 Mohawk, *Thinking in Indian*, 103.

124 Loftin, *Religion and Hopi Life*; Mohawk, *Thinking in Indian*.

125 Mohawk, *Thinking in Indian*, 92.

126 Bennett, *Vibrant Matter*, 122.

127 Suzuki with McConnell and Mason, *The Sacred Balance*, 158.

128 Latour, *On the Modern Cult of the Factish Gods*, 97; Latour, *We Have Never Been Modern*.

129 Bateson, *Steps to an Ecology of Mind*, 331.

130 See chapter 5 in Leduc, *Climate, Culture, Change*.

131 Bateson, *Steps to an Ecology of Mind*, 335.

CHAPTER FOUR

1 Cited in Greer, *Mohawk Saint*, 135.

2 Cited in Reguly, "In an Act of Atonement."

3 Greer, *Mohawk Saint*; Carpenter, *The Renewed, the Destroyed, and the Remade*.

4 Greer, *Mohawk Saint*, 52.

5 Patterson, *White Wampum*, 303.

6 Greer, *Mohawk Saint*, 204–5; Abley, "Reclaiming Kateri."

7 Saul, *A Fair Country*, 314.

8 Abley, "Reclaiming Kateri."

9 Mohawk, *Thinking in Indian*, 10.

10 Bonaparte, *A Lily among Thorns*. Also see Abley, "Reclaiming Kateri," where competing cultural approaches to Tekakwitha are discussed, including that of Bonaporte.

11 Lawrence, *"Real" Indians and Others*, 134.

12 Greer, *Mohawk Saint*. Other sources will be referenced over the following chapters that further clarify the view of saints as ancestors and the differences with Indigenous traditions.

13 Ibid., 199.

14 Ibid., 118.

15 Bateson, *Steps to an Ecology of Mind*, 335.

16 Patterson, *White Wampum*, 283.

17 Ibid., 270.

18 Cited in Hayes, *Historical Atlas of Canada*, 53.

19 Patterson, *White Wampum*, 269–70; Greer, *Mohawk Saint*, 135.

20 Parmenter, *The Edge of the Woods*, 128.

21 Ibid., 155; Abley, *Spoken Here*, 180.

22 Trigger, *The Children of Aataentsic*, 76; Engelbrecht, *Iroquoia*, 100.

23 Armstrong, *Saint Francis*, 24.

24 Le Goff, *The Medieval Imagination*, 54. For a review of influences on Saint Francis, also see Armstrong, *Saint Francis*, and Sorrell, *St. Francis of Assisi and Nature*.

25 Armstrong cited in Sorrell, *St. Francis of Assisi and Nature*, 24.

26 Marsden, *Sea-Road of the Saints*.

27 McNeill, *The Celtic Churches*.

28 Ibid., 73.

29 Campbell, *Primitive Mythology*, 470.

30 Markale, *The Cathedral of the Black Madonna*, 214; Jung and von Franz, *The Grail Legend*, 191.

31 Fanning, *Mystics of the Christian Tradition*.

32 Marsden, *Sea-Road of the Saints*, 48, 45.

33 Ibid., 314; White, Jr, "The Historical Roots of our Ecologic Crisis." Also see Taylor, *Dark Green Religion*, on the Western tradition of what he refers to as dark green religion.

34 Livingston, *The John A. Livingston Reader*, 73.

35 Rice, *The Rotinonshonni*, 216.

36 Ibid.

37 Cited in Sorrell, *St. Francis of Assisi and Nature*, 94; Underhill, *Mysticism*, 77.

38 Livingston, *Rogue Primate*, 117.

39 Mohawk, *Thinking in Indian*, 5.

40 Rice, *The Rotinonshonni*, 216.

41 Cited in Sagard, *Sagard's Long Journey*, xxix.

42 Cited in ibid., xv.

43 Sorrell, *St. Francis of Assisi and Nature*, 89–90; White, Jr, "The Historical Roots of our Ecologic Crisis"; Livingston, *The John A. Livingston Reader*.

44 Sagard, *Sagard's Long Journey*, 229.

45 Gibson cited in Livingston, *The John A. Livingston Reader*, xix.

46 Sorrell, *St. Francis of Assisi and Nature*.

47 Ibid., 38. It needs to be noted that Francis often failed to keep fasts and practise other austerities, and would write: "As a mother loves to see her child imbibe nourishment and expand, so God loves to see his children thrive on the nutriment he has furnished." Cited in ibid., 76.

48 Meek, "Between Faith and Folklore," 269; Bruce, *Prophecy, Miracles, Angels, and Heavenly Light?*

49 Greer, *Mohawk Saint*, 116–17.

50 Ibid., 144.

51 Greer, *Mohawk Saint*, 199.

52 Patterson, *White Wampum*, 269–70.

53 Greer, *Mohawk Saint*, 118, 142–3.

54 Ibid., 118.

55 Ibid., 143.

56 Ibid., 55.

57 Ibid., 199.

58 Blackburn, *Harvest of Souls*, 42; Dickason, *The Myth of the Savage*.

59 Greer, *The People of New France*, 85.

60 Greer, *Mohawk Saint*, 131.

61 Ibid., 166.

62 Cited in ibid., 131.

63 Ibid., 120.

64 Trigger, *The Children of Aataentsic*; Greer, *The Jesuit Relations*.

65 Blackburn, *Harvest of Souls*, 36.

66 Cited in Greer, *The Jesuit Relations*, 46.

67 Cited in Carpenter, *The Renewed, the Destroyed, and the Remade*, 12.

68 Greer, *Mohawk Saint*, 187; Blackburn, *Harvest of Souls*; Miller, *Skyscrapers Hide the Heavens*.

69 Trigger, *Natives and Newcomers*, 331.

70 Trigger, *Natives and Newcomers*; Eccles, *The French in North America*.

71 Delâge, *Bitter Feast*, 186.

72 Trigger, *Natives and Newcomers*, 259.

73 Rice, *The Rotinonshonni*, 267.

74 Ibid., 267.

75 Ibid., 16.

76 Thomas with Boyle, *Teachings from the Longhouse*, 17.

77 Ibid., 267.

78 Carpenter, *The Renewed, the Destroyed, and the Remade*; Miller, *Skyscrapers Hide the Heavens*; Delâge, *Bitter Feast*.

79 Parmenter, *The Edge of the Woods*, 72.

80 Ibid.

81 Thomas with Boyle, *Teachings from the Longhouse*.

82 Ibid., 127.

83 Cited in ibid., 28.

84 Ibid., 37. Also see Trigger, *The Children of Aataentsic*.

85 Trigger, *The Children of Aataentsic*. 82.

86 Greer, *Mohawk Saint*, 26–7.

87 Rice, *The Rotinonshonni*, 265.

88 Ibid., 26–7.

89 Ibid.

90 Ibid., 96.

91 Ibid., 57.

92 Ibid., 53.

93 Ibid.

94 Greer, *Mohawk Saint*, 53.

95 Cited in Underhill, *Mysticism*, 174.

96 Greer, *Mohawk Saint*, 120.

97 Underhill, *Mysticism*, 174.

98 Ibid.

99 Greer, *Mohawk Saint*, 124.

100 Woodworth and Leduc, "Recovering the Tribal Identity of Toronto."

101 Woodworth, *The Morning Star*, 132.

102 Ibid., 132–3.

103 Bobiwash, "The History of Native People in the Toronto Area," 19.

104 Ibid.

105 Miller, *Skyscrapers Hide the Heavens*, 254.

106 Lawrence, *"Real" Indians and Others*, 48–9.

107 Ibid., 55.

108 Bobiwash, "The History of Native People in the Toronto Area," 19.

109 Palmater, *Beyond Blood*, 195.

110 Ibid. Also see *Alfred, Peace, Power, Righteousness* for an analysis of these issues.

111 Palmater, *Beyond Blood*, 196.
112 Thomas with Boyle, *Teachings from the Longhouse*, 17.
113 Ibid., 143.
114 Ibid., 142.
115 Ibid., 144.
116 Ibid., 145. This challenge for the Haudenosaunee Good Mind is also discussed in terms of conscience and governance in Alfred, *Peace, Power, Righteousness*, 49. He writes: "In choosing between revitalizing indigenous forms of government and maintaining the European forms imposed on them, Native communities have a choice between two radically different kinds of social organization: One based on conscience and the authority of the good, and the other on coercion and authoritarianism."
117 Ibid., 144–5.
118 Ibid., 172.
119 Woodworth, *The Morning Star*, 129.
120 Longboat, *Owehna'shon*, 53.
121 Ibid.; Woodworth and Leduc, "Recovering the Tribal Identity of Toronto."
122 Woodworth, *The Morning Star*, 17.
123 See Rice, *The Rotinonshonni*, 265.
124 Woodworth, *The Morning Star*, 172.
125 Greer, *Mohawk Saint*, 22–3.
126 Ibid., 14.
127 Ibid., 170.
128 Berkes, "Epilogue," 348.
129 Saul, *A Fair Country*, 74.
130 Ibid.
131 Ophuls, *Plato's Revenge*, 177.
132 Ibid.
133 Ibid.
134 Ibid., 169.
135 Dickason, "From 'One Nation' in the Northeast to 'New Nation' in the Northwest," 19.
136 Ibid.; Martel, "When a Majority Becomes a Minority," 79.
137 See Anderson, *"Métis"*; and Anderson, "'I'm Métis, What's your Excuse?'"
138 Lawrence, *"Real" Indians and Others*, 135.
139 Saul, *A Fair Country*, 21.
140 Ibid., 138.

141 Ibid., 186.

142 Ibid., 138.

143 Ibid., 186, 138.

144 Berland, *North of Empire*, 219.

145 Harris, Diamond, Iyer, and Payne, *Don't Supersize Me!* For a broader analysis of our carbon-climate situation, also see Helm, *The Carbon Crunch*.

146 Ibid.

147 Ibid.

148 Anable, *Personal Mobility and Energy Demand*.

149 Smil, "The Long Slow Rise of Solar and Wind."

150 Ibid., 56–7.

151 Ibid. Also see Hansen, *Climate and Energy*.

152 Nikiforuk, *The Energy of Slaves*, 226–7.

153 Ophuls, *Plato's Revenge*, 162.

154 Suzuki with McConnell and Mason, *The Sacred Balance*, 182.

155 Ibid.

156 Bateson, *Steps to an Ecology of Mind*, 326.

157 Nikiforuk, *The Energy of Slaves*, 227.

158 Ophuls, *Plato's Revenge*, 187.

159 Thomas with Boyle, *Teachings from the Longhouse*, 135.

160 See Leduc and Crate, "Reflexive Shifts in Climate Research and Education"; Crate and Nuttall, eds., *Anthropology and Climate Change*.

161 Vanderheiden, *Atmospheric Justice*, 252, 243.

162 Ophuls, *Plato's Revenge*, 192. Also see Nikiforuk, *The Energy of Slaves*.

163 Letson and Higgins, *The Jesuit Mystique*, 111.

164 Ibid., 114.

165 Ibid., 115; Gutiérrez, *A Theology of Liberation*, 25.

166 Boff, *Ecology and Liberation*, 77. Also see Scharper, *Redeeming the Time*; Primavesi, *Sacred Gaia*.

167 Greer, *Mohawk Saint*, 111.

168 Rice, *The Rotinonshonni*, 265.

169 Greer, *Mohawk Saint*, 31. For other discussions on cultural braiding, third-space theory, or hybridization, see Morrison, *The Solidarity of Kin*; Lutz 'Pomo Wawa,' *Makuk*.

170 Ophuls, *Plato's Revenge*, 186.

171 Santmire, *The Travail of Nature*, 109.

172 Berry, *The Great Work*, 195.
173 Ibid., x.
174 Ibid., 195.
175 Ibid., x–xi.
176 Taylor, *Green Sisters*, 164.
177 Ibid.
178 Ibid, 113–14.
179 Cited in Sorrell, *St. Francis of Assisi and Nature*, 101.
180 Greer, *Mohawk Saint*, 18.
181 Lawrence, *"Real" Indians and Others*, 14–15.
182 Family genealogy done by Colleen Leduc.
183 Lawrence, *"Real" Indians and Others*, 135.

CHAPTER FIVE

1 Sioui, *Histories of Kanatha*, 162.
2 Ibid., 111.
3 Sioui, *Huron-Wendat*, 112–13.
4 A hundred and fifty years after Cartier, Lorette was primarily composed of Jesuit converts who survived the dispersal from the Wendat territory in southern Ontar:io, though there was a period in the 1670s when about fifty Mohawk also lived in the village. By 1695, Lorette's two hundred families were referred to by the Haudenosaunee "as 'nephews' of the governor of New France, suggesting that there was likely no longer a substantial League-affiliated Iroquois presence." See Parmenter, *The Edge of the Woods*, 386.
5 Cited in Kenton, *Indians of North America, Volume 2*, 297.
6 Sioui, *Histories of Kanatha*, 162.
7 Ibid., 133; Trigger, *The Children of Aataentsic*, 77.
8 Woodworth, *The Morning Star*.
9 Ibid., 104.
10 Sioui, *Histories of Kanatha*, 133.
11 Ibid.; Woodworth, *The Morning Star*; Rice, *The Rotinonshonni*; King, *The Truth about Stories*.
12 Sioui, *Histories of Kanatha*, 115.
13 Cited in Greer, *The Jesuit Relations*, 43.
14 Sioui, *Histories of Kanatha*, 97.
15 Ibid.

16 Ibid., 115.

17 Chenu, *Nature, Man, and Society in the Twelfth Century*, 103.

18 Cited in Skogan, *Mary of Canada*, 63.

19 Sioui, *Histories of Kanatha*, 294.

20 Ibid., 297–8.

21 Ibid., 298.

22 Cited in Greer, *The Jesuit Relations*, 41–3.

23 Sioui, *Histories of Kanatha*, 298.

24 Cited in ibid.

25 Sioui, *Huron-Wendat*, 136.

26 Ibid., 137.

27 Greer, *The Jesuit Relations*, 37.

28 Sioui, *Huron-Wendat*, 16; Trigger, *The Children of Aataentsic*;
    Engelbrecht, *Iroquoia*; Woodworth, *The Morning Star*; Rice, *The
    Rotinonshonni*; King, *The Truth about Stories*.

29 Cited in Greer, *The Jesuit Relations*, 42.

30 Woodworth, *The Morning Star*, 104.

31 Ibid. Also see Campbell, *Historical Atlas of World Mythology, Volume II*,
    145; Woodworth, *The Morning Star*; Sioui, *Huron-Wendat*.

32 Rice, *The Rotinonshonni*, 24.

33 Ibid., 22.

34 Ibid., 37.

35 Woodworth, *The Morning Star*, 104.

36 Rice, *The Rotinonshonni*, cited in Campbell, *Historical Atlas of World
    Mythology, Volume II*.

37 Ibid., 40.

38 Cited in ibid., 146.

39 Ibid., 156.

40 Rice, *The Rotinonshonni*, 40.

41 Woodworth, *The Morning Star*, 104; Sioui, *Huron-Wendat*, 16; Rice,
    The Rotinonshonni.

42 Cited in Greer, *The Jesuit Relations*, 42.

43 This island stretched "about fifty-six kilometres from east to west and
    thirty-two kilometres from north to south" and contained twenty-five
    villages by the time of Brûlé and Sagard. Many Laurentian Iroquois from
    Kanatha went westward to Wendake following the disease that arrived
    with Cartier's ship, with oral stories suggesting that they underwent a

dispersal not dissimilar to that of the colonial period of Saint Tekakwitha and the Wendat's return to the Kaniatarowanenneh. See Sioui, *Huron-Wendat*, 90; Trigger, *The Children of Aataentsic*; Parmenter, *The Edge of the Woods*.

44 Sioui, *Huron-Wendat*, 90.

45 Sioui, *Histories of Kanatha*, 133. Also cited in Greer, *The Jesuit Relations*, 42–3; Woodworth, *The Morning Star*; Rice, *The Rotinonshonni*.

46 Rice, *The Rotinonshonni*, 42.

47 Ibid., 16.

48 Ibid., 42.

49 Ibid.

50 Ibid., 18.

51 King, *The Truth about Stories*, 13.

52 Ibid., 15–16.

53 Ibid., 17–18.

54 Ibid., 28.

55 Cited in Greer, *The Jesuit Relations*, 41.

56 Ibid., 42.

57 Sagard, *Sagard's Long Journey*, 170.

58 King, *The Truth about Stories*, 24.

59 Ibid.

60 Frank, *Letting Stories Breathe*, 156.

61 Ibid., 24.

62 Ibid., 156.

63 White, Jr, "The Historical Roots of our Ecologic Crisis," 1203.

64 Ibid., 1205.

65 See Scharper, *Redeeming the Time*; Gardner, "Engaging Religion in the Quest for a Sustainable World."

66 Livingston, *The John A. Livingston Reader*, 305.

67 Ibid., 304.

68 Rice, *The Rotinonshonni*, 42; King, *The Truth about Stories*.

69 King, *The Truth about Stories*, 27.

70 White, Jr, "The Historical Roots of our Ecologic Crisis," 1206.

71 Woodworth, *The Morning Star*, 105.

72 Baring and Cashford, *The Myth of the Goddess*, 557.

73 Cited in Kenton, *Indians of North America*, 298.

74 Markale, *The Cathedral of the Black Madonna*, 43.
75 Chenu, *Nature, Man, and Society in the Twelfth Century*, 118.
76 Eliade, *The Sacred and the Profane*, 58–9.
77 Scott, *The Gothic Enterprise*, 132.
78 Ibid., 133.
79 Carabine, *John Scottus Eriugena*, 16.
80 Chenu, *Nature, Man, and Society in the Twelfth Century*, 4.
81 Favier, *The World of Chartres*, 132.
82 Prache, *Chartres Cathedral*, 61–2; Favier, *The World of Chartres*.
83 Cited in Woodworth, *The Morning Star*, 184–5.
84 Ibid., 110.
85 Cited in Greer, *The Jesuit Relations*, 46–7.
86 Ibid.
87 Cited in Sorrell, *St. Francis of Assisi and Nature*, 101.
88 Woodworth, *The Morning Star*, 110.
89 Baring and Cashford, *The Myth of the Goddess*, 579.
90 Newman, *Sister of Wisdom*, 102.
91 Hildegard, *Symphonia*, 127.
92 Newman introduction to ibid., 36. Contemporaries like Alan of Lille depicted Mary in various forms, including a Nature goddess who stated: "His working is one, whereas mine is many" (cited in Chenu, *Nature, Man, and Society in the Twelfth Century*, 47).
93 Rice, *The Rotinonshonni*, 53.
94 Cited in Kenton, *Indians of North America*, 298.
95 Patterson, *White Wampum*, 277.
96 Cited in Skogan, *Mary of Canada*, 59.
97 Lampman cited in Altmeyer, "Three Ideas of Nature in Canada," 100.
98 Gersh, *From Iamblichus to Eriugena*, 26.
99 Pseudo-Dionysius, *Pseudo-Dionysius*, 160.
100 Chenu, *Nature, Man, and Society in the Twelfth Century*, 24.
101 Carabine, *John Scottus Eriugena*, 16.
102 Carabine, *John Scottus Eriugena*; McNeill, *The Celtic Churches*.
103 Eriugena, *Periphyseon*, 119.
104 White, Jr, "The Historical Roots of our Ecologic Crisis," 1206.
105 Eriugena, *Periphyseon*, 275.
106 Gersh, *From Iamblichus to Eriugena*, 2, 23.

107 Serres, *Angels, a Modern Myth*, 25.

108 Pseudo-Dionysius, *Pseudo-Dionysius*, 187.

109 Adomnán of Iona, *Life of St Columba*, 220. Also see O'Reilly, "The Wisdom of the Scribe and the Fear of the Lord," 199; Bruce, *Prophecy, Miracles, Angels, and Heavenly Light?*

110 Bruce, *Prophecy, Miracles, Angels, and Heavenly Light?* 130.

111 Marsden, *Sea-Road of the Saints*, 132; Adomnán of Iona, *Life of St Columba*, 218.

112 Le Goff, *The Medieval Imagination*, 52–3.

113 Bechmann, *Trees and Man*, 262.

114 Markale, *The Cathedral of the Black Madonna*, 59; Baring and Cashford, *The Myth of the Goddess*, 595.

115 Gimpel, *The Medieval Machine*, cited in Scott, *The Gothic Enterprise*, 11.

116 Coates, *Nature*, 63.

117 Primavesi, *Sacred Gaia*, 126; Primavesi, *Gaia and Climate Change*.

118 Primavesi, *Sacred Gaia*, 123.

119 Ibid., 153.

120 Gersh, *From Iamblichus to Eriugena*, 283.

121 Eriugena, *Periphyseon*, 192.

122 Primavesi, *Sacred Gaia*, 126.

123 Sioui, *Huron-Wendat*, 114.

124 Ibid.

125 Berry, *The Great Work*, x.

126 Ibid., 32.

127 Teilhard de Chardin, *The Phenomenon of Man*, 301. Also see Teilhard de Chardin, *The Divine Milieu*.

128 Rice, *The Rotinonshonni*, 179.

129 Ibid., 40.

130 Favier, *The World of Chartres*, 132.

131 Newman introduction to Hildegard, *Symphonia*, 56.

132 Rogers, *Nature and the Crisis of Modernity*, 174.

133 Ibid.

134 Ibid.

135 Hulme, *Why We Disagree about Climate Change*, 355.

136 Demeritt, "The Construction of Global Warming and the Politics of Science," 325.

137 Chen, "Why Do People Misunderstand Climate Change?" 40; Bonner, "Ontology and Climate Change"; Hulme, *Why We Disagree about Climate Change*; Esbjörn-Hargens, "An Ontology of Climate Change."

138 Hulme, *Why We Disagree about Climate Change*. Also see Leduc, *Climate, Culture, Change.*

139 E.g., Hulme, *Why We Disagree about Climate Change*; Demeritt, "The Construction of Global Warming and the Politics of Science," 314.

140 Demeritt, "The Construction of Global Warming and the Politics of Science," 329.

141 Rogers, *Nature and the Crisis of Modernity*, 95.

142 Hulme, *Why We Disagree about Climate Change*, 355.

143 Ibid.

144 Latour, *On the Modern Cult of the Factish Gods*, 121.

145 Ibid., 121.

146 Ibid., 9.

147 Livingston, *Rogue Primate*, 136.

148 Latour, *On the Modern Cult of the Factish Gods*, 106.

149 See Sewall, "Beauty and the Brain," 271–2; Selhub and Logan, *Your Brain on Nature.*

150 Carr, *The Shallows*, 156.

151 Ibid., 196.

152 Innis, *Empire and Communications*, 196.

153 Hindman, *The Myth of Digital Democracy*, 18–19.

154 Latour, *On the Modern Cult of the Factish Gods*, 106.

155 Ibid.

156 Ibid., 107.

157 Selhub and Logan, *Your Brain on Nature*, 53; Sewall, "Beauty and the Brain"; Louv, *Last Child in the Woods.*

158 Sewall, "Beauty and the Brain," 281.

159 Latour, *On the Modern Cult of the Factish Gods*, 106–7.

160 Leopold, *A Sand County Almanac*, 239.

161 Livingston, *Rogue Primate*, 109.

162 Ibid.; also see Rogers, *Nature and the Crisis of Modernity*, 93.

163 Rogers, *Nature and the Crisis of Modernity*, 113.

164 Livingston, *Rogue Primate*, 107.

165 Rogers, *Nature and the Crisis of Modernity*, 93.

166 Ibid.

167 Latour, *On the Modern Cult of the Factish Gods*, 7.

168 Ibid., 50.

169 This debate between Latour and Serres is documented in Keller, *Face of the Deep*, 176.

170 Latour, *On the Modern Cult of the Factish Gods*, 123.

171 Ibid., 7.

172 Livingston, *Rogue Primate*, 117.

173 Ibid., 284.

174 Cited in Chenu, *Nature, Man, and Society in the Twelfth Century*, 103.

175 Catholic Encyclopaedia, "St. Lawrence."

176 Ibid.

177 Ibid., 217.

178 Primavesi, *Gaia and Climate Change*, 35.

179 Moogk, *La Nouvelle France*, 255.

180 Bergmann, "Theology in its Spatial Turn," 364.

181 Chenu, *Nature, Man, and Society in the Twelfth Century*, 47.

182 Sioui, *Histories of Kanatha*, 97.

183 Frank, *Letting Stories Breathe*, 153.

CHAPTER SIX

 1 Layard cited in Campbell, *The Mythic Image*, 463; Cipolla, *Labyrinth*; Favier, *The World of Chartres*; Chevalier and Gheerbrant, *Dictionary of Symbols*.

 2 Chenu, *Nature, Man, and Society in the Twelfth Century*, 18.

 3 This window is closer to the labyrinth and Royal Portal than the Black Madonna.

 4 Rice, *The Rotinonshonni*, 309–10.

 5 Ibid.; Woodworth, *The Morning Star*; Campbell, *Historical Atlas of World Mythology*.

 6 Thomas with Boyle, *Teachings from the Longhouse*, 23.

 7 Cited in Woodworth, *The Morning Star*, 126.

 8 Thomas with Boyle, *Teachings from the Longhouse*, 23.

 9 Ibid., 26.

10 Rice, *The Rotinonshonni*, 298; Woodworth, *The Morning Star*.

11 Campbell, *Historical Atlas of World Mythology*, 141; Woodworth, *The Morning Star*; Rice, *The Rotinonshonni*. The many duties of the Good Message go beyond the scope of this chapter, but a detailed telling of this story and duties can be found in Thomas with Boyle, *Teachings from the Longhouse*.

12 Woodworth, *The Morning Star*, 127; Rice, *The Rotinonshonni*.

13 Genesis, cited in Cohn, *Noah's Flood*, 12.

14 Rice, *The Rotinonshonni*, 48; Woodworth, *The Morning Star*.

15 Campbell, *Historical Atlas of World Mythology*, 160.

16 Ibid.

17 Woodworth, *The Morning Star*, 105–6.

18 Leduc, *Climate, Culture, Change*, 153.

19 Calvin, *A Brain for All Seasons*, 295.

20 Hulme, *Why We Disagree about Climate Change*, 341.

21 Chevalier and Gheerbrant, *Dictionary of Symbols*; Cipolla, *Labyrinth*.

22 Cipolla, *Labyrinth*, 30–1.

23 Woodworth, *The Morning Star*, 205–6; Woodworth and Leduc, "Recovering the Tribal Identity of Toronto."

24 Rice, *The Rotinonshonni*, 190.

25 Baring and Cashford, *The Myth of the Goddess*, 646.

26 Eliade, *The Sacred and the Profane*, 62.

27 Markale, *The Cathedral of the Black Madonna*, 192; Baring and Cashford, *The Myth of the Goddess*.

28 Sioui, *Huron-Wendat*, 24.

29 Rice, *The Rotinonshonni*, 189.

30 Sioui, *Huron-Wendat*, 90; Trigger, *The Children of Aataentsic*, 30.

31 Rice, *The Rotinonshonni*, 43; Woodworth, *The Morning Star*.

32 Rice, *The Rotinonshonni*, 43; Woodworth, *The Morning Star*.

33 Rice, *The Rotinonshonni*, 43–4.

34 Woodworth, *The Morning Star*, 185.

35 Ibid., 47–8.

36 Ibid., 48.

37 Ibid., 50.

38 Cited in Parmenter, *The Edge of the Woods*, xxxvi.

39 Woodworth, *The Morning Star*, 203.

40 Trigger, *The Children of Aataentsic*, 77–8.

41 Ibid.; Rice, *The Rotinonshonni*.
42 Campbell, *Historical Atlas of World Mythology*, 160.
43 Rice, *The Rotinonshonni*; Woodworth, *The Morning Star*.
44 Sioui, *Huron-Wendat*, 24–5.
45 Rice, *The Rotinonshonni*, 55.
46 Ibid., 61.
47 Ibid., 190.
48 Chevalier and Gheerbrant, *Dictionary of Symbols*.
49 Rice, *The Rotinonshonni*, 57.
50 Williamson, ed., *Toronto*; Root, Chant and Heidenreich, *Special Places*.
51 Macdougall, *Frozen Earth*, 14.
52 Cited in Campbell, *Historical Atlas of World Mythology*, 149.
53 Ibid.
54 City of Toronto, *High Park: Restoring a Jewel of Toronto's Park System*, 5; Root, Chant, and Heidenreich, *Special Places*.
55 Johnson, "The Indigenous Environmental History of Toronto," 60.
56 Eyles, *Ontario Rocks*; Williamson, ed., *Toronto*.
57 Judd and Speirs, *A Naturalist's Guide to Ontario*, 133.
58 Williamson, ed., *Toronto*, 20; Eyles, *Ontario Rocks*, 240.
59 Eyles, *Ontario Rocks*, 239; Williamson, ed., *Toronto*.
60 Trigger, *Natives and Newcomers*, 76.
61 Trigger, *The Children of Aataentsic*; Sioui, *Huron-Wendat*.
62 Calvin, *A Brain for All Seasons*, 18.
63 Woodworth, *The Morning Star*, 128.
64 Calvin, *A Brain for All Seasons*, 93. Also see Burroughs, *Climate Change in Prehistory*.
65 Calvin, *A Brain for All Seasons*, 93.
66 Hayden, *Shamans, Sorcerers and Saints*, 31–2. Also see Bellah, *Religion in Human Evolution*.
67 See Rogers, Timmerman, Leduc, and Dickinson, "The Why of the 'Hau'"; Godbout, with Caillé and Winkler, *The World of the Gift*.
68 Rice, *The Rotinonshonni*, 138.
69 Ibid., 48.
70 Ibid.
71 Ibid., 117.
72 Cohn, *Noah's Flood*, 16.

73 Eliade, *Patterns in Comparative Religion*, 211.

74 Cohn, *Noah's Flood*, 16; Eliade, *Patterns in Comparative Religion*, 211.

75 Cohn, *Noah's Flood*, 14.

76 Ibid., 1.

77 Sandars, *The Epic of Gilgamesh*, 108; Cohn, *Noah's Flood*; Hentsch, *Truth or Death*.

78 Sandars, *The Epic of Gilgamesh*, 108.

79 Flannery, *The Eternal Frontier*, 149–50.

80 Fagan, *The Long Summer*; Macdougall, *Frozen Earth*, 110–11. These cycling changes continued such that by 8,200 years ago another build-up of water entered the Atlantic through the Hudson Bay, once again leading to an ocean-mediated decrease in average global temperatures.

81 Wilson, *Before the Flood*, 8–9.

82 Ibid., xiv.

83 Johnston, *Ojibway Heritage*, 13.

84 Ibid.

85 Deloria, *God Is Red*, 72.

86 Wilson, *Before the Flood*, 12.

87 MacLennan and De Visser, *Rivers of Canada*, 13.

88 Suzuki with McConnell and Mason, *The Sacred Balance*, 103.

89 Deloria, *God Is Red*, 72.

90 Leduc, *Climate, Culture, Change*, 153.

91 Johnston, *Ojibway Heritage*, 14.

92 Johnston, *Honour Earth Mother*, xi.

93 Ibid., xiv. Also see McGregor, "Coming Full Circle."

94 Trigger, *The Children of Aataentsic*; Sioui, *Huron-Wendat*.

95 Sandars, *The Epic of Gilgamesh*, 106–7.

96 Ibid., 66.

97 Ibid., 69.

98 Ibid., 106–7.

99 Ibid.

100 Ibid, 111.

101 Hentsch, *Truth or Death*, 86; Cohn, *Noah's Flood*.

102 Thompson, *The Time Falling Bodies Take to Light*, 202–3.

103 Thompson, *Imaginary Landscape*, 80.

104 Thompson, *The Time Falling Bodies*, 202.

105 Ibid.

106 See Leduc, *Climate, Culture, Change*; Hulme, *Why We Disagree about Climate Change*.

107 Cheney and Weston, "Environmental Ethics as Environmental Etiquette," 134. See as well Schönfeld, "Introduction – Plan B: Global Ethics on Climate Change." I also began considering questions about climate and etiquette in a chapter written for Schönfeld, ed., *Global Ethics on Climate Change*.

108 Thompson, *The Time Falling Bodies*, 202.

109 Pseudo-Dionysius, *Pseudo-Dionysius*, 136. This mystic practice, which was a central influence in Europe right up to the Enlightenment, offers a longer and more circuitous route to wisdom. See O'Rourke, *Pseudo-Dionysius and the Metaphysics of Aquinas*, 58.

110 Chenu, *Nature, Man, and Society in the Twelfth Century*, 138.

111 Berry, *The Great Work*, 198.

112 Woodworth, *The Morning Star*, 177–8.

113 Cited in Hulme, *Why We Disagree about Climate Change*, 348.

114 Ibid., 352.

115 Victor, *Managing without Growth*, 2.

116 Ibid., 23.

117 Ibid., 168.

118 Rice, *The Rotinonshonni*, 88.

119 Victor, *Managing without Growth*, 149.

120 Fisher, *Radical Ecopsychology*, 85–6; Kidner, *Nature and Psyche*.

121 Rice, *The Rotinonshonni*, 138.

122 Ibid., 22.

123 Suzuki with McConnell and Mason, *The Sacred Balance*, 157–8.

124 Ibid., 158.

125 Ibid., 229–30.

126 Ibid.

127 Ibid., 68. Also see Thompson, *Imaginary Landscape*, 61, for the links he draws between these biological processes and myths about angels and fairies have interesting connections with the past two chapters.

128 Suzuki with McConnell and Mason, *The Sacred Balance*, 229–30.

129 Teilhard de Chardin, *Christianity and Evolution*, cited in Anonymous, *Meditations on the Tarot*, 535.

130 Anonymous, *Meditations on the Tarot*, 535.

131 Ibid., 35–6. These pages offer a clarification of the difference between a phenomenology of being and a phenomenology of love, an analysis that is subsequently connected to both the Neoplatonic cosmology that influenced Notre-Dame de Chartres and Teilhard de Chardin's Omega point.

132 Suzuki with McConnell and Mason, *The Sacred Balance*, 174.

133 Eliade, *The Forge and the Crucible*, 172–3.

134 Christianson, *Greenhouse*, 28; Roberts, *The End of Oil*, 33; Suzuki with McConnell and Mason, *The Sacred Balance*, 174.

135 Livingston, *Arctic Oil*, 110–11.

136 Berry, *The Great Work*, 158.

137 Bellah, *Religion in Human Evolution*, 158.

138 Hulme, *Why We Disagree about Climate Change*, 363; Thompson, *The Time Falling Bodies*, 210.

139 Hulme, *Why We Disagree about Climate Change*, 352.

140 Calvin, *A Brain for All Seasons*, 291.

141 Ophuls, *Plato's Revenge*, 138.

142 Victor, *Managing without Growth*, 2.

143 Ibid., 24–5.

144 Ibid., 183.

145 Ibid., 222.

146 Hulme, *Why We Disagree about Climate Change*, 336.

147 Klein, *This Changes Everything*, 54.

148 Hayden, *Shamans, Sorcerers and Saints*, 25.

149 Bateson, *Steps to an Ecology of Mind*, 332–3.

150 Hulme, *Why We Disagree about Climate Change*, 363.

151 Thompson, *The Time Falling Bodies*, 120.

152 Berry, *The Great Work*, 9.

153 Greer, *The Jesuit Relations*, 17, 14.

CHAPTER SEVEN

1 CBC News, *Global Warming Dials up Our Risks, UN Report Says*.

2 CBC, *The Current, IPCC Climate Change Report: Official Prophecy of Doom*.

3 Harrison, *The Dominion of the Dead*, 102–3.

4 Mowat in Livingston, *The John A. Livingston Reader*, xxi.

5 Gibson in Livingston, *The John A. Livingston Reader*, xv.

6 Heschel, *The Prophets, Volume II*, 63.

7 Woodworth, "Urban Place as an Expression of the Ancestors," 227.

8 Woodworth, *The Morning Star*; Rice, *The Rotinonshonni*; Longboat, *Owehna'shon*.

9 Longboat, *Owehna'shon*, 279.

10 Wilson, *Research Is Ceremony*, 137.

11 Woodworth, *The Morning Star*.

12 Wilson, *Research Is Ceremony*, 60–1.

13 Mohawk, *Thinking in Indian*, 6.

14 Chevalier and Gheerbrant. *Dictionary of Symbols*, 813.

15 Anonymous, *Meditations on the Tarot*, 389; Chevalier and Gheerbrant, *Dictionary of Symbols*, 813.

16 Campbell, *The Mythic Image*, 235; Chevalier and Gheerbrant. *Dictionary of Symbols*, 813–14.

17 Livingston, *The John A. Livingston Reader*, 19.

18 Ibid., 9.

19 Leopold, *A Sand County Almanac*, 197.

20 Fabiani, "The Greatest Environmentalist You've Never Heard of."

21 Woodworth, *The Morning Star*, 116; see 116–28 for more details.

22 Rice, *The Rotinonshonni*, 199.

23 Ibid., 202.

24 Longboat, *Owehna'shon*, 284. Also see Alfred, *Peace, Power, Righteousness*, which undertakes an analysis of colonial power structures using the frame of the Condolence Ceremony. Some of his ideas are brought up in other parts of the book (e.g., chapter 4), but I decided to keep this chapter closer to Thomas and his students.

25 Rice, *The Rotinonshonni*, 234.

26 Woodworth, *The Morning Star*, 124.

27 Ibid., 68.

28 Ibid., 76.

29 Mortimer-Sandilands, "Melancholy Natures, Queer Ecologies," 392.

30 Leopold cited in ibid., 331.

31 Ibid., 333.

32 Ibid., 342.

33 Ibid., 348–9.

34 Livingston, *The John A. Livingston Reader*, 8.

35 Leduc, *Climate, Culture, Change*. See chapter 4 for an analysis of Livingston's wasteland-theology thesis in relation to the north and the Inuit. On the one hand, he offers a valuable way of critiquing changing Western systems of colonial power, and, on the other, he seems to dismiss the value of Indigenous knowledge because their cultures have been changed by those systems of power.

36 Mortimer-Sandilands, "Melancholy Natures, Queer Ecologies," 333.

37 Ibid., 342.

38 Mowat in Livingston, *The John A. Livingston Reader*, xxi.

39 Fawcett and Russell, "Remembered: John Livingston," 236.

40 E.g., Leduc, "The Fallacy of Environmental Studies?"

41 Martin, "John Livingston, Naturalist 1923–2006."

42 Ibid.

43 Cited in ibid.

44 Cited in Suzuki, *The David Suzuki Reader*, 278.

45 Mortimer-Sandilands, "Melancholy Natures, Queer Ecologies," 354.

46 Hulme, *Why We Disagree about Climate Change*, 347.

47 Rice, *The Rotinonshonni*, 188.

48 Thomas with Boyle, *Teachings from the Longhouse*, 13.

49 Ibid., 189.

50 Longboat, *Owehna'shon*, 280.

51 See Greer, *Mohawk Saint*; Grim, *The Shaman*; Underhill, *Mysticism*; Hayden, *Shamans, Sorcerers and Saints*.

52 Greer, *Mohawk Saint*, 152.

53 Ibid., 155.

54 Ibid., 157.

55 Ibid., 159.

56 Abley, "Reclaiming Kateri."

57 Cited in Grim, *The Shaman*, 172–3.

58 Ibid., 146.

59 Greer, *Mohawk Saint*.

60 Ibid., 17.

61 Pseudo-Dionysius, *Pseudo-Dionysius*, 108.

62 Ibid., 213.

63 Lilburn, *Going Home*, 137.

64 Ibid., 128.

65 Ibid., 169.

66 Ibid., 173.

67 Livingston, *The John A. Livingston Reader*, 73.

68 Morton, *The Ecological Thought*, 77. As discussed in the Introduction, Timmerman, "The Human Dimensions of Global Change," also transforms environmental issues into mysteries that pull us in, thus highlighting the contextualizing of scientific knowledge within a kind of mystic sensibility.

69 Lilburn, *Going Home*, 177.

70 Ibid., 169–70.

71 Underhill, *Mysticism*, 400.

72 Ibid., 436.

73 Cited in ibid., 399.

74 Ibid., 228.

75 Ibid., 412.

76 Catherine of Siena, *The Dialogue*, 186–7.

77 Underhill, *Mysticism*, 400.

78 Longboat, *Owehna'shon*, 280.

79 Parmenter, *The Edge of the Woods*, 82.

80 Sioui, *Huron-Wendat*, 158.

81 Trigger, *The Children of Aataentsic*, 81.

82 Woodworth, *The Morning Star*, 167.

83 Sioui, *Huron-Wendat*, 158.

84 Ibid.

85 Trigger, *The Children of Aataentsic*, 81.

86 Woodworth, *The Morning Star*, 109, 114–15.

87 Danford, "Will the 'Real' False Face Please Stand up?" 267–8.

88 Husain, *The Goddess*, 91.

89 Livingston, *The John A. Livingston Reader*, 305.

90 Rice, *The Rotinonshonni*, 24.

91 Thompson, *The Time Falling Bodies Take to Light*, 9.

92 Anonymous, *Meditations on the Tarot*, 433.

93 Ibid., 109.

94 Danford, "Will the 'Real' False Face Stand up?" 263, 264.

95 Ibid., 263.

96 Woodworth, *The Morning Star*, 109.

97 Ibid., 167; Danford, "Will the 'Real' False Face Stand up?" 264.

98 Engelbrecht, *Iroquoia*, 5.

99 Danford, "Will the 'Real' False Face Stand up?" 257.

100 Ibid., 268.

101 Bellah, *Religion in Human Evolution*, 123. Also see Rappaport, *Ritual and Religion in the Making of Humanity*; Hayden, *Shamans, Sorcerers and Saints*.

102 Hayden, *Shamans, Sorcerers and Saints*, 32.

103 Chevalier and Gheerbrant. *Dictionary of Symbols*, 640.

104 Fenton, *The False Faces of the Iroquois*.

105 Woodworth, *The Morning Star*.

106 Rice, *The Rotinonshonni*, 16.

107 Danford, "Will the 'Real' False Face Stand up?" 257.

108 Ibid., 263.

109 Engelbrecht, *Iroquoia*, 5.

110 Danford, "Will the 'Real' False Face Stand up?" 266–7.

111 Ibid., 264.

112 Trigger, *The Children of Aataentsic*, 81.

113 Greer, *The Jesuit Relations*, 72.

114 Longboat, *Owehna'shon*, 290.

115 Eamon, *Science and the Secrets of Nature*, 29.

116 Harrison, *The Dominion of the Dead*, 158–9.

117 Mortimer-Sandilands, "Melancholy Natures, Queer Ecologies," 335.

118 Frank, *Letting Stories Breathe*, 159.

119 Livingston, *The John A. Livingston Reader*, 22.

120 Bennett, *Vibrant Matter*, 110.

121 Truth and Reconciliation Commission of Canada, *Honouring the Truth, Reconciling for the Future, Summary*. This final report of the commission came out after this book was already with the publisher, and its analysis and recommendations offer an important starting point for further considering what the healing being talked about here will entail. I briefly talk about some potential directions in the Epilogue, but clearly that is for me merely a start toward truly engaging this document with a spirited orenta.

122 Sinclair, "Opinion."

123 Ibid.

124 See chapter 2 for this discussion.

125 Woodworth, *The Morning Star*, 50.

126 Livingston, *The John A. Livingston Reader*, 152.

127 Woodworth, *The Morning Star*, 47.

128 Wilson, *Research Is Ceremony*, 136.

129 Thomas with Boyle, *Teachings from the Longhouse*, 140.

130 Mohawk, *Thinking in Indian*, 6.

EPILOGUE

1 Canadians for a New Partnership, http://www.cfnp.ca/.

2 Goar, "You Can Be Part of Aboriginal Reconciliation," A13.

3 Canadians for a New Partnership.

4 Woodworth, "Urban Place as an Expression of the Ancestors," 227.

5 Ibid., 228.

6 Ibid., 227.

7 First Story, https://firststoryblog.wordpress.com/aboutfirststory/.

8 Campbell, *Masks of God*, 47.

9 Ibid.

10 Truth and Reconciliation Commission of Canada, *Honouring the Truth*, 37–9. This centre opened its doors in 2015 at the University of Manitoba, and it is conceived by the commission as a place that will "encourage and engage in respectful dialogue on many issues that hinder or foster reconciliation." Its purpose has a great resonance with what Woodworth conceives for Toron:to on a local scale.

11 See Klein, *This Changes Everything*, 345. "The point of drawing this liquid map was clear to all present: of course all of these different [Indigenous] nations and [environment] groups would join together to fight the threat of an oil spill – they are all already united by water; by the lakes and rivers, streams and oceans that drain into one another."

12 Thomas with Boyle, *Teachings from the Longhouse*, 134.

13 Scharper, *For Earth's Sake*, 173.

14 Ibid, 172–4.

15 See Wilk, "Consuming Ourselves to Death," 266. Also see the discussion around Victor's ideas in chapter 6.

16 Pope Francis, *Encyclical Letter Laudato Si' of the Holy Father Francis on Care for our Common Home*, 152–3.

17 Scharper, "Pope Francis's Important Ecology Lesson."

18 Niehardt, *Black Elk Speaks*, 20–47.

19 Ibid.

Bibliography

Abley, Mark. "Reclaiming Kateri." Canada's History, April–May 2014.
– *Spoken Here: Travels among Threatened Languages*. Boston and New York: Houghton Mifflin 2003.
Abram, D. *Becoming Animal: An Earthly Cosmology*. New York: Vintage 2010.
– *The Spell of the Sensuous: Perception and Language in a More-Than-Human World*. New York: Pantheon Books 1996.
Adomnán of Iona and Richard Sharpe. *Life of St Columba* [Vita S. Columbae]. Penguin Classics. Harmondsworth, UK, and New York: Penguin Books 1995.
Alfred, Taiaiake. *Peace, Power, Righteousness: An Indigenous Manifesto*, 2nd ed. Don Mills, ON: Oxford University Press 2008.
Altmeyer, George. "Three Ideas of Nature in Canada." In Chad Gaffield and Pam Gaffield, eds., *Consuming Canada: Readings in Environmental History*. 96–118. Toronto: Copp Clark 1995.
Anable, J. *Personal Mobility and Energy Demand*. Presentation at UK Energy Research Centre Summer School, London, 2010.
Anderson, Chris. "'I'm Métis, What's Your Excuse?' On the Optics and the Ethics of the Misrecognition of Métis in Canada." *Aboriginal Policy Studies*, 1(2) (2011): 161–5.
– *"Métis": Race, Recognition, and the Struggle for Indigenous Peoplehood*. Vancouver and Toronto: UBC Press 2014.
Anonymous (trans. Robert Powell). *Meditations on the Tarot: A Journey into Christian Hermeticism*. New York: J.P. Tarcher/Putnam 2002.
Archer, George H. "I Remember." *Globe and Mail*, s7, 2006.

Armstrong, Edward A. *Saint Francis: Nature Mystic*. Berkeley: University of California Press 1973.

Arthur, Eric Ross, and Stephen A. Otto. *Toronto: No Mean City*. Toronto: University of Toronto Press 1986.

Asfeldt, M., I. Urberg, and B. Henderson. "Wolves, Ptarmigan, and Lake Trout: Critical Elements of a Northern Canadian Place-Conscious Pedagogy." *Canadian Journal of Environmental Education*, 14 (2009): 33–41.

Atleo, E.R. Umeek. *Tsawalk: A Nuu-chah-nulth Worldview*. Vancouver: UBC Press 2004.

Bai, H. "Reanimating the Universe: Environmental Education and Philosophical Animism." In M. McKenzie et al., eds., *Fields of Green: Restorying Culture, Environment and Education*. 135–52. New York: Hampton Press 2009.

Baring, A., and J. Cashford. *The Myth of the Goddess: The Evolution of an Image*. London: Viking Arkana 1991.

Baskerville, Peter A. *Ontario: Image, Identity and Power*. Don Mills, ON: Oxford University Press 2002.

Bassett, John M. *Elizabeth Simcoe: First Lady of Upper Canada*. Don Mills, ON: Fitzhenry and Whiteside 1974.

Bateson, G. *Steps to an Ecology of Mind*. New York: Ballantine Books 1972.

Bechmann, Roland. *Trees and Man: The Forest in the Middle Ages* [Des arbres et des hommes]. 1st ed. New York: Paragon House 1990.

Behringer, Wolfgang. *Witches and Witch-Hunts: A Global History*. Cambridge, UK, and Malden, MA: Polity Press 2004.

Bellah, R.N. *Religion in Human Evolution: From the Paleolithic to the Axial Age*. Cambridge and London: Harvard University Press 2011.

Benn, Carl. "Colonial Transformations." In R.F. Williamson, ed., *Toronto: An Illustrated History of Its First 12,000 Years*. 53–72. Toronto: James Lorimer 2008.

Bennett, Jane. *Vibrant Matter: A Political Ecology of Things*. Durham, NC, and London: Duke University Press 2010.

Berchem, F.R. *The Yonge Street Story 1793–1860: An Account from Letters, Diaries and Newspapers*. Toronto: Natural Heritage/Natural History 1997.

Bergmann, Sigurd. "Theology in Its Spatial Turn: Space, Place and Built Environments Challenging and Changing the Images of God." *Religion Compass*, 1/3 (2007): 353–79.

Berkes, Fikret. "Epilogue: Making Sense of Arctic Environmental Change?" In I. Krupnik and D. Jolly, eds., *The Earth Is Faster Now: Indigenous Observations of Arctic Environmental Change*. 334–49. Fairbanks: Arcus 2002.

Berland, Jody. *North of Empire: Essays on the Cultural Technologies of Space*. Durham, NC, and London: Duke University Press 2009.

Berman, Morris. *The Reenchantment of the World*. Toronto and New York: Bantam 1984.

Berry, Thomas. *The Great Work: Our Way into the Future*. New York: Bell Tower 1999.

Blackburn, C. *Harvest of Souls: The Jesuit Missions and Colonialism in North America, 1632–1650*. Montreal and Kingston, ON: McGill-Queen's University Press 2000.

Bobiwash, A. Rodney. "The History of Native People in the Toronto Area: An Overview." In F. Sanderson and H. Howard-Bobiwash, eds., *The Meeting Place: Aboriginal Life in Toronto*. 5–24. Toronto: Native Canadian Centre of Toronto 1997.

Boff, Leonardo. *Ecology and Liberation: A New Paradigm*. Maryknoll, NY: Orbis Books 1995.

Bonaparte, Darren. *A Lily Among Thorns*. North Charleston, SC: BookSurge Publishing 2009.

Bonnell, J., and M. Fortin. *Ashbridges Bay*. Don Valley Historical Mapping Project. 2009. http://maps.library.utoronto.ca/dvhmp/ashbridges-bay.html (accessed 20 January 2014).

Bonner C. "Ontology and Climate Change." *Forum on Public Policy*, 2012.

Boudreau, Julie-Anne, Roger Keil, and Douglas Young. *Changing Toronto: Governing Urban Neoliberalism*. Toronto: University of Toronto Press 2009.

Bruce, James. *Prophecy, Miracles, Angels, and Heavenly Light?: The Eschatology, Pneumatology, and Missiology of Adomnán's Life of Columba*. Carlisle, UK, and Waynesboro, GA: Paternoster 2004.

Bruce, James P., and Stewart J. Cohen. "Impacts of Climate Change in Canada." In H. Cohen and A.J. Weaver, eds., *Hard Choices: Climate*

Change in Canada. 73–88. Waterloo, ON: Wilfrid Laurier University Press 2004.

Burroughs, W.J. *Climate Change in Prehistory: The End of the Reign of Chaos.* Cambridge and New York: Cambridge University Press 2005.

Calvin, William H. *A Brain for All Seasons: Human Evolution and Abrupt Climate Change.* Chicago: University of Chicago Press 2002.

Campbell, Joseph. *Historical Atlas of World Mythology, Volume II: The Way of the Seeded Earth; Part 2: Mythologies of the Primitive Planters: The Northern Americas.* New York: Harper and Row 1988.

– *Masks of God: Creative Mythology.* Harmondsworth, UK, and New York: Penguin Books 1976.

– *Masks of God: Primitive Mythology.* Harmondsworth, UK, and New York: Penguin Books 1976.

– *The Mythic Image.* Princeton, NJ: Princeton University Press 1981.

Campbell, Mora. "Technology and Temporal Ambiguity." In E. Higgs, A. Light, and D. Strong, eds., *Technology and the Good Life?* 256–70. Chicago: University of Chicago Press 2000.

CBC. *The Current.* "IPCC Climate Change Report: Official Prophecy of Doom." 31 March 2014. http://www.cbc.ca/thecurrent/episode/2014/03/31/ipcc-climate-change-report-official-prophecy-of-doom/ (accessed 29 May 2014).

CBC News. "Global Warming Dials up Our Risks, UN Report Says." 31 March 2014. http://www.cbc.ca/news/technology/global-warming-dials-up-our-risks-un-report-says-1.2592307 (accessed 29 May 2014).

Canadians for a New Partnership. http://www.cfnp.ca/ (accessed 27 November 2014).

Carabine, Deirdre. *John Scottus Eriugena.* New York: Oxford University Press 2000.

Carpenter, Roger M. *The Renewed, the Destroyed, and the Remade: The Three Thought Worlds of the Huron and the Iroquois, 1609–1650.* East Lansing: Michigan State University Press 2004.

Carr, Nicholas. *The Shallows: What the Internet Is Doing to Our Brains.* New York and London: W.W. Norton 2010.

Carrothers, G., S. Kline, and J. Livingston, *FESKIT: Essays into Environmental Studies: Being Some Interpretations and Amplifications of the FES Curriculum Model.* Toronto: FES, York University, 1968/1988.

Carson, Rachel. *Silent Spring.* Boston and New York: Houghton Mifflin 2002 (1962).

Catherine of Siena. *The Dialogue*. Trans. S. Noffke. In *The Classics of Western Spirituality*. New York and Mahwah, NJ: Paulist Press 1980.

Catholic Encyclopaedia. "St. Lawrence." http://www.newadvent.org/cathen/09089a.htm (accessed 10 May 2014).

Cenovus. "Rising to the Challenges." 2014. http://www.cenovus.com/news/rising-to-the-challenges.html (accessed 8 April 2014).

Chastko P.A. *Developing Alberta's Oil Sands: From Karl Clark to Kyoto*. Calgary, AB: University of Calgary Press 2004.

Chen, I. "Why Do People Misunderstand Climate Change? Heuristics, Mental Models and Ontological Assumptions." *Climatic Change*, 108 (2011): 31–46.

Cheney, J., and A. Weston. "Environmental Ethics as Environmental Etiquette." *Environmental Ethics*, 21 (1999): 115–34.

Chenu, Marie-Dominique. *Nature, Man, and Society in the Twelfth Century: Essays on New Theological Perspectives in the Latin West*. Selected, edited, and translated by Jerome Taylor and Lester K. Little. Chicago: University of Chicago Press 1968.

Chevalier, Jean, and Alain Gheerbrant. *Dictionary of Symbols*. Trans. J. Buchanan-Brown. London, New York, and Toronto: Penguin Books 1994.

Christianson, Gale E. *Greenhouse: The 200-Year Story of Global Warming*. New York: Walker 1999.

Cipolla, Geatano. *Labyrinth: Studies on an Archetype*. New York, Ottawa, and Toronto: Legas 1987.

City of Toronto. *Change Is in the Air: Climate Change, Clean Air and Sustainable Energy Action Plan: Moving from Framework to Action*. Toronto: Toronto Environment Office, June 2007.

– *High Park: Restoring a Jewel of Toronto's Park System*. Toronto: Forestry, Parks and Recreation Division 2002.

– *High Park Woodland and Savannah Management Plan*. Toronto: Urban Forestry Services, Parks and Recreation Division 2002.

Coates, Pete. *Nature: Western Attitudes since Ancient Times*. Berkeley: University of California Press 2005.

Code, Lorraine. *Ecological Thinking: The Politics of Epistemic Location*. New York: Oxford University Press 2006.

Cohn, Norman. *Noah's Flood: The Genesis Story in Western Thought*. New Haven, CT, and London: Yale University Press 1996.

Crate S.A., and M. Nuttall, eds. *Anthropology and Climate Change: From Encounters to Actions*. Walnut Creek, CA: Left Coast Press 2009.

Cronon, W. "The Trouble with Wilderness; or, Getting Back to the Wrong Nature." In *The Introductory Reader in Human Geography*, 167–78. Malden, MA: Blackwell Publishing 2007 (1992).

Cruikshank J. *Do Glaciers Listen?: Local Knowledge, Colonial Encounters, and Social Imagination.* Vancouver and Seattle: UBC Press and University of Washington Press 2005.

Curthoys, L.P. "Finding a Place of One's Own: Reflections on Teaching in and with Place." *Canadian Journal of Environmental Education*, 12 (2007): 68–79.

Danford, Joanne. "Will the 'Real' False Face Please Stand Up?" *Canadian Journal of Native Studies*, 2 (1989): 253–72.

de Chardin, Pierre Teilhard. *The Divine Milieu.* New York: Perennial 2001 (1960).

– *The Phenomenon of Man.* New York: Perennial 2002 (1959).

Delâge, Denys. *Bitter Feast: Amerindians and Europeans in Northeastern North America, 1600–64.* Vancouver: UBC Press 1993.

Deloria, V. *God Is Red: A Native View of Religion.* Golden, CO: North American Press 1994.

Demeritt, D. "The Construction of Global Warming and the Politics of Science." *Annals of the Association of American Geographers*, 91(2) (2001): 307–37.

– "Science Studies, Climate Change and the Prospects for Constructivist Critique." *Economy and Society*, 35(3) (2006): 453–79.

Desfor, G., and R. Kiel. *Nature and the City: Making Environmental Policy in Toronto and Los Angeles.* Tucson: University of Arizona Press 2004.

Dickason, Olive Patricia. "From 'One Nation' in the Northeast to 'New Nation' in the Northwest: A Look at the Emergence of the Métis." In J. Peterson and J. Brown, eds., *The New Peoples: Being and Becoming Métis in North America.* 19–36. Winnipeg: University of Manitoba Press 1985.

– *The Myth of the Savage: And the Beginnings of French Colonialism in the Americas.* Edmonton: University of Alberta Press 1984.

Dillon, John. *The Heirs of Plato: A Study of the Old Academy (347–274 BC).* Oxford: Clarendon Press 2003.

Eamon, William. *Science and the Secrets of Nature: Books of Secrets in Medieval and Early Modern Culture.* Princeton, NJ: Princeton University Press 1994.

Eccles, W.J. *The French in North America, 1500–1765*. East Lansing: Michigan State University Press 1998.

Eliade, Mircea. *The Forge and the Crucible*. Trans. S. Corrin. Chicago: University of Chicago Press 1978.

– *The Myth of the Eternal Return: Or, Cosmos and History*. Bollingen Series. Princeton, NJ: Princeton University Press 1971.

– *Patterns in Comparative Religion*. London and New York: Sheed and Ward 1958.

– *The Sacred and the Profane: The Nature of Religion*. A Harvest Book. 1st American. New York: Harcourt, Brace 1959.

Engelbrecht, W. *Iroquoia: The Development of a Native World*. Syracuse, NY: Syracuse University Press 2002.

Eriugena, Johannes Scotus. *Periphyseon on the Division of Nature* [De divisione naturae]. The Library of Liberal Arts. 1st ed. Vol. 157. Indianapolis, IN: Bobbs-Merrill 1976.

Esbjörn-Hargens, S. "An Ontology of Climate Change: Integral Pluralism and the Enactment of Multiple Objects." *Journal of Integral Theory and Practice*, 5(1) (2010): 143–74.

Esbjörn-Hargens, S., and M.E. Zimmerman. *Integral Ecology: Uniting Multiple Perspectives on the Natural World*. Boston and London: Integral Books 2009.

Evernden, Neil. *The Natural Alien: Humankind and Environment*. 2nd ed. Toronto: University of Toronto Press 1993.

Eyles, N. *Ontario Rocks: Three Billion Years of Environmental Change*. Markham, ON: Fitzhenry and Whiteside 2002.

Fabiani, Louise. "The Greatest Environmentalist You've Never Heard of." Toronto *Star*, 28 April 2007.

Fagan, Brian. *The Long Summer: How Climate Changed Civilization*. New York: Basic Books 2004.

Fanning, Steven. *Mystics of the Christian Tradition*. London and New York: Routledge 2001.

Favier, Jean. *The World of Chartres*. Trans. F. Garvie. London and New York: Thames and Hudson 1990.

Fawcett, Leesa, and Connie Russell. "Remembered: John Livingston." *Canadian Journal of Environmental Education*, 11 (2006): 236–40.

Fawcett, Leesa, Constance L. Russell, and Anne C. Bell. "Guiding Our Environmental Praxis: Teaching and Learning for Sustainable Environmental Education." In W.L. Filho, ed., *Teaching Sustainability at*

Universities – Towards Curriculum Greening. 223–38. London: Peter
   Lang Scientific Publishers 2002.
Fenton, William. *The False Faces of the Iroquois.* Norman: University
   of Oklahoma Press 1987.
First Story. https://firststoryblog.wordpress.com/aboutfirststory/ (accessed
   27 November 2014).
Fisher, Andy. *Radical Ecopsychology: Psychology in the Service of Life.*
   Albany: State University of New York Press 2002.
Flannery, Tim F. *The Eternal Frontier: An Ecological History of North
   America and Its Peoples.* London: Heineman 2001.
Ford, M.P. *Beyond the Modern University: Toward a Constructive
   Postmodern University.* Westport, CT, and London: Praeger 2002.
Frank, Arthur W. *Letting Stories Breathe: A Socio-Narratology.* Chicago:
   University of Chicago Press 2010.
Freeman, Victoria. "Indigenous Hauntings in Settler-Colonial Spaces: The
   Activism of Indigenous Ancestors in the City of Toronto." In C.E. Boyd
   and C. Thrush, eds., *Phantom Past, Indigenous Presence: Native Ghosts
   in North American Culture and History.* 209–49. Lincoln: University of
   Nebraska Press 2011.
Frye, Northrop. *The Bush Garden: Essays on the Canadian Imagination.*
   Toronto: Anansi 1971.
Fryer, Mary B., and Christopher Dracott. *John Graves Simcoe (1752–
   1806): A Biography.* Toronto: Dundurn Press 1998.
Gardner, G. "Engaging Religion in the Quest for a Sustainable World."
   In L. Starke, ed., *State of the World 2003.* 152–75. New York: W.W.
   Norton 2003.
Gersh, S.E. *From Iamblichus to Eriugena: An Investigation of the
   Prehistory and Evolution of the Pseudo-Dionysian Tradition* [Studien
   Zur Problemgeschichte Der Antiken Und Mittelalterlichen Philosophie].
   Vol. 8. Leiden: Brill 1978.
Gibson, Graeme. "Appreciation." In John A. Livingston, *The John A.
   Livingston Reader.* Toronto: McClelland and Stewart 2007.
Gimpel, Jean. *The Medieval Machine: The Industrial Revolution of the
   Middle Ages.* New York: Penguin Books 1977.
Glickman, Susan. *The Picturesque and the Sublime: A Poetics of the
   Canadian Landscape.* Montreal and Kingston, ON: McGill-Queen's
   University Press 1998.

Goar, C. "You Can Be Part of Aboriginal Reconciliation." Toronto *Star*, 17 November 2014, A13.

Godbout, J., with A. Caillé and D. Winkler. *The World of the Gift*. Montreal and Kingston, ON: McGill-Queen's University Press 1998.

Grady, W. *The Great Lakes: A Natural History of a Changing Region*. Vancouver: Greystone Books 2007.

– *Toronto the Wild: Field Notes of an Urban Naturalist*. Toronto: MacFarlane Walter and Ross 1995.

Greer, Allan. *The Jesuit Relations: Natives and Missionaries in Seventeenth-Century North America*. Boston: Bedford/St Martin's 2000.

– *Mohawk Saint: Catherine Tekakwitha and the Jesuits*. Oxford and New York: Oxford University Press 2005.

– *The People of New France*. Toronto: University of Toronto Press 1997.

Grim, John A. *The Shaman: Patterns of Religious Healing among the Ojibway Indians*. Norman: University of Oklahoma Press 1983.

Gruenewald, D.A. "The Best of Both Worlds: A Critical Pedagogy of Place." *Educational Researcher*, 32(4) (2003): 3–12.

Gutiérrez, Gustavo. *A Theology of Liberation: History, Politics, and Salvation* [Teología de la liberación]. Maryknoll, NY: Orbis Books 1988.

Hansen, James. *Climate and Energy: Fundamental Facts, Responsibilities and Opportunities*. Testimony to the United States Senate Committee on Foreign Relations. 13 March 2014.

Haraway, Donna. "Cyborgs and Symbionts: Living Together in the New World Order." In C.H. Gray, ed., *The Cyborg Handbook*. New York: Routledge 1995.

– *How Like a Leaf* (an interview with T.N. Goodeve). New York and London: Routledge 1999.

– "Situated Knowledges: The Science Question in Feminism and the Privilege of Partial Perspective." *Feminist Studies*, 14(3) (1988): 575–99.

Harper, S. "Harper's Index: Stephen Harper Introduces the Tar Sands Issue." 2006. http://www.dominionpaper.ca/articles/1491 (accessed 10 April 2013).

Harris, Cole. *The Reluctant Land: Society, Space, and Environment in Canada before Confederation*. Vancouver and Toronto: UBC Press 2008.

Harris, J., R. Diamond, M. Iyer, and C. Payne. *Don't Supersize Me! Toward a Policy of Consumption-Based Energy Efficiency*. University of California Institute 2006.

Harrison, Robert P. *Gardens: An Essay on the Human Condition.* Chicago: University of Chicago Press 2008.

Harrison, Robert Pogue. *The Dominion of the Dead.* Chicago: University of Chicago Press 2003.

Hayden, Brian. *Shamans, Sorcerers and Saints: A Prehistory of Religion.* Washington: Smithsonian Books 2003.

Hayes, Derek. *Historical Atlas of Canada: Canada's History Illustrated with Original Maps.* Vancouver and Toronto: Douglas and McIntyre 2006.

– *Historical Atlas of Toronto.* Vancouver and Toronto: Douglas and McIntyre 2009.

Helm, Dieter. *The Carbon Crunch: How We're Getting Climate Change Wrong – and How to Fix It.* New Haven, CT, and London: Yale University Press 2012.

Hennepin, Louis. *A New Discovery of a Vast Country in America.* Toronto: Coles Publishing 1974.

Hentsch, Thierry. *Truth or Death: The Quest for Immortality in the Western Narrative Tradition.* Vancouver: Talonbooks 2004.

Heschel, Abraham J. *The Prophets, Volume II.* New York: Harper Torchbooks 1975.

Hessing, M., M. Howlett, and T. Summerville. *Canadian Natural Resource and Environmental Policy, 2nd Edition.* Vancouver and Toronto: UBC Press 2005.

Hildegard. Ed. Barbara Newman. *Symphonia: A Critical Edition of the Symphonia Armonie Celestium Revelationum [Symphony of the Harmony of Celestial Revelations].* 2nd ed. Ithaca, NY: Cornell University Press 1998.

Hillman, James. *The Soul's Code: In Search of Character and Calling.* 1st ed. New York: Random House 1996.

Hindman, Matthew. *The Myth of Digital Democracy.* Princeton, NJ, and Oxford: Princeton University Press 2009.

Homer-Dixon T. "The Tar Sands Disaster." New York *Times*, 31 March 2013. http://www.nytimes.com/2013/04/01/opinion/the-tar-sands-disaster.html?_r=0 (accessed 19 April 2013).

Homer-Dixon, T.F. *Environment, Scarcity, and Violence.* Princeton, NJ: Princeton University Press 1999.

– *The Ingenuity Gap.* Toronto: Vintage Canada 2001.

Hough, Michael. *Cities and Natural Process*. London and New York: Routledge 1995.

Hudema M. "Tar Sands Spills." *Greenpeace*, 19 September 2013. http://www.greenpeace.org/canada/en/Blog/tar-sands-spills-up-to-152899671-litres-and-s/blog/46700/ (accessed 30 September 2013).

Hulme, M. *Why We Disagree about Climate Change: Understanding Controversy, Inaction and Opportunity*. New York: Cambridge University Press 2009.

Husain, Shahrukh. *The Goddess: Power, Sexuality, and the Feminine Divine*. Ann Arbor: University of Michigan Press 2003.

Indigenous Environmental Network. "First Nations Responds to Suncor Tar Sands Tailings Breach." 26 March 2013. http://www.ienearth.org/first-nations-responds-to-suncor-tar-sands-tailings-breach/ (accessed 10 April 2014).

Innis, Harold A. *Empire and Communications*. Lanham, MD: Rowman and Littlefield 2007 (1950).

– *The Fur Trade in Canada: An Introduction to Canadian Economic History*. Toronto: University of Toronto Press 1999 (1930).

– *Staples, Markets, and Cultural Change*. Ed. D. Drache. Montreal and Kingston, ON: McGill-Queen's University Press 1995.

Intergovernmental Panel on Climate Change. *Climate Change 2013: The Physical Science Basis – Summary for Policymakers. Fifth Assessment Report*. Geneva: IPCC Secretariat/World Meteorological Organization 2013.

Issar, A.S., and M. Zohar. *Climate Change – Environment and Civilization in the Middle East*. Berlin: Springer 2004.

Jasen, Patricia Jane. *Wild Things: Nature, Culture, and Tourism in Ontario, 1790–1914*. Toronto: University of Toronto Press 1995.

Johnson, Jon. "The Indigenous Environmental History of Toronto, the Meeting Place." In L. Anders Sandberg et al., eds., *Urban Explorations: Environmental Histories of the Toronto Region*. 59–72. Hamilton, ON: L.R. Institute for Canadian History, McMaster University 2013.

Johnston, Basil H. *Honour Earth Mother: Mino-Audjaudauh Mizzu-Kummik-Quae*. Wiarton, ON: Kegedonce Press 2003.

– *Ojibway Heritage*. Toronto: McClelland and Stewart 1981.

Judd, William W., and J.M. Speirs. *A Naturalist's Guide to Ontario*. Toronto: University of Toronto Press 1964.

Judson, Clara Ingram. *St. Lawrence Seaway*. Chicago: Follett 1959.

Jung, Emma, and Marie-Louise von Franz. *The Grail Legend*. Princeton, NJ: Princeton University Press 1998.

Keller, Catherine. *Face of the Deep: A Theology of Becoming*. London and New York: Routledge 2003.

Kennedy, C., et al. "Methodology for Inventorying Greenhouse Gas Emissions from Global Cities." *Energy Policy* (2009).

Kenton, Edna. *Indians of North America, Volume 2*. New York: Harcourt 1927.

Kidner D. *Nature and Psyche: Radical Environmentalism and Politics of Subjectivity*. New York: SUNY Press 2001.

King, S.J., and I.L. Stefanovic. "Children and Nature in the City." In I.L. Stefanovic and S.B. Scharper, eds., *The Natural City: Re-Envisioning the Built Environment*. 322–42. Toronto: University of Toronto Press 2012.

King, Thomas. *The Truth about Stories: A Native Narrative*. Toronto: Anansi Press 2003.

Klein, Naomi. *This Changes Everything: Capitalism vs. The Climate*. Toronto: Alfred A. Knopf Canada 2014.

Latour, B. *We Have Never Been Modern*. Cambridge, MA: Harvard University Press 1993.

Latour, Bruno. *On the Modern Cult of the Factish Gods*. Durham, NC, and London: Duke University Press 2010.

Lawrence, Bonita. *"Real" Indians and Others: Mixed-Blood Urban Native Peoples and Indigenous Nationhood*. Vancouver and Toronto: UBC Press 2004.

Leduc, Adrienne. *Antoine: Coureur de Bois*. Vancouver: Mazarin Publications 1996.

Leduc, T., and D. Morley. *Five Decades of FES at York: The Praxis of Environmental Studies*. Toronto: ABL Group 2015.

Leduc T., and S. Crate. "Reflexive Shifts in Climate Research and Education: Towards Re-localizing Our Lives." *Nature and Culture*, 8(2) (2013): 134–61.

Leduc, T. "Ancestral Climate Wisdom: Return to a Thoughtful Etiquette." In M. Schönfeld, ed., *Global Ethics on Climate Change: The Planetary Crisis and Philosophic Alternatives*. 107–19. London and New York: Routledge 2013.

– "Climates of Ontological Change: Past Wisdom in Current Binds?" *Wiley Interdisciplinary Reviews: Climate Change*, 5 (2014): 247–260.

Leduc, Timothy B. *Climate, Culture, Change: Inuit and Western Dialogues with a Warming North*. Ottawa: University of Ottawa Press 2010.

– "A Climate for Wisdom?" *Tikkun Magazine*, 26 (2011): 3.

– "The Fallacy of Environmental Studies? An Interdisciplinary Foray thru Canada's Academic Programs." *Environments*, 37(2) (2010): 1–28.

– "Inuit Economic Adaptations for a Changing Global Climate." *Ecological Economics*, 60(1) (2006): 27–35.

– "A Thanksgiving Species." Centre for Humans and Nature, "What Does It Mean to Be Human?" 2012. http://www.humansandnature.org/what-does-it-mean-to-be-human-question-2.php.

Legget, Robert Ferguson. *The Seaway*. Toronto: Clarke, Irwin 1979.

Le Goff, Jacques. *The Medieval Imagination*. Chicago: University of Chicago Press 1988.

Leopold A. *A Sand County Almanac: With Essays on Conservation from Round River*. New York: Ballantine Books 1970 (1949).

Letson, Douglas Richard, and Michael W. Higgins. *The Jesuit Mystique*. Chicago: Jesuit Way 1995.

Lilburn, Tim. *Going Home: Essays*. Toronto: Anansi Press 2008.

Livingston, John A. *Arctic Oil*. Toronto: CBC Merchandising 1981.

– *John Livingston: The Natural History of a Point of View*. Toronto, ON: CBC, *The Nature of Things*, 1998.

– *Rogue Primate*. Toronto: Key Porter Books 1994.

– *The John A. Livingston Reader: The Fallacy of Wildlife Conservation and One Cosmic Instant*. Toronto: McClelland and Stewart 2007.

Loftin, John D. *Religion and Hopi Life, 2nd Edition*. Bloomington and Indianapolis: Indiana University Press 2003.

Longboat, Roronhiakwen Daniel. *Owehna'shon (The Haudenosaunee Archipelago): The Nature and Necessity of Bio-Cultural Restoration and Revitalization*. Toronto: York University 2008.

Louv, Richard. *Last Child in the Woods: Saving Our Children from Nature-Deficit Disorder*. New York: Algonquin Books 2008.

Luibhéid, Colm, and Paul Rorem, eds. *Pseudo-Dionysius: The Complete Works. Classics of Western Spirituality* [*Works*]. Vol. 54. New York: Paulist Press 1987.

Lutz, John S. *'Pomo Wawa,' Makuk: A New History of Aboriginal-White Relations*. Vancouver: UBC Press 2008.

Macdougall, Doug. *Frozen Earth: The Once and Future Story of Ice Ages.* Berkeley: University of California Press 2004.

MacLennan, Hugh. *Seven Rivers of Canada.* Toronto: Macmillan 1961.

MacLennan, Hugh, and John De Visser. *Rivers of Canada.* Toronto: Macmillan 1974.

Markale, Jean. *The Cathedral of the Black Madonna: The Druids and the Mysteries of Chartres* [Chartres et l'énigme des Druides]. 1st US ed. Rochester, VT: Inner Traditions 2004.

Marsden, John. *Sea-Road of the Saints: Celtic Holy Men in the Hebrides.* Edinburgh: Floris Books 1995.

Martel, Gilles. "When a Majority Becomes a Minority: The French-Speaking Métis in the Canadian West." In D.R. Louder and E. Waddell, eds., and F. Philip, trans., *French America: Mobility, Identity, and Minority Experience across the Continent.* 69–99. Baton Rouge and London: Louisiana State University Press 1993.

Martin, Sandra. "John Livingston, Naturalist 1923–2006." *Globe and Mail,* 28 January 2006.

May, John Bentley. *Emerald City: Toronto Visited.* Toronto: Viking 1994.

McGregor, D. "Coming Full Circle: Indigenous Knowledge, Environment and Our Future." *American Indian Quarterly,* 28(3/4) (2004): 385–410.

McIntosh, Robert M. *Earliest Toronto.* Renfrew, ON: General Store Publishing House 2006.

McKay, Paul. *Electric Empire: The Inside Story of Ontario Hydro.* Toronto: Between the Lines 1983.

McNeill, John Robert. *Something New under the Sun: An Environmental History of the Twentieth-Century World.* 1st ed. New York: W.W. Norton 2000.

McNeill, John Thomas. *The Celtic Churches: A History a.d. 200 to 1200.* Chicago: University of Chicago Press 1974.

Meek, Donald E. "Between Faith and Folklore: Twentieth-Century Interpretations and Images of Columba." In D. Broun and T.O. Clancy, eds., *Spes Scotorum Hope of Scots: Saint Columba, Iona and Scotland.* 253–70. Edinburgh: T&T Clark 1999.

Merchant, C. *The Death of Nature: Women, Ecology, and the Scientific Revolution.* New York: Harper and Row 1989.

M'Gonigle, Michael, and J. Starke. *Planet U: Sustaining the World, Reinventing the University.* Gabriola Island, BC: New Society Publishers 2006.

Miller, J.R. *Skyscrapers Hide the Heavens: A History of Indian-White Relations in Canada.* 3rd ed. Toronto: University of Toronto Press 2000.

Mills, H. "The Living Machine: An Interview with Helen Mills of the Lost Rivers Project." In W. Reeves and C. Palassio, eds., *HTO: Toronto's Water.* 212–21. Toronto: Coach House Books 2008.

Mohawk, J. *Thinking in Indian: A John Mohawk Reader* (J. Barreiro, ed.). Golden, CO: Fulcrum 2010.

– *Utopian Legacies: A History of Conquest and Oppression in the Western World.* Santa Fe, NM: Clear Light 1999.

Mokuku T. "Lehae La Rona: Epistemological Interrogation to Broaden our Conception of Environment and Sustainability." *Canadian Journal of Environmental Education,* 17 (2012): 159–72.

Moogk, Peter N. *La Nouvelle France: The Making of French Canada: A Cultural History.* East Lansing: Michigan State University Press 2000.

Morissonneau, Christian. "The Ungovernable People: French-Canadian Mobility and Identity." In D.R. Louder and E. Waddell, eds., and F. Philip, trans., *French America: Mobility, Identity, and Minority Experience across the Continent.* 15–32. Baton Rouge and London: Louisiana State University Press 1993.

Morito, Bruce. *Thinking Ecologically: Environmental Thought, Values and Policy.* Halifax: Fernwood 2002.

Morrison, Kenneth. *The Solidarity of Kin.* Albany, NY: SUNY Press 2002.

Mortimer-Sandilands, Catriona. "Melancholy Natures, Queer Ecologies." In C. Mortimer-Sandilands and B. Erickson, eds., *Queer Ecologies: Sex, Nature, Politics, Desire.* 331–58. Bloomington: Indiana University Press 2010.

Morton, Timothy. *The Ecological Thought.* Cambridge, MA: Harvard University Press 2010.

Newman, Barbara. *Sister of Wisdom: St. Hildegard's Theology of the Feminine.* Berkeley: University of California Press 1987.

Niehardt, John G. *Black Elk Speaks.* Lincoln: University of Nebraska Press 1968.

Nikiforuk A. *The Energy of Slaves: Oil and the New Servitude.* Vancouver: Greystone Books 2012.

– *Tar Sands: Dirty Oil and the Future of a Continent.* Vancouver: Greystone Books 2008.

Ophuls, William. *Plato's Revenge: Politics in the Age of Ecology.* Cambridge, MA, and London: MIT Press 2011.

O'Reilly, Jennifer. "The Wisdom of the Scribe and the Fear of the Lord." In D. Broun and T.O. Clancy, eds., *Spes Scotorum Hope of Scots: Saint Columba, Iona and Scotland.* 159–211. Edinburgh: T&T Clark 1999.

O'Rourke, Fran. *Pseudo-Dionysius and the Metaphysics of Aquinas.* Leiden and New York: E.J. Brill 1992.

Orr, David W. *Earth in Mind: On Education, Environment, and the Human Prospect.* Washington, Covelo, CA, and London: Island Press 2004 (1992).

Ottawa *Citizen.* "The Lost Villages," 28 June 2008. http://www.canada.com/ottawacitizen/news/story.html?id=b888ee7c-b7b1-4cod-b52f-451271952ba4 (accessed 3 April 2013).

Otto, Rudolf. *The Idea of the Holy: An Inquiry into the Non-Rational Factor in the Idea of the Divine and Its Relation to the Rational.* London: Oxford University Press 1924.

Oxford Canadian Dictionary. Don Mills, ON: Oxford University Press Canada 2004.

Palmater, Pamela D. *Beyond Blood: Rethinking Indigenous Identity.* Saskatoon: Purich 2011.

Parham, Claire Puccia. *From Great Wilderness to Seaway Towns: A Comparative History of Cornwall, Ontario and Massena, New York, 1784–2001.* Albany: State University of New York Press 2004.

Parmenter, Jon. *The Edge of the Woods: Iroquoia, 1534–1701.* East Lansing: Michigan State University Press 2010.

Patterson, Frances Taylor. *White Wampum: The Story of Kateri Tekakwitha.* New York: Longmans, Green 1934.

Petrides, George A., and J. Wehr. *Eastern Trees: Peterson Field Guides.* Boston and New York: Houghton Mifflin 1998.

Plumwood, V. *Environmental Culture: The Ecological Crisis of Reason.* London and New York: Routledge 2002.

Podruchny, Carolyn. *Making the Voyageur World: Travellers and Traders in the North American Fur Trade.* Toronto: University of Toronto Press 2006.

Pope Francis. *Encyclical Letter Laudato Si' of the Holy Father Francis on Care for our Common Home.* http://w2.vatican.va/content/dam/francesco/pdf/encyclicals/documents/papa-francesco_20150524_enciclica-laudato-si_en.pdf. 2015 (accessed 10 June 2015).

Prache, Anne. *Chartres Cathedral: Image of the Heavenly Jerusalem.* Trans. Janice Abbott. Paris: CNRS Editions 1993.

Primavesi, A. *Exploring Earthiness: The Reality and Perception of Being Human Today*. Eugene, OR: Cascade Books 2013.

– *Gaia and Climate Change: A Theology of Gift Events*. London and New York: Routledge 2009.

– *Sacred Gaia: Holistic Theology and Earth System Science*. London and New York: Routledge 2000.

Rappaport, R.A. *Ritual and Religion in the Making of Humanity*. Cambridge and New York: Cambridge University Press 1999.

Razack, Sherene H. "Colonization: The Good, the Bad, and the Ugly." In A. Baldwin, L. Cameron, and A. Kobayashi, eds., *Re-Thinking the Great White North: Race, Nature, and the Historical Geographies of Whiteness in Canada*. 264–71. Vancouver and Toronto: UBC Press 2011.

Reguly, Eric. "In an Act of Atonement, Vatican Makes Kateri Tekakwitha the First Native Canadian Saint." *Globe and Mail*, 21 October 2012.

Rice, Brian. *The Rotinonshonni: A Traditional Iroquoian History through the Eyes of Teharonhia:wako and Sawiskera*. New York: Syracuse University Press 2013.

Roberts, Paul. *The End of Oil: On the Edge of a Perilous New World*. Boston: Houghton Mifflin 2004.

Roberts, Wayne. "Whose Land? The Displacement of the Mississauga Has Left Toronto with a Major Cultural Deficit." *NOW*, 32(45) (11–18 July 2013): 18.

Robertson, Heather. *Walking into Wilderness: The Toronto Carrying Place and Nine Mile Portage*. Winnipeg: Heartland 2010.

Robinson, John. "Being Undisciplined: Transgressions and Intersections in Academia and Beyond." *Futures*, 40 (2008): 70–86.

Robinson, Percy J. *Toronto during the French Régime*. Toronto: University of Toronto Press 1965.

Rogers, R.A., P. Timmerman, T. Leduc, and M. Dickinson. "The Why of the 'Hau': Scarcity, Gifts, and Environmentalism." *Ecological Economics*, 51 (2004): 177–89.

Rogers, Raymond A. *Nature and the Crisis of Modernity: A Critique of Contemporary Discourse on Managing the Earth*. Montreal: Black Rose Books 1994.

– *Solving History: The Challenge of Environmental Activism*. Montreal: Black Rose Books 1998.

Root, B.I., D A. Chant, and C.E. Heidenreich. *Special Places: The Changing Ecosystems of the Toronto Region*. Vancouver and Toronto: UBC Press 1999.

Sachs, Wolfgang. "Environment." In W. Sachs, ed., *The Development Dictionary*. 26–37. London: Zed Books 1992.

Sagard, Gabriel. *Sagard's Long Journey to the Country of the Hurons*. New York: Greenwood 1968.

Sandars, N.K. *The Epic of Gilgamesh* (An English Version with an Introduction by N.K. Sandars). New York: Penguin Books 1972.

Sandberg, Anders, G.R. Wekerle, and L. Gilbert. *The Oak Ridges Moraine Battles: Development, Sprawl, and Nature Conservation in the Toronto Region*. Toronto: University of Toronto Press 2013.

Santmire, Paul. *The Travail of Nature: The Ambiguous Ecological Promise of Christian Theology*. Philadelphia: Fortress Press 1985.

Saul, John Ralston. *On Equilibrium*. Toronto: Viking Books 2001.

– *A Fair Country: Telling Truths about Canada*. Toronto: Penguin Canada 2009.

– *Reflections of a Siamese Twin: Canada at the End of the Twentieth Century*. Toronto: Viking 1997.

Scharper, Stephen B. *For Earth's Sake: Toward a Compassionate Ecology* (ed. S. Appolloni). Toronto: Novalis 2013.

– "Pope Francis's Important Ecology Lesson." Toronto *Star*, 21 June 2015 (http://www.thestar.com/opinion/commentary/2015/06/21/pope-francis-important-ecology-lesson.html) (accessed 28 June 2015).

– *Redeeming the Time: A Political Theology of the Environment*. New York: Continuum 1997.

Schmalz, P. *The Ojibwa of Southern Ontario*. Toronto: University of Toronto Press 1991.

Schmidt, J., and F. Remiz. "High Park Waterways: Forward to the Past." In W. Reeves and C. Palassio, eds., *HTO: Toronto's Water*. 284–91. Toronto: Coach House Books 2008.

Schönfeld, M., ed. *Global Ethics on Climate Change: The Planetary Crisis and Philosophic Alternatives*. London and New York: Routledge 2013.

– "Introduction – Plan B: Global Ethics on Climate Change." *Journal of Global Ethics*, 7(2) (2011): 129–36.

Scott, Robert A. *The Gothic Enterprise: A Guide to Understanding the Medieval Cathedral*. Berkeley: University of California Press 2003.

Selhub, Eva M., and Alan C. Logan. *Your Brain on Nature: The Science of Nature's Influence on Your Health, Happiness, and Vitality.* Mississauga, ON: John Wiley and Sons Canada 2012.

Senior, Elinor Kyte. *From Royal Township to Industrial City: Cornwall, 1784–1984.* Belleville, ON: Mika 1983.

Serres, Michel, and Philippa Hurd. *Angels: A Modern Myth* [Légende des anges]. Paris: Flammarion 1995.

Seton, Ernest Thompson. *Wild Animals I Have Known.* Toronto: McClelland and Stewart 1981 (1898).

Sewall, Laura. "Beauty and the Brain." In P.H. Kahn, Jr, and P.H. Hasbach, eds., *Ecopsychology: Science, Totems, and the Technological Species.* 265–84. Cambridge, MA: MIT Press 2012.

Shepard, P. "Ecology and Man: A Viewpoint." In P. Shepard and D. McKinley, eds., *The Subversive Science: Essays Toward an Ecology of Man.* 1–10. Boston: Houghton Mifflin 1967.

– *The Only World We've Got: A Paul Shepard Reader.* San Francisco: Sierra Club Books 1996.

Sheridan, J., and R.D. Longboat. "The Haudenosaunee Imagination and the Ecology of the Sacred." *Space and Culture*, 9(4) (2006): 365–81.

Shiva, V. *Monocultures of the Mind.* London: Zed Books 1993.

Simcoe, Elizabeth. *The Diary of Mrs. John Graves Simcoe*, ed. J.R. Roberson. Toronto: William Briggs 1973 (1911).

Sinclair, Justice Murray. "Opinion: Reconciliation not Opportunity to 'Get Over It.'" CBC News, 18 April 2014. http://www.cbc.ca/news/aboriginal/ reconciliation-not-opportunity-to-get-over-it-justice-murray-sinclair-1 .2614352 (accessed 10 August 2014).

Sioui, Georges. *Histories of Kanatha: Seen and Told.* Ottawa: University of Ottawa Press 2008.

– *Huron-Wendat: The Heritage of the Circle.* Vancouver: UBC Press 1999.

Skogan, Joan. *Mary of Canada: The Virgin Mary in Canadian Culture, Spirituality, History, and Geography.* Banff, AB: Banff Centre Press 2003.

Smil, Vaclav. "The Long Slow Rise of Solar and Wind." *Scientific American*, ScientificAmerican.com, January 2014.

Smith, Linda Tuhiwai. *Decolonizing Methodologies: Research and Indigenous Peoples.* London and New York: Zed Books; Dunedin, NZ, and New York: University of Otago Press 1999.

Sorrell, Roger D. *St. Francis of Assisi and Nature: Tradition and Innovation in Western Christian Attitudes toward the Environment.* New York and Oxford: Oxford University Press 1988.

Soulé, M.E., and D. Press, "What is Environmental Studies?" *BioScience,* 48(5) (1998): 397–405.

Sousa, E. "Re-Inhabiting Taddle Creek." In W. Reeves and C. Palassio, eds., HTO: *Toronto's Water.* 234–47. Toronto: Coach House Books 2008.

Sperling, D., and D. Gordon. *Two Billion Cars: Driving toward Sustainability.* Oxford: Oxford University Press 2009.

Spitzer, Leo. "Milieu and Ambiance: An Essay in Historical Semantics." *Philosophy and Phenomenological Research,* 3(2) (1942): 169–218.

Storck, Peter L. *Journey to the Ice Age: Discovering an Ancient World.* Vancouver: UBC Press 2004.

Suzuki, David. *The David Suzuki Reader: A Lifetime of Ideas from a Leading Activist and Thinker.* Vancouver: Greystone Books 2003.

Suzuki, David, with Amanda McConnell and Adrienne Mason. *The Sacred Balance: Rediscovering Our Place in Nature.* Vancouver: Greystone Books 2007.

Swift, Jamie, and Keith Stewart. *Hydro: The Decline and Fall of Ontario's Electric Empire.* Toronto: Between the Lines 2004.

Taylor, Bron. *Dark Green Religion: Nature Spirituality and the Planetary Future.* Berkeley: University of California Press 2010.

Taylor, Sarah McFarland. *Green Sisters: A Spiritual Ecology.* Cambridge, MA: Harvard University Press 2007.

Thomas, Chief Jacob, with Terry Boyle. *Teachings from the Longhouse.* Toronto: Stoddart 1994.

Thompson, W.I. *Imaginary Landscape: Making Worlds of Myth and Science.* New York: St Martin's Press 1989.

– *The Time Falling Bodies Take to Light: Mythology, Sexuality and the Origins of Culture.* New York: St Martin's Press 1981.

Thoreau, Henry D. *Walden.* New Haven, CT, and London: Yale University Press 2006 (1854).

Timmerman, Peter. "The Human Dimensions of Global Change." In T. Munn and P. Timmerman, eds., *Encyclopedia of Global Environmental Change.* Vol. 5. 1–9. Toronto: John Wiley and Sons 2002.

Trigger, B. *The Children of Aataentsic: A History of the Huron People to 1660.* Montreal and Kingston, ON: McGill-Queen's University Press 1987.

Trigger, Bruce G. *Natives and Newcomers: Canada's "Heroic Age" Reconsidered*. Montreal and Kingston, ON: McGill-Queen's University Press 1985.

Truth and Reconciliation Commission of Canada. *Honouring the Truth, Reconciling for the Future: Summary of the Final Report of the Truth and Reconciliation Commission of Canada*. 2015.

Underhill, Evelyn. *Mysticism: A Study in the Nature and Development of Man's Spiritual Consciousness*. New York: E.P. Dutton 1961.

Vanderburg, W.H. *Living in the Labyrinth of Technology*. Toronto: University of Toronto Press 2005.

Vanderheiden, Steve. *Atmospheric Justice: A Political Theory of Climate Change*. Oxford: Oxford University Press 2008.

Victor, Peter. *Managing without Growth: Slower by Design, Not Disaster*. Cheltenham, UK, and Northampton, MA: Edward Elgar 2008.

Wadland, John. *Ernest Thompson Seton: Man in Nature and the Progressive Era, 1880–1915*. New York: Arno 1978.

White, Jr, Lynn. "The Historical Roots of our Ecologic Crisis." *Science*, 155(3767) (1967): 1203–7.

Wilbert J. *Mindful of Famine: Religious Climatology of the Warao Indians*. Cambridge, MA: Harvard University Press 1997.

Wilk, Richard. "Consuming Ourselves to Death: The Anthropology of Consumer Culture and Climate Change." In S.A. Crate and M. Nuttall, eds., *Anthropology and Climate Change: From Encounters to Actions*. 265–76. Walnut Creek, CA: Left Coast Press 2009.

Williams, Raymond. *Problems in Materialism and Culture*. London and New York: Verso 1980.

Williamson, R.F., ed. *Toronto: An Illustrated History of Its First 12,000 Years*. Toronto: James Lorimer 2008.

Williamson, R.F., and R.I. MacDonald. "A Resource like No Other: Understanding the 11,000-year Relationship between People and Water." In W. Reeves and C. Palassio, eds., *HTO: Toronto's Water*. 42–51. Toronto: Coach House Books 2008.

Wilson, Ian. *Before the Flood: The Biblical Flood as a Real Event and How It Changed the Course of Civilization*. London: St Martin's Press 2004.

Wilson, Shawn. *Research Is Ceremony: Indigenous Research Methods*. Halifax and Winnipeg: Fernwood 2008.

Winfield, M. "Opinion: Lac Mégantic Disaster Caused by a Series of Failures." Montreal *Gazette*, 26 July 2003. http://www.montrealgazette. com/news/Opinion+M%C3%A9gantic+disaster+caused+series+failures/ 8712833/story.html (accessed 20 August 2013).

Woodworth, William Raweno:kwas. *Indigenous Ancestry and Environmental Education in Toron:to.* Presentation to Graduate Environmental Education class, Humber River, Toronto, 2013.

– *Indigenous Ancestry and Environmental Education in Toron:to.* Presentation to Graduate Environmental Education class, Humber River, Toronto, 2012.

– "The Morning Star: It Is Bright Tawennawetah Teyohswathe." PhD dissertation, California Institute of Integral Studies 2001.

– "Urban Place as an Expression of the Ancestors." In I.L. Stefanovic and S.B. Scharper *The Natural City: Re-Envisioning the Built Environment.* 223–30. Toronto: University of Toronto Press 2012.

Woodworth, William Raweno:kwas, and Tim Leduc. "Recovering the Tribal Identity of Toronto: Ancestral Conversations." Locating Compassion in Land Ethics Conference. Toronto, 23–25 March 2012.

Index